D0900204

Marmosets and Tamarins

The support of the
**Fundação Biodiversitas, Belo Horizonte, Brazil,**
is gratefully acknowledged by the Editor.

# Marmosets and Tamarins

## *Systematics, Behaviour, and Ecology*

---

*Edited by*

ANTHONY B. RYLANDS

*Universidade Federal de Minas Gerais, Brazil*

Oxford New York Tokyo

OXFORD UNIVERSITY PRESS

1993

*Oxford University Press, Walton Street, Oxford OX2 6DP*
*Oxford New York Toronto*
*Delhi Bombay Calcutta Madras Karachi*
*Kuala Lumpur Singapore Hong Kong Tokyo*
*Nairobi Dar es Salaam Cape Town*
*Melbourne Auckland Madrid*
*and associated companies in*
*Berlin Ibadan*

*Oxford is a trade mark of Oxford University Press*

*Published in the United States*
*by Oxford University Press Inc., New York*



*A catalogue record for this book is available from the British Library*

*Library of Congress Cataloging in Publication Data*
*Marmosets and tamarins : systematics, behaviour, and ecology / edited by Anthony B. Rylands.*
*Includes bibliographical references and index.*
*1. Marmosets—Classification—Congresses. 2. Tamarins—Classification—Congresses. 3. Marmosets—Behavior—Congresses. 4. Tamarins—Behavior—Congresses. 5. Marmosets—Ecology—Congresses. 6. Tamarins—Ecology—Congresses. I. Rylands, Anthony B.*
*QL737.P92M38 1993 599.8'2—dc20 92-27771*
*ISBN 0 19 854022 1*

*Typeset by Graphicraft Typesetters Ltd, Hong Kong*
*Printed in Great Britain by*
*Biddles Ltd, Guildford and King's Lynn*

# Acknowledgements

The book is based on a symposium held during the XIIth Congress of the International Primatological Society, Brasília, 24–29 July 1988, entitled 'Adaptive Unity in the Callitrichidae: a Systematic Comparison of Species Differences in the Family'. The chapters arising from this meeting include 2 (Snowdon), 4 (Epple *et al.*), 5 (Abbott *et al.*), 6 (Dixson), 8 (Caine), 10 (Yamamoto), 11 (Soini), 12 (Rylands and Faria), and 13 (Garber). I am most grateful to Milton Thiago de Mello (University of Brasília), organizer of the Congress. I would also like to register my gratitude to Warwick Kerr (in 1976 Director of the National Institute for Amazon Research – INPA, Manaus) who gave me the opportunity to study marmosets, and to Tim Clutton-Brock (Large Animal Research Group, Cambridge University) and Peter Jewell (Research Group in Mammalian Ecology and Reproduction, Cambridge University) for showing me how. Russell Mittermeier (Conservation International, Washington DC), Adelmar Coimbra-Filho (Rio de Janeiro Primate Centre – CPRJ), Célio Valle (Federal University of Minas Gerais), and Jeremy Mallinson (Jersey Wildlife Preservation Trust, Jersey), and not least my wife, Miriam, have been a constant source of inspiration and encouragement during my 15 years working with these animals in Brazil. Lastly, I extend my thanks to the other contributors for their goodwill, enthusiasm, and patience.

*Belo Horizonte, Brazil* A.B.R.
June 1992

# Contents

## Part III: Ecology

# Contributors

*D.H. Abbott*

MRC/AFRC Comparative Physiology Research Group, Institute of Zoology, Zoological Society of London, Regent's Park, London NW1 4RY, UK
Present address for correspondence: Wisconsin Regional Primate Research Center, 1223 Capitol Court, Madison, WI 53715–1299, USA

*J. Barrett*

MRC/AFRC Comparative Physiology Research Group, Institute of Zoology, Zoological Society of London, Regent's Park, London NW1 4RY, UK

*A.M. Belcher*

Monell Chemical Senses Center, 3500 Market Street, Philadelphia, PA 19104, USA and Deutsches Primatenzentrum, Kellnerweg 4, D-3400 Göttingen, Germany

*Nancy G. Caine*

Department of Psychology, Bucknell University, Lewisburg, PA 17837, USA
Present address: Psychology Program, California State University, San Marcos, San Marcos CA 92096-0001, USA

*Adelmar F. Coimbra-Filho*

Centro de Primatologia do Rio de Janeiro, Fundação Estadual de Engenharia do Meio Ambiente, Rua Fonseca Teles 121/1624, 20.940 Rio de Janeiro RJ, Brazil

*Kurt Darms*

Institut für Anthropologie, Universität Göttingen, Bürgerstrasse 50, D-3400 Göttingen, Germany

*A.F. Dixson*

Centre de Primatologie, Centre International de Recherches Médicales de Franceville, B.P. 769, Franceville, Gabon

*G. Epple*

Monell Chemical Senses Center, 3500 Market Street, Philadelphia, PA 19104, USA and Deutsches Primatenzentrum, Kellnerweg 4, D-3400 Göttingen, Germany

*Doris S. de Faria*

Departamento de Biologia Animal, Instituto de Ciências Biológicas, Universidade de Brasília, 70.910 Brasília DF, Brazil

*Stephen F. Ferrari*

Departamento de Zoologia, Museu Paraense Emílio Goeldi, Caixa Postal 399, 66.040 Belém PA, Brazil

*Paul A. Garber*

Department of Anthropology, University of Illinois, Urbana, IL 61801, USA

*L.M. George*

MRC/AFRC Comparative Physiology Research Group, Institute of Zoology, Zoological Society of London, Regent's Park, London NW1 4RY, UK

*K.L. Greenfield*

Monell Chemical Senses Center, 3500 Market Street, Philadelphia, PA 19104, USA

*Mary L. Harrison*

Department of Biological Sciences, Kent State University, Stark Campus, 6000 Frank Avenue, Canton, OH 44720, USA

*I. Küderling*

Deutsches Primatenzentrum, Kellnerweg 4, D-3400 Göttingen, Germany

*Russell A. Mittermeier*

Conservation International, 1015 18th Street N.W., Suite 1000, Washington, DC 20036, USA and Department of Anatomical Science, Health Sciences Center, State University of New York, Stony Brook, NY 11794, USA

*Alcides Pissinatti*

Centro de Primatologia do Rio de Janeiro, Fundação Estadual de Engenharia do Meio Ambiente, Rua Fonseca Teles 121/1624, 20.940 Rio de Janeiro RJ, Brazil

*Hartmut Rothe*

Institut für Anthropologie, Universität Göttingen, Bürgerstrasse 50, D-3400 Göttingen, Germany

*Anthony B. Rylands*

Departamento de Zoologia, Instituto de Ciências Biológicas, Universidade Federal de Minas Gerais, 31.270 Belo Horizonte MG, Brazil

*L. Scolnick*

Monell Chemical Senses Center, 3500 Market Street, Philadelphia, PA 19104, USA

*Mary A. Simek*

Department of Anthropology, University of Tennessee, Knoxville, TN 37996, USA

*A.B. Smith III*

Monell Chemical Senses Center, 3500 Market Street, Philadelphia, PA 19104, USA and Department of Chemistry, University of Pennsylvania, Philadelphia, PA 19104, USA

*Charles T. Snowdon*

Department of Psychology, University of Wisconsin, 1202 West Johnson Street, Madison, WI 53706, USA

*Pekka Soini*

Ministerio de Agricultura, Region Agraria XXII, Iquitos, Peru and Apartado 341, Iquitos, Loreto, Peru

*Suzette D. Tardif*

Marmoset Research Center, Oak Ridge Associated Universities, Oak Ridge, TN 37831, USA
Present address for correspondence: Department of Anthropology, University of Tennessee, 252 South Stadium Hall, Knoxville, TN 37996, USA

*Maria Emília Yamamoto*

Setor de Psicobiologia, Centro de Biociências, Universidade Federal do Rio Grande do Norte, Caixa Postal 1511, 59.000 Natal RN, Brazil

*U. Zeller*

Zentrum Anatomie, Universität Göttingen, Kreuzbergring 36, D-3400 Göttingen, Germany

# Introduction

*Anthony B. Rylands*

In 1978, Devra Kleiman, introducing her book the *Biology and Conservation of the Callitrichidae*, pointed out that the 'ultimate indication of our ignorance of this family is the tendency to refer to all forms as 'the marmoset', regardless of genus or species . . . The assumption that species are alike arises only because we know so little of the differences.' Kleiman's book pointed to many of these differences, and was a milestone in our understanding of these animals which had been steadily growing during the late 1960s and 1970s but has increased so dramatically during the last decade.

To date 52 species and subspecies of marmoset and tamarin monkeys have been described, some very recently. Interest in these animals, which include the Amazonian pygmy marmoset, the smallest of all the monkeys, the marmosets and tamarins, and the colourful lion tamarins of south-east Brazil, has focused on many aspects. Not least is their remarkable social life, involving reproductive privileges restricted to one female in the group, who physiologically and behaviourally inhibits breeding by other females, variable mating systems with documented cases in the wild of monogamy and polyandry as well as polygyny, the active participation of the males along with other group members in rearing the twin offspring, and the use of scent glands, variably concentrated in three epidermal regions, for communication. The marmosets *Cebuella* and *Callithrix* show an extraordinary tree-gouging and gum-eating behaviour, unique amongst the primates. Callitrichids occur in very diverse habitats, from Panama to northeastern Paraguay and southern Brazil, and although some are highly adaptable in their ability to survive in small and degraded patches of forests otherwise destroyed by man, others, notably the lion tamarins, are on the brink of extinction. Interest in this case results from the need to understand biological aspects related to their breeding and management in captivity, as well as their geographical distributions and the behavioural ecology of wild populations for the management of, in a number of species, already critically small populations.

This book provides reviews of our current knowledge of some of the main aspects which have excited considerable research interest over the last decade. These include the geographical distributions and the still contested taxonomy of this highly unusual group of primates, along with comparative reviews of such topics as vocalizations, hybridization, scent-marking, reproductive inhibition, social organization and mating systems, infant care and development, and their behaviour and ecology in the wild. The underlying theme in all cases is to examine the extent of generic/specific variation in the topics considered, a comparative approach which is fundamental for our understanding of the adaptive functions of the social organization and ecology of callitrichids in particular and mammals in general.

The first part deals with taxonomy, geographic distributions and conservation status. Rylands, Coimbra-Filho, and Mittermeier (Chapter 1) review the divergent opinions regarding callitrichid taxonomy, at the family/subfamily level as well as regarding species and subspecies, in the latter case affecting most especially the marmosets, *Callithrix*, and lion tamarins, *Leontopithecus*. With respect to species and subspecies, opinions are divided into two extremes, that of Hershkovitz (1977) who emphasizes subspecies, and that of Vivo (1988) who, dealing only with *Callithrix*, considered all to be species. Rylands, Coimbra-Filho, and Mittermeier may be considered mediators of these divergent positions; for example, they currently regard all non-Amazonian marmosets as species, while following Hershkovitz's classification of the Amazonian forms. Further taxonomic (and distributional) studies are required, and it is my feeling that in the case of the marmosets, a thorough revision of the two forms with the largest distributions (*C. argentata* and *C. penicillata*) will clarify many of the problems and even misunderstandings. In this chapter, geographical distributions are also described in some detail, providing an update on the information available since (and enabled by) the remarkable synthesis of Hershkovitz (1977). In some cases this has resulted in a redefinition of the distributional limits, and also contributes to some of the arguments regarding callitrichid taxonomy. The elucidation of geographic distributions is a matter of some urgency, considering that the release of captive animals far outside their natural range is an ever increasing problem, and the destruction of their natural forests, especially in northern Colombia and south-east Brazil, means that the original or natural distributional limits are probably already impossible to determine in a number of cases. Finally, a brief description is given of the conservation status of each form, along with a provisional listing of the conservation units within their geographic ranges. Due to an almost complete lack of information on the populations of most of the callitrichids, decisions on conservation status are based almost entirely on the size of the geographic range and a gross overview of the state of destruction/development of the regions involved. Only the lion tamarins, *Leontopithecus*, and the endangered cotton-top tamarin, *Saguinus oedipus*, have received special studies of their population status in both the wild and captivity.

Resolution of the taxonomic status of the callitrichids has to date depended on morphological studies, along with considerations on pelage colour and geographic distribution patterns. Here, Snowdon (Chapter 2) offers a fresh and evidently very valuable addition to the study of behavioural aspects, specifically vocalizations. Structural analyses of the long-calls permit the construction of phylogenetic relationships. To my knowledge no other behavioural aspects have been examined in this way, and Snowdon's contribution indicates the potential of similar comparative analyses of other communication systems (visual and olfactory). Finally in this section, Coimbra-Filho, Pissinatti, and Rylands report on captive hybrids and some natural hybrid zones in the marmosets of south-east Brazil. The authors emphasize that, although requiring many generations, controlled hybrid experiments have considerable potential for elucidating taxonomic problems, as well as potential usefulness in biomedical research. Likewise, the recent discovery of a number of natural hybrid zones opens up a gamut of questions regarding their extent and stability, and the behaviour of the populations and groups involved compared with non-hybrid populations.

The second section deals with a number of developmental, reproductive, and behavioural aspects which not only show subtle and sometimes marked differences between genera, and in some cases species, but are also fundamental for an understanding of the evolution of their social and reproductive system. Epple and co-workers (Chapter 4) discuss their longterm studies of epidermal scent glands and scent-marking behaviour, which they regard as the most elaborately developed of any of the simian primates. They have worked mainly on three tamarins, *Saguinus fuscicollis*, *S. oedipus*, and *S. leucopus*, and argue that chemical messages, far from being the result of single substances, are made up of complexes of volatile as well as non-volatile components from several sources (glandular secretion, urine, genital discharge, and products of bacterial action) and represent a wide variety of encoded messages. They describe the behavioural studies which have elucidated some of the information carried in these scent marks, but little research has yet been carried out in wild groups, and studies of the contexts, frequencies, and locations of scent marks in natural situations are fundamental for an understanding of the functions of these complex communication systems. Studies of the three tamarin species have demonstrated similarities and differences (quantitative and qualitative), which as the authors indicate point the way to taxonomic considerations along the lines of Snowdon (Chapter 2).

Scent marks have been identified as one of the means by which dominant females inhibit reproductive activity in subordinates, a remarkable characteristic of this family, first reported by Abbott and John Hearn in 1978. This is reviewed by Abbott, Barrett, and George (Chapter 5), who discuss the research carried out on four species of three genera (*Callithrix jacchus*, *Saguinus oedipus*, *S. fuscicollis*, and *Leontopithecus rosalia*). This phenomenon was, until recently, thought to be a trait of all the callitrichids, but recent physiological and field research has indicated that not only do the mechanisms vary between genera and species (at least the two tamarins), but also that physiological suppression of ovulation is absent altogether in the lion tamarins. The authors also discuss the functional aspects of this trait and possible reasons for the species differences (see below).

The mating system, in the 1970s thought to be universally monogamous, has been found to be highly variable, with monogamy occurring alongside cases of polyandry and polygyny, evidently even within populations, and depending on the group composition. Dixson (Chapter 6) and Rothe and Darms (Chapter 7), provide reviews of different approaches to the study of mating systems in these animals, both however arguing for the need to combine captive and field data. Dixson discusses the importance of understanding sexual behaviour and its degree of dependence on hormonal mechanisms, of studies of the morphological features of the reproductive tract, and finally of the great potential of DNA 'fingerprinting' techniques to provide detailed genealogies, not only for recognizing the breeding male(s) but also for an understanding of such parameters as dispersal and reproductive success, fundamental for explanations of the evolution of their social system. Rothe and Darms emphasize detailed studies of the social structure and behavioural interactions of captive groups. They review some of the results of the long-term studies of their common marmosets (*C. jacchus*) maintained in large and complex cages, discussing particularly sociodemography, the roles and interrelationships of individuals in differing social and familial contexts, and individual mating strategies. They regard monogamy as overwhelmingly typical of callitrichids (at least marmosets and tamarins). They point out that social contact is a fundamental driving force and that the detailed study of functional units within groups (as opposed to the generalizations resulting from descriptions of sociographic units) will provide many of the answers regarding callitrichid social structure. Rothe and Darms argue against the importance of such population parameters as immigration/emigration (breeding opportunities) and predation because of the rarity of such events in the wild. This contrasts with the views of Caine (Chapter 8), who despite obtaining the majority of her data from the study of captive groups of *Saguinus labiatus*, is more comfortable in extrapolating her findings to fit in with the conclusions from a number of studies in the wild. She emphasizes the co-operative nature of callitrichid social groups, and argues that the role of predation is 'among the most important, if not the most important selection pressure influencing the social behaviour and group structure of tamarin species'. As pointed out by Caine, sentinel behaviour and mobbing, like food calls and food sharing, combine with infant care in the repertoire of co-operative behaviours shown by callitrichids. In Chapter 9 Tardif and co-workers compare a number of parameters related to infant care in callitrichids and link them to differences in social organization and ecology. They emphasize the energetic costs of carrying infants and argue that species differences in the use of food resources (food types and their abundance and dispersion) may explain differences between genera.

Finally in this section, Yamamoto (Chapter 10) reviews the behavioural ontogeny of callitrichid offspring, concentrating particularly on the roles of different group members and the effect of group composition on infant care patterns. Yamamoto agrees with Tardif and co-workers in her conclusion that the overall patterns of behavioural ontogeny in the infant callitrichids studied to date are quite similar. Like Caine (Chapter 8), however, she argues that flexibility is an important adaptive feature in this group, notable in the relative constancy of the amount of infant care provided despite variable group composition (one-parent and two-parent families, for example).

The first four chapters in the third section provide comparative reviews of ecological aspects of wild groups of each of the four genera. Soini (Chapter 11) examines differences in feeding and ranging behaviour of sympatric *Cebuella* and *Saguinus*. Rylands and Faria (Chapter 12) discuss variation in habitat and diet (particularly gum-feeding), and their effects on home range and group size in the gum-feeding marmosets, *Callithrix*. Garber (Chapter 13) reviews the tamarins, *Saguinus*, examining particularly three contrasting modes of animal prey foraging. Divergent insect foraging behaviour is the key to permitting sympatry between the moustached tamarins (*S. mystax*, *S. imperator*, and *S. labiatus*) and the saddle-back tamarins (*S. fuscicollis*), but Garber concludes that habitat utilization, diet, and positional behaviour are extremely similar for all tamarins, and emphasizes unity rather than variety with regard to the 11 species he discusses.

Despite the rather limited data available on the ecology of lion tamarins, *Leontopithecus*, Rylands (Chapter 14) indicates that the evidence to date suggests differences in home range size (large), feeding behaviour (specialized animal prey foragers), sleeping sites (holes in trees), and reproduction (generally breeding once a year rather than twice; no physiological suppression of subordinate females, see Abbott, Chapter 5) in comparison to all other callitrichids. These differences will hold the clue to understanding why lion tamarins are now so rare. In this chapter, Rylands also points to some differences between the four lion tamarin species due to contrasting habitats (coastal/inland, seasonal/aseasonal rainfall).

The patterns described for *Saguinus* by Garber also apply in large part to the similar-sized *Callithrix*. One significant feature separating the two, however, is the latter's ability to specialize on plant exudates (gums) during fruit shortage. This increases the range of habitats they can occupy and also affects home range size, and probably patterns of reproduction and dispersal, and hence group size, composition, and stability. This aspect is reviewed by Ferrari (Chapter 15 and Ferrari and Lopes Ferrari 1989), who also discusses other aspects including feeding ecology, size effects (all are small, but *Leontopithecus* is approximately six times larger than *Cebuella*), patterns of sympatry, the formation of mixed species groups, and habitat utilization. In the light of our current knowledge of the differences, principally in feeding ecology, of the four callitrichid genera, he also discusses and speculates on the zoogeography and evolutionary history of the group.

In this book, a number of different features of the extraordinary ways of life shown by callitrichids are foregrounded, and a number of the authors speculate on the selective forces behind the evolution of their complex social and reproductive behaviour. The behavioural, physiological, and ecological differences ( and similarities) between the species and genera which are now emerging are beginning to provide us with hypotheses as well as answers to questions concerning the adaptive functions of their mating systems, reproductive inhibition, paternal and group care of infants, and high reproductive rate. As emphasized by a number of the contributors, key features requiring long-term population studies include infant survival and reproductive success, predation, immigration and emigration, and most importantly a better understanding of their habitat, particularly their food resources. Studies of key resources and habitat size, quality, and availability will provide the answers regarding the evolution of the breeding strategies of these animals, and will, I believe, explain the differences between lion tamarins and the remaining species (Abbott *et al.*, Rylands) and between *Cebuella* and the other three genera (Soini, Ferrari). As argued convincingly by Ferrari, the differences between tamarins, *Saguinus*, and marmosets, *Callithrix*, lie fundamentally in the latter's specialization for gum feeding. Important for both genera, however, is the availability and quality of second-growth patches (size, distribution, age, floristic composition etc.), prime sources of small fruits, animal prey, and sleeping sites, important for protection from predators, and the limiting factor for breeding opportunities, and hence the determinant for dispersal patterns and group size and composition. This raises the very important question of the environment in which the callitrichid social system evolved. In a forest untouched by human activity, secondary-growth patches and edge habitat, for example, are available as an unpredictable and dispersed mosaic of treefall gaps, river edges, etc. Forest cutting and timber extraction have completely changed this pattern, and what we see today, at least in the majority of field studies, probably reflects the flexibility of callitrichid social organization (stressed by various authors) to highly modified habitats, and not the scenario in which their social system evolved in the past.

# Part I

## *Systematics*

# 1

# Systematics, geographic distribution, and some notes on the conservation status of the Callitrichidae

*Anthony B. Rylands, Adelmar F. Coimbra-Filho, and Russell A. Mittermeier*

## Introduction

The phylogeny and systematics of the New World marmosets and tamarins is still under considerable discussion, even at the family level (Rosenberger 1981; Ford 1986*a*; Vivo 1991), although the remarkable synthesis of Hershkovitz (1977) remains the baseline for all current research and investigation on the taxonomy and distributions of these animals. Sixteen years on, some new forms have been described and there is still controversy regarding the taxonomic status of, most particularly, the marmosets, genus *Callithrix*. Considerable new information is also available regarding geographic distributions. In this chapter, we provide accounts of the current opinions regarding taxonomic status along with a revision of the distributions and some notes on the conservation status of each of the 50 or so marmosets and tamarins recognized to date.

## Families, subfamilies, tribes, subtribes, and genera

The longstanding arrangement for the extant platyrrhine primates, adopted by Pocock in 1925 and maintained by Hill (1957, 1959, 1960, 1962), Napier and Napier (1967), Simons (1972), and Thorington (1976) gives just two families, the Callitrichidae (marmosets and tamarins) and Cebidae (the remainder), with the enigmatic Goeldi's monkey, *Callimico goeldii*, as a subfamily of the former or the latter (Cabrera 1957; Simpson 1945). Some authors, however, have given *Callimico* its own family, Callimiconidae (for example, Dollman 1933; Hershkovitz 1977). *Callimico* has both of the principal features used in the past to distinguish cebids from callitrichids —clawed digits *and* three molar teeth (see Rosenberger 1981)—and its alignment with either of the two families resulted from the relative importance (character weighting) of these features and also whether the claws were considered true and primitive or rather specialized nails (see Garber 1980*a*,*b*; Ford 1986*b*).

Thorington (1976) argued that the callitrichids, excepting *Callimico*, are not sufficiently diverse to warrant subdivision, and implied the existence of only two subfamilies (Callitrichinae and Callimiconinae). He recognized, however, only two unambiguous subfamily groupings in the Cebidae—the Pitheciinae (*Pithecia*, *Cacajao*, and *Chiropotes*) and the Atelinae (*Ateles*, *Lagothrix*, and *Brachyteles*). Excepting the family status of *Callimico*, Hershkovitz (1977) also followed this scheme, with all the remaining extant genera in their own subfamilies.

Rosenberger (1981), however, has provided an alternative arrangement, based on testable suppositions regarding phylogenetic relationships and 'their common possession of presumably "derived" rather than "primitive" characteristics' (p.21) rather than the presence or absence of specific characters. Rosenberger (1980, 1981, 1984; see also Rosenberger and Strier 1989; Rosenberger *et al.* 1990) argued for a phylogenetic dichotomy separating two major families, the Cebidae Bonaparte, 1831 (redefined) and Atelidae Gray, 1825. The former includes the subfamilies Callitrichinae Thomas, 1903 (marmosets, tamarins, and *Callimico*) and Cebinae Bonaparte, 1831 (*Cebus* and *Saimiri*). The Atelidae also includes two subfamilies, Atelinae Gray, 1825 (*Lagothrix*, *Ateles*, *Alouatta*, and *Brachyteles*) and Pitheciinae Gray, 1949 (*Aotus*, *Callicebus*, *Pithecia*, *Cacajao*, and *Chiropotes*).

Rosenberger (1981) defined the Callitrichinae as a monophyletic unit characterized by such features as clawed digits (except for the hallux), second molar reduction, reduced hallux and pollex, and a modified form of incisal occlusion, which he argued are derived and indicative of a divergent adaptive patterns. Rosenberger (1984; Rosenberger *et al.* 1990) erected two tribes within the subfamily: Callitrichini Thomas, 1903, with two subtribes, Callitrichina Thomas, 1903 (*Callithrix*, *Cebuella*, and *Leontopithecus*) and Leontocebina Miranda Ribeiro, 1940 (*Saguinus*); and the monospecific tribe Callimiconini Thomas, 1913 (*Callimico*). These are distinguished by the presence of two rather than three molars, a tricuspid rather than four-cusped molar, and the birth of twins rather than singletons, respectively. This demonstrates Rosenberger's conviction that *Callimico* is most closely related to the marmosets (Rosenberger 1984), and also that *Leontopithecus* is more closely related to *Callithrix* than to *Saguinus*. Natori (1989) also argued that *Leontopithecus* is the closest genealogical relative of *Callithrix-Cebuella* on the basis of two shared apomorphies: the septa in the tympanic cavity, and a reduced hypocone on $M^1$. The Callitrichina have more V-shaped lower jaws, triangular upper molars, and moderate- to high-crowned incisors (considered derived traits), whereas *Saguinus* (Leontocebina) has anteriorly wide jaws, non-triangular upper molars, and low-crowned incisors. This revolutionary view is not shared by other recent authors such as Cronin and Sarich (1978) and Ford (1986*a*) who argued for an early splitting off of *Leontopithecus* from the ancestral stock, followed by *Saguinus* and subsequently the marmosets (see also Hill 1957; Byrd 1981; Boer 1974). Rosenberger and Coimbra-Filho (1984) pointed out that the common view, pairing *Saguinus* and *Leontopithecus* (tamarins) on the one hand and *Cebuella* and *Callithrix* (marmosets) on the other, is based on differences in the morphology of the anterior dentition (see Coimbra-Filho and Mittermeier 1978) with the tamarins having the 'long-tusked' lower dentition and the marmosets being 'short-tusked' (a misnomer, being more correctly a feature independent of canine size but reflecting the projection of the lower canines above the tooth row, with the incisors being larger in the marmosets). Rosenberger and Coimbra-Filho (1984) argued that the 'tamarin' anterior dentition is primitive for the Callitrichinae and for platyrrhines in general and therefore invalid as a cladistic common denominator, and concluded that '*Leontopithecus* presents an acceptable model of a "basically callitrichine" dentition being transformed in the direction of *Callithrix*' (p.168).

Currently, the most stable point of platyrrhine taxonomy is at the generic level, with the widespread recognition of the classification adopted by Hershkovitz (1977). There is only one case we know of where this has been disputed. Rosenberger (1984) questioned the generic status of the pygmy marmoset, *Cebuella*. According to him, *Cebuella* and some (not stated which) of the eastern Brazilian marmosets (the '*Jacchus*' group of Hershkovitz 1977) share a number of morphological traits (staggering and orientation of lower incisors, symphyseal shape, jaw shape, condylar height) which are not present in the Amazonian *Callithrix humeralifer* and *C. argentata*. On this basis, he proposed a single genus (*Callithrix*) with three subgenera, representing *Cebuella*, the Amazonian *Callithrix* ('*Argentata*' Group), and the eastern Brazilian *Callithrix* ('*Jacchus*' group), respectively (see also Rosenberger and Coimbra-Filho 1984), although this was not upheld in the classification presented in Rosenberger *et al.* (1990).

## Species, subspecies, and species groups

Whereas the generic classification of platyrrhine primates of Hershkovitz (1977) has been widely accepted, this is not so for the arrangement of the callitrichid (callitrichine) species and subspecies. Four recent taxonomies are shown in Table 1.1. There are divergent opinions regarding the taxonomic status of the south-east Brazilian *Callithrix*. Hershkovitz (1977) placed them all as subspecies of *C. jacchus*, whereas Coimbra-Filho and Mittermeier (Coimbra-Filho 1971, 1984, 1990; Coimbra-Filho and Mittermeier 1973*b*; Mittermeier and Coimbra-Filho 1981; Mittermeier *et al.* 1988*b*), and Natori (1986) defended their species status, and Vivo

**Table 1.1** Four recent taxonomies for callitrichids (callitrichines). Hershkovitz (1977, 1979, 1982) and Mittermeier *et al.* (1988) considered all forms, Vivo (1991) reviewed only the genus *Callithrix*, and Coimbra-Filho's (1990) review was restricted to the Brazilian taxa

| Hershkovitz (1977, 1979, 1982) | Mittermeier *et al.* (1988*b*) |
|---|---|
| *Cebuella* | *Cebuella* |
| *C. pygmaea* | *C. pygmaea* |
| *Callithrix* | *Callithrix* |
| *C. argentata argentata* | *C. argentata argentata* |
| *C. argentata leucippe* | *C. argentata leucippe* |
| *C. argentata melanura* | *C. argentata melanura* |
| | *C. emiliae* |
| *C. humeralifer humeralifer* | *C. humeralifer humeralifer* |
| *C. humeralifer intermedius* | *C. humeralifer intermedius* |
| *C. humeralifer chrysoleuca* | *C. humeralifer chrysoleuca* |
| *C. jacchus jacchus* | *C. jacchus* |
| *C. jacchus penicillata* | *C. penicillata* |
| *C. jacchus geoffroyi* | *C. geoffroyi* |
| | *C. kuhli* |
| *C. jacchus aurita* | *C. aurita* |
| *C. jacchus flaviceps* | *C. flaviceps* |
| *Saguinus* | *Saguinus* |
| *S. nigricollis nigricollis* | *S. nigricollis nigricollis* |
| *S. nigricollis graellsi* | *S. nigricollis graellsi* |
| *S. nigricollis hernandezi* | *S. nigricollis hernandezi* |
| *S. fuscicollis fuscicollis* | *S. fuscicollis fuscicollis* |
| *S. fuscicollis fuscus* | *S. fuscicollis fuscus* |
| *S. fuscicollis avilapiresi* | *S. fuscicollis avilapiresi* |
| *S. fuscicollis cruzlimai* | *S. fuscicollis cruzlimai* |
| *S. fuscicollis illigeri* | *S. fuscicollis illigeri* |
| *S. fuscicollis leucogenys* | *S. fuscicollis leucogenys* |
| *S. fuscicollis nigrifrons* | *S. fuscicollis nigrifrons* |
| *S. fuscicollis lagonotus* | *S. fuscicollis lagonotus* |
| *S. fuscicollis weddelli* | *S. fuscicollis weddelli* |
| *S. fuscicollis primitivus* | *S. fuscicollis primitivus* |
| *S. fuscicollis melanoleucus* | *S. fuscicollis melanoleucus* |
| *S. fuscicollis acrensis* | *S. fuscicollis acrensis* |
| *S. fuscicollis crandalli* | *S. fuscicollis crandalli* |
| *S. fuscicollis tripartitus* | *S. tripartitus* |
| *S. mystax mystax* | *S. mystax mystax* |
| *S. mystax pileatus* | *S. mystax pileatus* |
| *S. mystax pluto* | *S. mystax pluto* |
| *S. labiatus labiatus* | *S. labiatus labiatus* |
| *S. labiatus thomasi* | *S. labiatus thomasi* |
| *S. imperator imperator* | *S. imperator imperator* |
| *S. imperator subgrisescens* | *S. imperator subgrisescens* |
| *S. inustus* | *S. inustus* |
| *S. midas midas* | *S. midas midas* |
| *S. midas niger* | *S. midas niger* |
| *S. bicolor bicolor* | *S. bicolor bicolor* |
| *S. bicolor ochraceus* | *S. bicolor ochraceus* |
| *S. bicolor martinsi* | *S. bicolor martinsi* |
| *S. leucopus* | *S. leucopus* |
| *S. oedipus oedipus* | *S. oedipus* |
| *S. oedipus geoffroyi* | *S. geoffroyi* |
| *Leontopithecus* | *Leontopithecus* |
| *L. rosalia rosalia* | *L. rosalia* |
| *L. rosalia chrysomelas* | *L. chrysomelas* |
| *L. rosalia chrysopygus* | *L. chrysopygus* |

| Vivo (1988, 1991) (only the genus *Callithrix*) | Coimbra-Filho (1990) (only Brazilian taxa) |
|---|---|
| *Callithrix* | *Cebuella* |
| *C. argentata* | *C. pygmaea* |
| *C. leucippe* | *Callithrix* |
| *C. melanura* | *C. argentata argentata* |
| *C. emiliae* | *C. argentata leucippe* |
| *C. humeralifera* | *C. argentata melanura* |
| *C. intermedia* | *C. emiliae* |
| *C. chrysoleuca* | *C. humeralifera* |
| *C. jacchus* | *C. intermedia* |
| *C. penicillata* | *C. chrysoleuca* |
| *C. geoffroyi* | *C. jacchus* |
| *C. aurita* | *C. penicillata* |
| *C. flaviceps* | *C. geoffroyi* |
| | *C. kuhli* |
| | *C. aurita aurita* |
| | *C. aurita flaviceps* |
| | *Saguinus* |
| | *S. nigricollis nigricollis* |
| | *S. fuscicollis fuscicollis* |
| | *S. fuscicollis fuscus* |
| | *S. fuscicollis avilapiresi* |
| | *S. fuscicollis cruzlimai* |
| | *S. fuscicollis weddelli* |
| | *S. fuscicollis primitivus* |
| | *S. melanoleucus melanoleucus* |
| | *S. melanoleucus acrensis* |
| | *S. melanoleucus crandalli* |
| | *S. mystax mystax* |
| | *S. mystax pileatus* |
| | *S. mystax pluto* |
| | *S. labiatus labiatus* |
| | *S. labiatus thomasi* |
| | *S. imperator imperator* |
| | *S. imperator subgrisescens* |
| | *S. inustus* |
| | *S. midas midas* |
| | *S. midas niger* |
| | *S. bicolor bicolor* |
| | *S. bicolor ochraceus* |
| | *S. bicolor martinsi* |
| | *Leontopithecus* |
| | *L. rosalia* |
| | *L. chrysomelas* |
| | *L. chrysopygus chrysopygus* |
| | *L. chrysopygus caissara* |

(1988, 1991) argued that *all* the *Callithrix* forms are species. Mittermeier *et al.* (1988*b*) followed Hershkovitz (1977) regarding the Amazonian forms, *C. humeralifer* and *C. argentata*. Mittermeier *et al.* (1988*b*), Coimbra-Filho (1984, 1990), Rylands 1989*b*, and Natori (1990) also recognize *C. kuhli* from southern Bahia, regarded by Hershkovitz (1977) as a hybrid, and by Vivo (1991) as indistinguishable from *C. penicillata*. Coimbra-Filho (1990) recommended that the three *C. humeralifer* subspecies be considered species following Boer (1974) and Vivo (1991). The *Saguinus* species and subspecies recognized by Hershkovitz (1977, 1979, 1982) have been generally accepted, although Thorington (1988) and Coimbra-Filho (1990) questioned the status of some of the *S. fuscicollis* subspecies, and Mittermeier and Coimbra-Filho (1981) and Mittermeier *et al.* (1988*b*) (see also Hanihara and Natori 1987) have argued for the validity of *S. geoffroyi* (placed by Hershkovitz 1977 as a subspecies of *S. oedipus*). Finally, whereas Hershkovitz (1977) recognized only one species of *Leontopithecus*, Rosenberger and Coimbra-Filho (1984) (see also Mittermeier *et al.* 1988*b*) regarded them as distinct species. The discovery of the black-faced lion tamarin, *L. caissara*, by Lorini and Persson in 1990, is also causing some controversy, with Coimbra-Filho (1990) regarding it as only subspecific to the black lion tamarin, *L. chrysopygus*.

The taxonomic arrangement of the callitrichids (callitrichines) at the species/subspecies level is an academic question which involves the artificial definition of discrete thresholds of divergence in terms of their morphology and reproduction, linked directly to the reconstruction of their phylogeny. Perhaps the most useful approach is that adopted by Natori and Hanihara and co-workers, who use a comparative statistical method for craniodental morphology which permits an objective classification which they link with geographic distance and from the starting point of the differences observed between sympatric species. Thus, for example, if two forms are more different to each other than two sympatric forms, they should also be considered species. Analogous differences can be employed for *Callithrix* and *Leontopithecus*.

In this review, we maintain, for the present, the taxonomic arrangement presented in Mittermeier *et al.* (1988*b*), although in the treatments of each species we discuss some modifications at the specific and subspecific levels. As mentioned above, Mittermeier *et al.* (1988*b*) have argued that the *Callithrix* of the Atlantic forest of north-east and south-east Brazil should be considered valid species, and not subspecies of *C. jacchus*. Here we maintain this classification, although Coimbra-Filho (1990) argued that *C. flaviceps* should more correctly be placed as a subspecific form of *C. aurita*. We also consider that the Amazonian, *C. humeralifer intermedius*, first described by Hershkovitz (1977), should more correctly be considered a subspecies of *C. argentata*, both in terms of its pelage colouration and patterns and its distribution. Mittermeier *et al.* (1988*b*) recognize the form

*C. emiliae* on the basis of Vivo (1985). *C. emiliae* was regarded by Hershkovitz (1977) as merely a dark form of *C. a. argentata*. However, Vivo (1985) was considering a marmoset from Rondônia, which although darker and more brownish, he synonymized with the paler, more silvery *C. emiliae* first described from southern Pará (illustrated in Cruz Lima 1945), the distribution therefore being disjunct and separated by *C. argentata melanura*. We suggest, therefore, that Vivo's Rondônia marmoset should be considered distinct: at present it lacks a scientific name, true *C. emiliae* occurring in southern Pará. Martins *et al.* (1988) have also indicated the existence of another *C. argentata* form in southern Pará, between the Rios Xingu and Irirí and on the left bank of the Rio Irirí. We recognize here a new marmoset from the Rio Madeira, *C. nigriceps* Ferrari and Lopes 1992*b*. Ferrari and Lopes (1992*b*) gave this marmoset species status following Vivo (1991), although we predict that Hershkovitz, if recognizing the form, would place it as a subspecies of *C. argentata*. Finally, we must mention two undescribed forms aligned to *C. humeralifer*.

Besides the promotion of *S. fuscicollis tripartitus* to species status (following Thorington 1988), we also indicate that the pale form from the western Amazon in Brazil, *melanoleucus*, should be considered a valid species, with two subspecies *crandalli* and *acrensis* (following Carvalho 1957*b*; Coimbra-Filho 1990). Natori and Hanihara (1992), following detailed studies of the dental morphology in the genus *Saguinus*, found that the two subspecies of *S. midas* were as different from each other as *S. labiatus* is from *S. mystax*, and *S. fuscicollis* from *S. nigricollis*, and on this basis we suggest that *S. m. midas* and *S. m. niger*, separated from each other by the Rio Amazonas, be considered distinct species. Regarding the phylogeny of this genus, Natori (1988) concluded that *S. bicolor* is the closest relative of *S. midas*. This is entirely compatible with their distributions, and argues against Hershkovitz's alignment of *S. bicolor* with *S. oedipus* and *S. leucopus*. Finally, as a result of a re-evaluation of the known distributions of the three *S. mystax* subspecies, we predict that the Bolivian *S. mystax* mentioned by Izawa and Bejarano (1981) may well be a fourth distinct subspecies. These authors also referred to a different form of *S. labiatus*, although callitrichid (callitrichine) distributions in northern Bolivia are evidently complex and they did not rule out the possibility of it being a hybrid with *S. i. imperator*.

The differing opinions regarding the alpha taxonomy of the callitrichids (callitrichines) involve almost exclusively their status as species or subspecies. In only two cases is there argument regarding the validity of the form even as a subspecies (*C. kuhli* and *C. emiliae*—the latter is regarded by Hershkovitz as a dark form of *C. a. argentata*). For *Callithrix*, extremes are adopted by Hershkovitz (1977) (just three species) and Vivo (1991) (12 species; *C. kuhli* not valid). Mittermeier *et al.* (1988*b*) sit in the middle, accepting just three species in Amazonia (*C. emiliae* valid), but six species from east and south-east Brazil (*C. kuhli* valid). Coimbra-Filho (1990) moved his position slightly more towards that of Vivo (1991) by considering the *C. humeralifer* subspecies of Hershkovitz (1977) as species, following Boer (1974) who identified differing diploid chromosome numbers for *C. humeralifer* and *C. chrysoleucos* sic. (*C. h. intermedius* Hershkovitz 1977 was also given species status by default but is here recognized as being more closely aligned to *C. argentata*). Nothing is known of variation in chromosome numbers of *C. argentata*, Boer (1974) examining only *C. a. argentata*, and Coimbra-Filho (1990) maintained the arrangement of Mittermeier *et al.* (1988*b*), following Hershkovitz (1977). The only Hershkovitzian element remaining in Coimbra-Filho's (1990) *Callithrix* classification, therefore, is that of *C. argentata*.

If we consider that (1) *chrysoleuca* is a subspecies of *C. humeralifer*, (2) a further two '*humeralifer*' forms have yet to be described, (3) *C. h. intermedius* is a subspecies of *C. argentata*, (4) there is a further undescribed *C. argentata* from the Rio Xingu; (5) the '*C. emiliae*' from Rondônia is a distinct form which should be renamed as a *C. argentata* subspecies; and (6) that *C. nigriceps* could well be considered a *C. argentata* subspecies, we have the possible arrangement of four subspecies of *C. humeralifer* and seven subspecies of *C. argentata*. True *C. emiliae* from southern Pará is apparently sympatric with an undescribed *C. argentata* ssp. mentioned by Martins *et al.* (1988) and therefore considered a valid species. This poses the question of why have the Amazonian *Callithrix* subspeciated to such an extent whereas the Atlantic forest *Callithrix* have not? Alternatively, why should there be six species in the Atlantic forest, and only three in the vast area of the Amazon, south of the Rio Amazonas and east of the Rio Madeira? The answers would require an understanding of the mechanisms (principally rivers in the Amazon, but also topography and vegetation types, seasonality, structure, and floristic composition) as well as the periods (history) of isolation of the respective populations (see Hershkovitz 1977; Prance 1983; Whitmore and Prance 1987). Vivo (1991) in considering all marmoset forms to be species, provides at least a consistent treatment (why should *C. argentata* have numerous subspecies and *C. humeralifer* and the Atlantic forest marmosets not?) but implies a uniform adaptive radiation, with similar speciation rates for all the *Callithrix* lineages (two or three species groups, see below). Intuitively this would seem unlikely.

With regard to an understanding of the evolution of callitrichids and their present-day distribution, perhaps more important—as a first approximation—is the concept of species groups and their affinities. These are shown in Table 1.2. We follow Hershkovitz's groupings in this sense except: (1) the Atlantic forest species could usefully be divided into two groups, '*Aurita* group' (*C. aurita* and *C. flaviceps*) and '*Jacchus* group' (*C. penicillata*, *C. jacchus*, C. *kuhli*, and *C. geoffroyi*) which are phenotypically and ecologically distinct (see Natori 1986; Rylands and Faria, this volume); and

**Table 1.2** Species groups of *Callithrix* and *Saguinus* of Hershkovitz (1977)

| |
|---|
| **Marmosets** |
| *Callithrix jacchus* group—true marmosets |
| *C. jacchus* |
| *Callithrix argentata* group—bare-ear and tassel-ear marmosets |
| *C. argentata, C. humeralifer* |
| **Hairy-face tamarin section** |
| *Saguinus nigricollis* group—white-mouth tamarins |
| *S. nigricollis, S. fuscicollis* |
| *Saguinus mystax* group—moustached tamarins |
| *S. mystax, S. labiatus, S. imperator* |
| *Saguinus midas* group—midas tamarin group |
| *S. midas* |
| **Mottled-face tamarin section** |
| *S. inustus* |
| **Bare-face tamarin section** |
| *Saguinus bicolor* group—Brazilian bare-faced tamarins |
| *S. bicolor* |
| *Saguinus oedipus* group—Colombian and Panamanian bare-face tamarins |
| *S. oedipus, S. geoffroyi, S. leucopus* |

(2) following Natori and Hanahari (1992, see also Hanihara and Natori 1989), *S. m. midas* and *S. m. niger* (as species or subspecies) should more correctly be grouped with *S. bicolor*, excluding the former from the hairy-face section (*S. fuscicollis*, *S. nigricollis*, *S. labiatus S. mystax*, and *S. imperator*) and the latter from the bare-face tamarin section of (*S. oedipus*, *S. geoffroyi*, and *S. leucopus*) of Hershkovitz (1977) (see Nogami and Natori 1986; Natori 1988; Hanihara and Natori 1987, 1989; Natori and Hanihara 1992 for a discussion of the phylogeny of these species based on cranial and dental morphology). The enigmatic and solitary *S. inustus* (mottled-face tamarin section) is the only remaining species occurring north of the Rio Amazonas and Rio Japurá, between the Colombian/Central American species and west of *S. bicolor/S. midas* from the middle and lower Amazon, and its morphological/evolutionary affinities are unknown.

## Distributions

Considerable new evidence has been obtained since the publication of Hershkovitz's (1977) evaluation and descriptions of the distributions of callitrichids (callitrichines). Here we refine some of Hershkovitz's distributions in the light of new, if fragmentary, data obtained since the publication of his book.

In some cases, consideration of the known and supposed distributions have influenced our opinions regarding taxonomic status. This is the case, for example with *C. emiliae*, where Vivo's (1985) alignment of populations in Rondônia with those in southern Pará would seem unlikely because

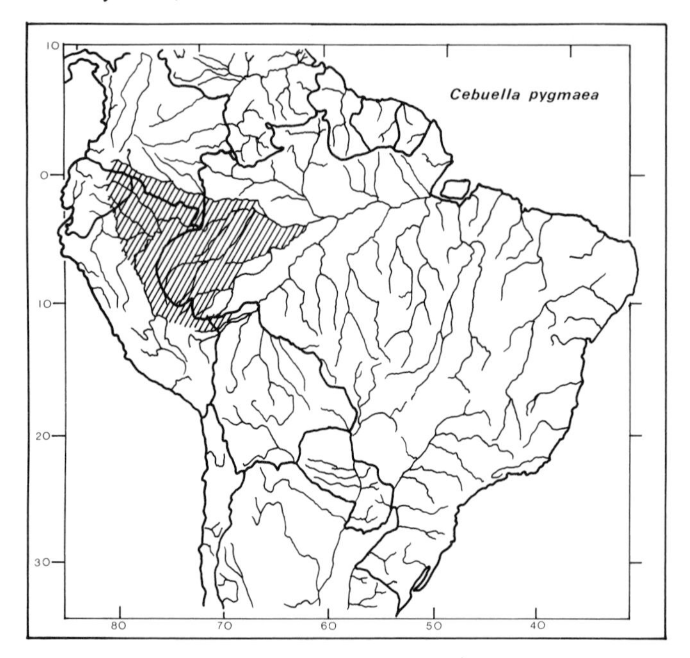


**Fig. 1.1** Distribution of *Cebuella pygmaea*.

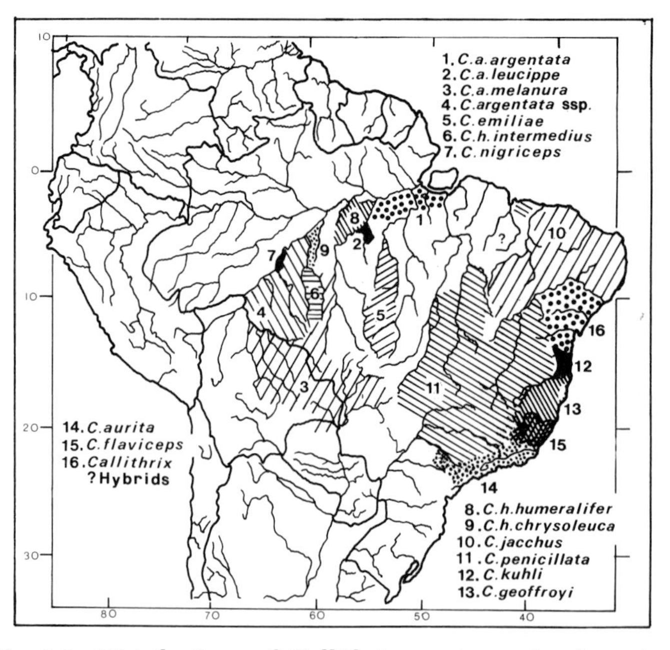


**Fig. 1.2** Distributions of *Callithrix* species and subspecies.

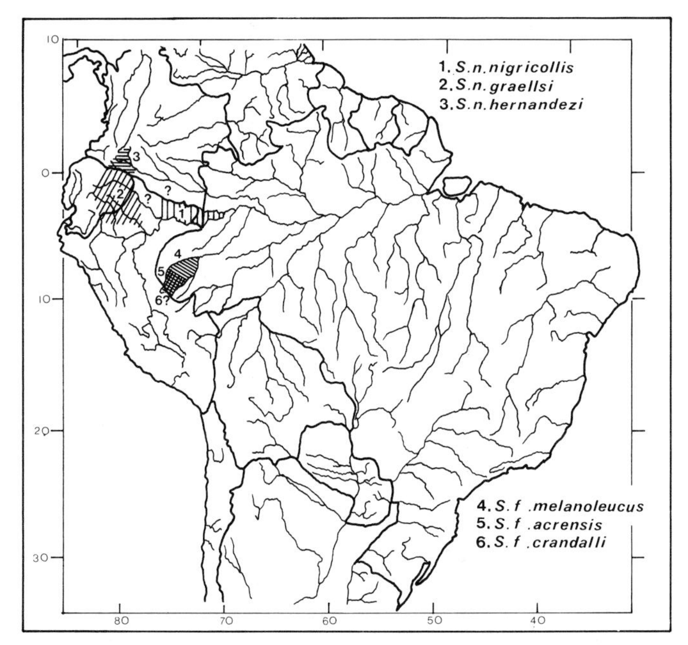


**Fig. 1.3** Distributions of the *Saguinus nigricollis* subspecies and three *Saguinus fuscicollis* subspecies.

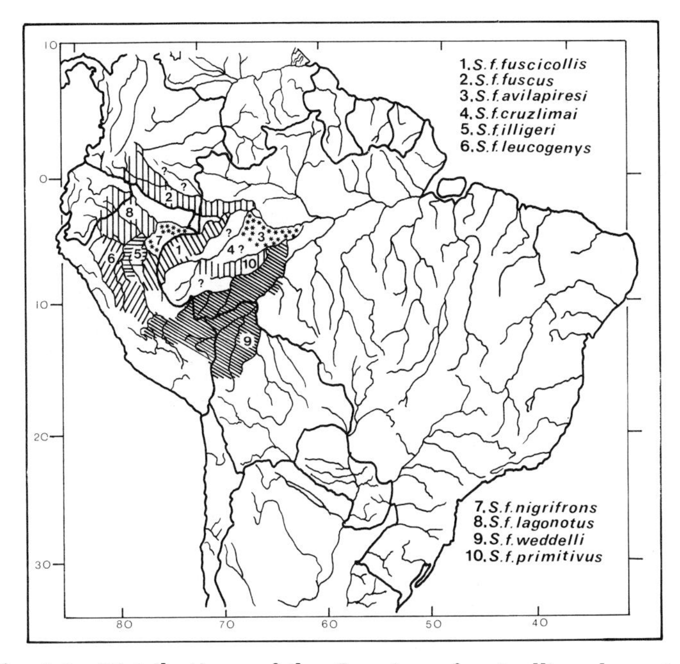


**Fig. 1.4** Distributions of the *Saguinus fuscicollis* subspecies.

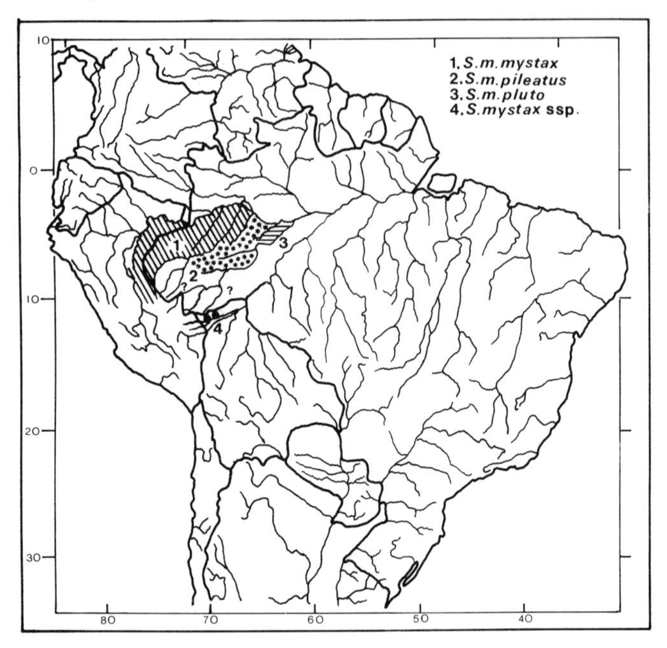


**Fig. 1.5** Distributions of the *Saguinus mystax* subspecies.

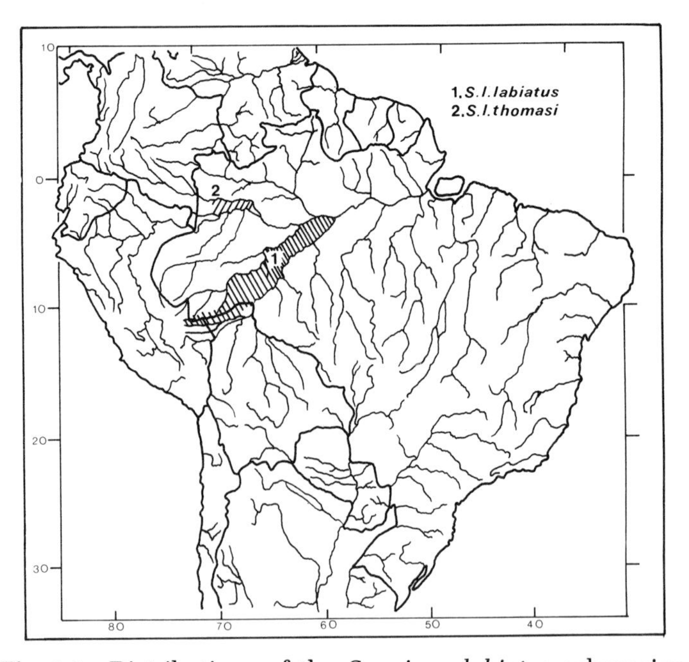


**Fig. 1.6** Distributions of the *Saguinus labiatus* subspecies.

**Fig. 1.7** Distributions of the *Saguinus imperator* subspecies.

**Fig. 1.8** Distributions of *Saguinus inustus* and the *Saguinus bicolor* subspecies.

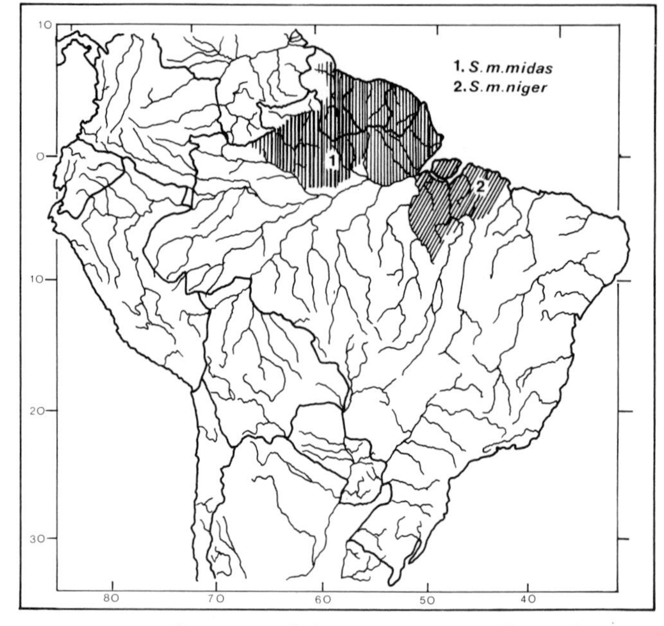


**Fig. 1.9** Distributions of the *Saguinus midas* subspecies.

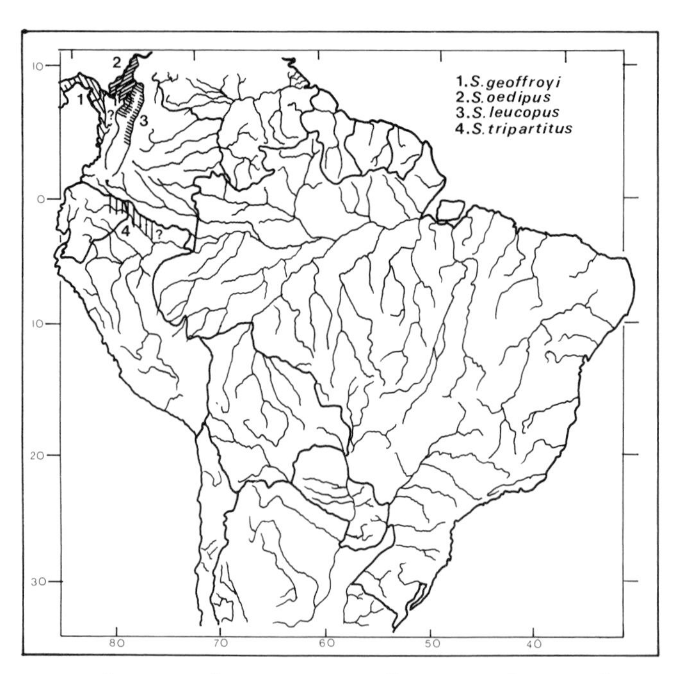


**Fig. 1.10** Distributions of *Saguinus geoffroyi*, *S.oedipus*, *S.leucopus*, and *S.tripartitus*.

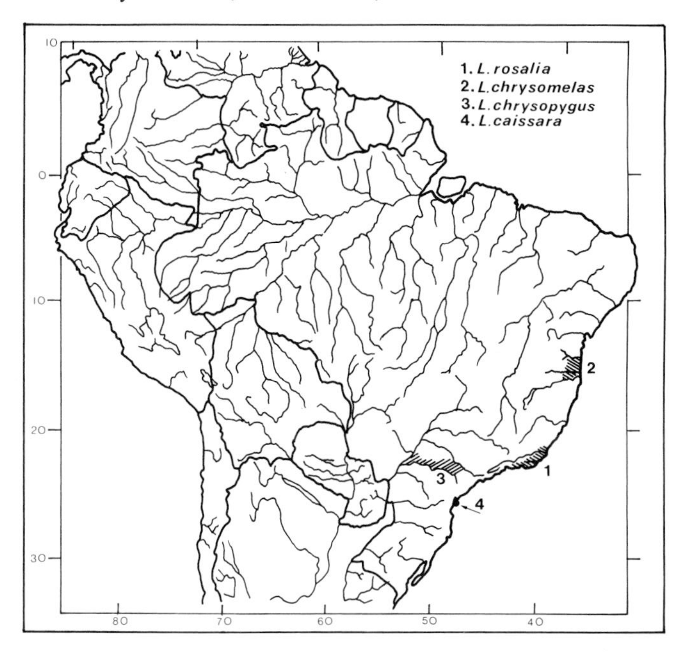


**Fig. 1.11** Distributions of the four lion tamarin species, *Leontopithecus*.

they are separated by *C. a. melanura*. There is no evidence that *C. humeralifer intermedius* is sympatric with *C. a. melanura*, as was indicated by Hershkovitz (1977), and its geographic range is surrounded by *C. argentata* forms, which contributes to our belief that it is more likely a *C. argentata* subspecies. Unfortunately little new evidence is available regarding the distributions of the saddleback tamarins, except for some minor changes in those of *S. f. nigrifrons* and *S. f. fuscicollis*. Izawa and Bejarano (1981) have demonstrated a complex situation regarding the moustached tamarin group in northern Bolivia. We have restricted the distribution of *Saguinus mystax pluto* on the basis of the only three known localities, and for this reason indicate that the resulting distant populations in northern Bolivia may well be a new subspecific form.

We emphasize that the callithrichid (callitrichine) distributions are still poorly understood, and based in many cases on very few locality records. The widespread destruction of the Brazilian Atlantic forest means that the historic ranges of the *Callithrix* taxa occurring there may never be known (see for example, Coimbra-Filho *et al.* 1991).

## Conservation status

Detailed information on the conservation status of the marmosets and tamarins is largely unavailable. Specific information and realistic appraisals

**Table 1.3** Endangered and vulnerable taxa of Neotropical callitrichids (callitrichines)

**Endangered**

*Callithrix argentata leucippe*

*Callithrix aurita*

*Callithrix flaviceps*

*Saguinus bicolor bicolor*

*Saguinus oedipus*

*Leontopithecus rosalia*

*Leontopithecus chrysomelas*

*Leontopithecus chrysopygus*

*Leontopithecus caissara*

**Vulnerable**
(restricted distributions in rapidly developing areas)

*C. argentata* ssp. A ('*C. emiliae*' of Vivo 1985)
*C. argentata* ssp. B (Rio Irirí basin)

*Callithrix humeralifer humeralifer*
*Callithrix humeralifer intermedius*
*Callithrix humeralifer chrysoleuca*
*Callithrix humeralifer* ssp. A
*Callithrix humeralifer* ssp. B

*Callithrix kuhli*

*Callithrix geoffroyi*

*Saguinus fuscicollis melanoleucus*
*Saguinus fuscicollis acrensis*
*Saguinus fuscicollis crandalli*[a]

*Saguinus tripartitus*

*Saguinus imperator imperator*
*Saguinus imperator subgrisescens*

*Saguinus leucopus*

*Saguinus geoffroyi*

[a] Distribution unknown but possibly in the frontier region of Peru and the state of Acre, Brazil (Hershkovitz 1977).

are generally available only for those forms considered critically endangered and/or subjected to field studies. In this paper, we provide brief comments and indicate the principle references for each taxon. Table 1.3 summarizes our conclusions regarding the endangered and vulnerable taxa considered in this paper. Nine species and subspecies are undoubtedly endangered through habitat destruction, and a further 17 we consider vulnerable, having very restricted distributions in areas which are currently undergoing widespread development.

For discussions of the principal factors endangering wild primate populations in South America, we refer the reader to some recent reviews: Bolivia—Brown and Rumiz (1986); Brazilian Amazonia—Ayres (1983), Mittermeier and Coimbra-Filho (1977), Mittermeier *et al.* (1978), Peres (1987), Rylands and Mittermeier (1982); Brazil, Atlantic forest—Mittermeier *et al.* (1981, 1982), Oliver and Santos (1991), Santos *et al.* (1987), Torres de Assumpção (1983*a*); Brazilian callitrichids—Coimbra-Filho (1984); Colombia—Hernandez-Camacho and Cooper (1976); Hernandez-Camacho and Defler (1985, 1989); General—Mittermeier *et al.* (1986), Marsh and Mittermeier (1987), Mittermeier and Cheney (1987); French Guiana—Mittermeier *et al.* (1978), Roussilhon (1988); Neotropics—Mittermeier and Coimbra-Filho (1983), Mittermeier *et al.* (1989); Peru—Soini (1982*a*), Soini *et al.* (1989); Suriname—Baal *et al.* (1988), Mittermeier and van Roosmalen (1982); Mittermeier *et al.* (1978).

With regard to their occurrence in conservation units, it should be remembered that in many cases the presence of a species/subspecies in a protected area is presumed, according to its geographic distribution, rather than proven. The principal references used include: Bolivia—Brown and Rumiz (1986); Brazil—Brazil, MINTER-Ibama (1989); Brazilian Amazonia—Rylands (1985*a*, 1991), Rylands and Bernardes (1989), Rylands and Mittermeier (1982); Brazil, Atlantic forest—Coimbra-Filho (1984), Mendes (1991*a*), Mittermeier *et al.* (1981, 1982), Oliver and Santos (1991); Colombia—Colombia, MA-INDERENA (1989); Hernandez–Camacho and Defler (1985, 1989); Peru—Soini (1982*a*); Paraguay—Stallings (1985); Suriname—Baal *et al.* 1988; Mittermeier and van Roosmalen (1982); All countries—IUCN (1982*b*). The abbreviations given after the names of the Brazilian conservation units refer to the state, as follows: AC, Acre; AL, Alagoas; AP, Amapá; AM, Amazonas; BA, Bahia; CE, Ceará; ES, Espírito Santo; GO, Goiás; MG, Minas Gerais; MT, Mato Grosso; PA, Pará; PB, Paraíba; PI, Piauí; RJ, Rio de Janeiro; RN, Rio Grande do Norte; RO, Rondônia; RR, Roraima; SP, São Paulo; TO, Tocantins.

## Accounts of species

## Genus CEBUELLA Gray 1866

### ***Cebuella pygmaea*** (Spix 1823)

Pygmy marmoset (Engl.), mico-leãozinho (Br.), sagui-leãozinho (Br.), sagui-pigmeu (Br.), titi-pielroja (Co.), leonzito (Co.) (*v*. Hernandez-Camacho and Defler 1989), leoncito (Pe.) (*v*. Burgos 1974).

**Type locality** Tabatinga, Rio Solimões, Amazonas, Brazil (*v*. Hershkovitz 1977).

**Distribution** The exact distribution of this species is not well known, but evidence to date suggests that it is confined to the upper Amazon basin, west of the Rio Purus and Japurá in Brazil, to the Andes in southern Colombia, south of the Río Caquetá, and Ecuador on the upper reaches of the Ríos Napo and Putumayo, to the Río Pastaza, lower Río Huallaga, and Río Ucayali in northern Peru (Napier 1976; Hershkovitz 1977), and to the region of Cobija in northwestern Bolivia (Heltne *et al.* 1976). In Bolivia, Izawa (1979) and Izawa and Bejarano (1981) confined it to the north and west of the Ríos Orthon and Manuripi, northern tributaries of the Río Madre de Dios. Brown and Rumiz (1986), however, doubt that it occurs as far south as the Río Manuripi, limiting its distribution to the north of the Río Tahuamanu. Its presence in northern Bolivia indicates that it should occur in parts of eastern Acre, including the Ríos Acre and uppermost Abunã, not indicated by Hershkovitz (1977). The Río Abunã is a tributary of the Rio Madeira, and the easternmost limits in this region remain undetermined. The southernmost locality reported so far is the Manu National Park, approximately 12°S (Soini 1988).

Hernandez-Camacho and Cooper (1976) recorded that it is well known south of the Río Caquetá in Colombia, although reports of its occurrence further north, in the upper Río Guaviare region, remain to be confirmed. The only evidence to date is a captive specimen reportedly obtained from Cano Morrocoy on the south bank of the Río Guaviare. Izawa (1975) reported that it is absent from the Río Peneya, north of the Río Caquetá, and during further surveys was unable to confirm its presence anywhere else north of the Río Caquetá, although in a later publication (Izawa 1979) indicated that it may occur on the Río Orteguaza, a northern tributary of the uppermost reaches. The easternmost localities in Brazil include the Paraná do Aranapu and Paraná do Jarauá on the lower Rio Japurá (Ayres 1986), Tefé and Lago Ipixuna, on the south bank of the Río Solimões (Hershkovitz 1977) and specimens in the Zoological Collection of the São Paulo Museum, São Paulo, from Pauiní, left bank of the Rio Purus, this latter confirming the supposition of Hershkovitz (1977). Surveys on the Rio Bauana, lower Rio Tefé, in 1980, indicated that it is scarce if not absent from the region (Rylands and Clutton-Brock, unpublished data) and, likewise, a study of the primate community at Porto da Castanha on the Lago de Tefé nearby by Johns (1986) also failed to confirm its presence. This, with the lack of specimens from the area between the Rios Juruá and Purus, argue that its distribution may not be uniform through these parts.

**Status** This species is generally considered common (Mittermeier *et al.* 1978; Coimbra-Filho 1984; Hernandez-Camacho and Defler 1985, 1989; Rylands and Mittermeier 1982; Soini 1982*a*; Soini *et al.* 1989). Its small size means that it is not hunted for food and the principal reason for its inclusion on the CITES Appendix I in 1977–1979 (now relegated to Appendix II; see Mack and Mittermeier 1984) was the international trade, particularly from the area of Iquitos, Colombia. It is largely restricted to river-edge forest (Soini 1988; this volume) and therefore its abundance in a given locality is dependent on the availability of suitable habitat. A number of authors, however, have indicated its occurrence in secondary forest (Moynihan 1976*a*; Hernandez-Camacho and Cooper 1976). It may be particularly abundant in areas moderately affected by agricultural activities and hunting (Soini 1982*a*, 1988) and is capable of existing in isolated forest patches near human settlements (Hernandez-Camacho and Defler 1985).

The following conservation units are within its known geographical distribution:

Bolivia
- Manuripi Heath Nature Reserve (1 884 375 ha)

Brazil
- Serra do Divisor National Park (605 000 ha) AC
- Abufarí Biological Reserve (288 000 ha) (left bank of Rio Purus) AM
- Rio Acre Ecological Station (72 000 ha) AC
- Juamí-Japurá Ecological Station (745 830 ha) AM
- Jutaí-Solimões Ecological Reserve (284 285 ha) AM
- Mamirauá State Ecological Station (1 124 000 ha) AM

Colombia
- Amacayacu National Natural Park (293 500 ha)
- La Paya National Natural Park (422 000 ha)
- Cahuinari National Natural Park (575 500 ha)

Peru
- Manu National Park (1 532 806 ha)
- Tingo Maria National Park (18 000 ha)
- Pacaya-Samiria National Reserve (1 478 800 ha)
- Tambopata Natural Wildlife Reserve (5500 ha).

## Genus Callithrix Erxleben 1777

### ***Callithrix argentata argentata*** (Linnaeus 1771)

Silvery marmoset (Engl.), sagui-branco-de-cauda-preta (Br.) (*v*. Hershkovitz 1977).

**Type locality** Pará, Brazil, restricted by Carvalho (1965) to Cametá, left bank of lower Río Tocantins (*v*. Hershkovitz 1977).

**Distribution** *C. a. argentata* occurs south of the Rio Amazonas, in relatively flat, lowland forest, between the mouth of the Rio Tocantins in the east and the Rios Tapajós and Cuparí (an eastern tributary) in the west (Ferrari and Lopes Ferrari 1990*a*), extending south to the Rio Irirí as far as the lower Rio Curuá (Hershkovitz 1977). The southernmost record listed by Hershkovitz (1977) is the type locality, Maloca, upper Rio Curuá, of an individual with a 'blackish crown and grayish brown back' described by Thomas (1920) as *Hapale emiliae*, and illustrated by Cruz Lima (1945) as *Callithrix emiliae*. Cruz Lima (1945) pointed out that *C. emiliae* 'bears a closer resemblance to the form *melanura* of Mato Grosso than to the typical form of the marginal river zone of the Amazon'. Avila-Pires (1986) also argued for the validity of the form *Callithrix argentata emiliae* on the basis of specimens obtained further south, on the Rio Peixoto de Azevedo (see *C. a. melanura* and *C. a. emiliae* below). This restricts the range of *C. a. argentata* to the north of the mouth of the Rio Irirí, with the southern limits being somewhere between the Rios Cuparí and Irirí as indicated by Hershkovitz (1977; see also Martins *et al.* 1988).

**Status**

Endemic to the Brazilian Amazon

The species was found to be common near the mouth of the Rio Tapajós (Mittermeier, unpublished data: Mittermeier and Coimbra-Filho 1977). It occurs in *terra firme* primary forests and in extensive areas of secondary growth forest (Belterra and Fordlândia, east bank of the Rio Tapajós) and has also been observed in forest patches in Amazonian white-sand savanna at Alter do Chão, south of Santarém, Rio Tapajós. Otherwise nothing is known of the conservation status of this subspecies.

There are no conservation units within its geographical distribution.

## *Callithrix argentata leucippe* (Thomas 1922)

Golden-white bare-ear marmoset (Engl.) (*v.* Hershkovitz 1977), sagui-branco (Br).

**Type locality** Pimental, right bank of the Rio Tapajós, below mouth of Rio Jamanxim, Pará, Brazil (*v.* Hershkovitz 1977).

**Taxonomy and distribution** *C. a. leucippe*, regarded by Carvalho (1959*a*) as a subspecies of *C. chrysoleuca*, was placed as a subspecies of *C. argentata* by Hershkovitz (1966*a*). It is confined to the small area between the Rios Cuparí and Tapajós, south of the Rio Amazonas, extending south to the Rio Jamanxim in the state of Pará (Hershkovitz 1977).

**Status**

1988 IUCN Red List of Threatened Animals—Vulnerable

1989 List of Brazilian Fauna Threatened with Extinction

Endemic to the Brazilian Amazon
This subspecies was first observed in the wild by Mittermeier in 1973, and is potentially threatened because of its extremely small distribution and the extensive destruction of the area cut by the Transamazônica (BR230) and Cuiabá–Santarém (BR163) highways (Mittermeier and Coimbra-Filho 1977: Mittermeier *et al.* 1978; Rylands and Mittermeier 1982). Vivo (1979, 1988) observed *C. a. leucippe* at Pimental during surveys in the Amazônia National Park. Vivo found it to be most common in degraded and secondary forest. The buffer zone of the Amazônia National Park, proposed in its management plan (Brazil, MA-IBDF and FBCN 1979), would include populations of this marmoset. Mittermeier (1973) suggested the creation of the Cupari Biological Reserve for its protection, but to date there are no conservation units within its geographical range.

## *Callithrix argentata melanura* (Humboldt 1812)

Black-tailed marmoset (Engl.), sauim (Br.), sagui (Br.), sagui-do-cerrado (Br.) (*v.* Vivo 1988).

**Type locality** Brazil, restricted to Cuyabá (=Cuiabá) by Allen (1916) (*v.* Hershkovitz 1977).

**Taxonomy and distribution** The most widespread of the *C. argentata* group, and the only *Callithrix* occurring naturally outside of Brazil, this southern subspecies extends from Brazil into Bolivia and Paraguay. Hershkovitz (1977) indicated the Rio Tacuarí in Brazil and the headwaters of the Río Mamoré in Bolivia as the southern limit of its distribution, but Stallings and Mittermeier (1983) and Stallings (1985) recorded it also from the northeastern Paraguayan *chaco*, extending its southern limit to approximately 20° S. In Bolivia it occurs east of the Río Mamoré, in the Departments of Beni and Santa Cruz (Brown and Rumiz 1986). According to Hershkovitz (1977), in Brazil it is confined to the east of the Rio Madeira, from the mouth of the Rio Aripuanã extending south to beyond the Rio Guaporé and west to the Rio Roosevelt (Hershkovitz 1977). It also occurs east of the Rio Aripuaná, north at least to 10° S (Rylands 1981), and probably west to the Rio Juruena, or the Rio Teles Pires, where Ávila-Pires (1986) predicted that it would meet the range of the form *C. a. emiliae*.

However, there is some confusion as to the identity of the marmoset to the east of the Rio Madeira, in the states of Amazonas and Rondônia. Vivo (1985, 1988) indicated that it was not the form *melanura*, but *Callithrix emiliae* (Vivo regarded all *Callithrix* taxa as species), according to its similarity with the marmosets obtained from the Rio Curuá (slightly darker, but otherwise conforming to the description of Thomas 1920) in Pará, here recognized as true *C. emiliae*. The *C. emiliae* of Vivo, however, occurs to the west of the distribution of *C. a. melanura* and, if aligned with *C. emiliae* of Thomas (1920), Cabrera (1957), and Ávila-Pires (1986), would indicate a disjunct distribution, being separated by typical *C. a. melanura* between the Rios Aripuanã and Juruena (or Teles Pires). The Rondônia marmosets, as described by Vivo (1985, 1988), are distinct from *melanura* in their paler colouration and the lack of the distinct pale thigh stripe, with a hip patch which is either indistinct or absent. Vivo (1985, 1988) ascribes the northernmost Foz do Rio Castanho specimen (near the junction of the Rios Roosevelt, Guariba, and Aripuanã in the state of Amazonas) of Hershkovitz (locality 197b, p.569, 1977) to *C. emiliae*, but makes no mention of two other localities indicated by Hershkovitz: 214b, mouth of the Rio Jiparaná, upper Rio Madeira; and 214c, Urupá, Rio Jiparaná. Vivo's implication is that these are also *C. emiliae*, and that the range of this form is between the left bank of the Rios Aripuanã–Roosevelt and the right bank of the Rio Madeira, extending south to the Rio Guaporé. It is almost certain that these Rondônia marmosets should be considered a new species or subspecies, similar to, but distinct from, both *emiliae* and *melanura*.

**Status** *C. a. melanura* has a widespread distribution and, although occurring in areas with considerable agricultural development, is not believed to be vulnerable or threatened.

A large part of the distribution of the form described by Vivo (1988) as *C. emiliae*, considered as a pale form of *C. a. melanura* by Hershkovitz (1977), is within the state of Rondônia which is suffering widespread forest destruction through the Polonoroeste Colonization Programme. Mahar (1989) demonstrated that 20.6 per cent (50 421 km$^2$) of the forests in the state of Rondônia were cleared between 1980 and 1988 (see also Lisboa *et al.* 1987). The uncertainty regarding the extent of the distribution of this form indicates that it should be considered vulnerable until information is available to the contrary. It may occur in the following conservation units:

Brazil
- Pacáas Novos National Park (?) (764 801 ha) RO
- Guaporé Biological Reserve (?) (600 000 ha) RO
- Jaru Biological Reserve (268 150 ha) RO
- Samuel State Park (20 865 ha) RO

*C. a. melanura* occurs in the following conservation units:

Bolivia
- Amboró National Park (?) (98 640 ha)

Brazil
- Pantanal Matogrossense National Park (135 000 ha)
- Chapada dos Guimarães National Park (33 000 ha)

Taiamá Ecological Station (14 325 ha)
Serra das Araras Ecological Station (28 700 ha)
Iquê Ecological Station (?) (200 000 ha)
Paraguay
Defensores del Chaco National Park (780 000 ha)

## ***Callithrix emiliae*** (Thomas 1920)

Snethlage's marmoset (Engl.), sagui (Br).

**Type locality** Maloca, upper Rio Curuá, upper Rio Irirí, Rio Xingú, Pará, Brazil.

**Taxonomy and distribution** This marmoset, named by Thomas (1920) as *Hapale emiliae*, was considered a dark form of *C. a. argentata* by Hershkovitz (1977). It was recognized by Cabrera (1957) and Hill (1957), however, as a distinct subspecies of *C. argentata*, and Ávila-Pires (1986) also argued its validity on the basis of three skins obtained from the Rio Peixoto de Azevedo, well to the south of the type locality. Cabrera (1957) describes its distribution as the south of the state of Pará, possibly entering contiguous parts of the state of Mato Grosso. Ávila-Pires (1986) was more exact, indicating that it occurs south from the Rio Irirí (*C. a. argentata* occurring to the north—confirmed by Martins *et al.* 1988), at least as far south as the southern (left) margin of the Rio Peixoto de Azevedo, an eastern tributary of the Rio Teles Pires. More recently, Martins *et al.* (1988) recorded its presence on the left bank of the Rio Irirí, south from its mouth, interestingly evidently sympatric with what they refer to as a *C. argentata* ssp. (undescribed), and therefore arguing for its specific status. The southern limits would evidently not be beyond the headwaters and upper Rio Paraguai, approximately 14° 30′ S, where *C. a. melanura* has been registered for a number of localities (Hershkovitz 1977; Vivo 1988). Ávila-Pires suggested that the Rio Teles Pires marks the western limit of its range. Martins *et al.* (1988) indicated that *C. emiliae* is limited to the west (left) bank of the lower Rio Irirí, with the undescribed *C. argentata* subspecies mentioned above being the only form occurring between the Rios Irirí and Xingú. These authors also indicated that no marmoset species occur east of the Rio Xingú above the mouth of the mouth of the Rio Irirí.

The distribution (and taxonomic status) of *C. emiliae* has been confused somewhat by its alignment with a similar, if slightly darker, form in the state of Rondônia by Vivo (1988). This is discussed under *C. a. melanura*, and we suggest that the two will need to be considered distinct because of widely separated distributions, with the Rondônia marmoset requiring a new name and, if valid, the southern Pará/northern Mato Grosso marmoset, first collected by Emilia Snethlage in 1914, retaining rights to the name *emiliae*. Assuming the two are distinct, then the validity of *C. emiliae* will need to reappraised on the basis of individuals from southern Pará alone.

**Status**

Endemic to the Brazilian Amazon

This marmoset if known from only two localities, but it evidently has a relatively wide distribution, and, although not protected in any conservation units, there is no reason to suspect that its status is threatened or vulnerable at the present time.

## *Callithrix nigriceps* Ferrari and Lopes 1992

The black-headed marmoset (Engl.), sagui-de-cabeça-preta (Br.) (*v*. Ferrari and Lopes 1992*b*).

**Type locality** Lago dos Reis (7° 31′S, 62° 52′W, = Lago Paraiso), 17 km east of Humaitá, Amazonas, Brazil, on the Trans-Amazon highway BR-230 (right or east bank of the Madeira River).

**Taxonomy and distribution** Ferrari and Lopes (1992*b*) align this marmoset with the bare-eared 'argentata' group, but they describe it as a full species following the taxonomy of Vivo (1988). It is darker than the form *emiliae*, decribed by Vivo (1985) from adjacent Rondônia, but differs in the pigmentation of the face and ears, pheomelanization of the forelimbs, mantle and ventrum, a brown rather than grey dorsum, an orange/russet colouration of the posterior limbs, and pale hips and upper thighs. Although known from only two localities separated by little more than 50 km, the paratype locality being Calama, (8° 03′ S, 62° 53′ W), Rondônia, Brazil, (right or east bank of Madeira River), east of Jiparaná river, this marmoset is believed to occur between the Rio dos Marmelos in the north and east, the Rio Madeira in the west and the Rio Jiparaná in the south, in the state of Rondônia, Brazil. Ferrari and Lopes (1992*b*) argued that it is unlikely to extend further west than the Rios Madeira and Jiparaná, nor east to the Rios Aripuanã and Roosevelt. The southeastern limits are defined by an area of savanna vegetation at the headwaters of the Rio dos Marmelos and along the middle Rio Jiparaná.

**Status**

Endemic to Brazil

Ferrari and Lopes (1992*b*) indicated that its natural range is little more than 10 000 km$^2$, one of the smallest of any Amazonian primate species, and potentially one of the most precarious. The area is currently undergoing rapid colonization, with access by asphaulted highway from Rondônia, and is traversed by the Trans-Amazon. Principle threats include widespread logging, goldmining, and cattle-ranching. The authors pointed out that while marmosets are well able to adapt to habitat disturbance in the short term, continued deforestation will eventually have deleterious effects on the population as a whole.

## ***Callithrix humeralifer humeralifer*** (E. Geoffroy 1812)

Santarém marmoset (Engl.), black-and-white tassel-ear marmoset (Engl.) (*v.* Hershkovitz 1977), sagui-de-Santarém (Br.) (*v.* Vivo 1988).

**Type locality** Brazil, restricted to Paricatuba, left bank of the Rio Tapajóz, near mouth, Pará, Brazil (*v.* Hershkovitz 1966a).

**Taxonomy and distribution** This marmoset is considered by Hershkovitz (1977) to be the nominal subspecies of three, *C. h. humeralifer*, *C. h. intermedius*, and *C. h. chrysoleuca*, but which Vivo (1988, 1991) considered valid species. Boer (1974) regarded *C. humeralifer* and *C. chrysoleucos* (sic.) as separate species on the basis of differing chromosome numbers; $2n = 44$ for *C. humeralifer* and $2n = 46$ for *C. chrysoleucos*. Following this, Coimbra-Filho (1990) placed them as separate species. However, Hershkovitz (1977, p.576) indicated that this was based on misidentification of '*C. chrysoleucos*' by Bender and Mettler (1960), responsible for the chromosome number cited by Boer (1974). Hershkovitz (1977, pp.598–599) argued, convincingly, that the individual karyotyped by Bender and Mettler was in fact either *C. jacchus* or *S. fuscicollis melanoleucus*. Although the authorship of the specific name *humeralifer* is usually ascribed to E. Geoffroy (1812), Vivo (1988) argued that Humboldt published the description of *Simia humeralifera* previously, although in the same year. Vivo (1988) also pointed out that *Callithrix*, being a feminine noun , demands a feminine adjective ending, thus he uses *humeralifera* instead of *humeralifer*.

According to current knowledge, it occurs south of the Rio Amazonas, between the Rio Tapajós in the east, south as far as Vila Braga, and west to the Rio Canumã (=Cunumã) (Hershkovitz 1977). The southern limit is not known, but it does not extend below 10° S between the Rios Aripuanã and Juruena, as was indicated by Hershkovitz (1977). The marmoset occurring between the Rios Aripuanã and Juruena is *C. a. melanura* (Rylands 1981) and the southern limits to the range of *C. humeralifer* are possibly in the region of the Serra do Sucundurí and the headwaters of such as the Rio Sucundurí, nearer to 8° S. However, the southernmost locality for the Santarém marmoset is Vila Braga, 4° 25′ S, and so the possibility remains that it extends no further south, and either *C. a. melanura* extends further north or there may be an as yet underscribed marmoset in the intervening area. As pointed out by Hershkovitz (1977), the interfluve between the lower Rios Tapajós and Madeira has an extraordinarily complex network of rivers and canals where further, probably subspecific, forms of *C. humeralifer* in isolated enclaves may well be discovered in the near future. A pale orange-brown marmoset, with typical *C. humeralifer* ear tufts, obtained from the Rio Arapiuns and in the collection of the Belém Primate Centre, would fit the category of a true *intermedius* (see below). Mittermeier *et al.* (1988*b*, p.20) provided a photograph of this undescribed form. There are also very dark-coloured marmosets in the region of Maués (M. Schwarz, personal communication). Both of these have yet to be described.

**Status**
1988 IUCN Red List of Threatened Animals—Vulnerable
1989 List of Brazilian Fauna Threatened with Extinction
Endemic to the Brazilian Amazon
*C. h. humeralifer* is considered to be vulnerable because of its proximity to a number of expanding urban centres as well as the mainstream of the Rio Amazonas, and the resulting forest destruction (Coimbra-Filho 1982, 1984; IUCN 1982*a*). Its distribution, smaller even than depicted by Hershkovitz (1977, p.569) is also cut by the Transamazônica highway (BR230). It occurs in the Amazônia National Park (994 000 ha), Pará, where population surveys were carried out by Ayres and Milton (1981) and Branch (1983).

## ***Callithrix humeralifer intermedius*** Hershkovitz 1977

Tassel-ear marmoset (Engl.) (*v*. Hershkovitz 1977), sagui (Br.) (*v*. Vivo 1988).

**Holotype locality** Near mouth of Rio Guariba, left bank of Rio Aripuanã, southeastern Amazonas, Brazil (*v*. Hershkovitz 1977).

**Taxonomy and distribution** This marmoset, first described by Hershkovitz (1977), has to date been considered an intermediate colour form between the pale *C. humeralifer chrysoleuca* and the dark *C. h. humeralifer*, hence its classification as *C. h. intermedius*. Vivo (1988) and Coimbra-Filho (1990) upgraded it to specific status, *C. intermedia* (the latter simply as a result of giving *C. h. humeralifer* and *C. h. chrysoleuca* species status). However, following Ávila-Pires (1986), two features indicate that it is quite clearly a form more closely aligned to *C. argentata*, and we argue that it be more correctly referred to as *C. argentata intermedia*. It is notably similar to *C. a. melanura* in such aspects as the distinct pale thigh stripe, similarly coloured hindquarters, dark crown (just a little paler than *C. a. melanura*), and the lack of an ear-tuft (it has a rudimentary tuft from only behind the pinna and not the well-developed tuft from within and around the pinna as in *C. humeralifer*).The face is variably depigmented (some individuals have quite dark-greyish faces), the forequarters are paler, and varying parts of the tail are pale rather than black, when compared to *C. a. melanura*. Some *C. a. melanura* individuals have considerable depigmentation around the nose and mouth. Apart from the entirely black tail, the black-and-white photograph illustrating *C. a. melanius* in Hershkovitz (1975, p.159; 1977, p.571) could easily be misconstrued for *C. h. intermedius* (including the slight ear-tufts). In fact the '*C. a. melanius*' illustrated is considerably paler than typical *C. a. melanura*, with pelage of the lower dorsal region contrasting with the black tail. J.B. Carroll (personal communication) pointed out that in *C. a. melanura* the dark pelage of the lower dorsum is continuous with the tail. The second feature is that Hershkovitz (1977) indicated sympatry between *C. h. intermedius* and *C. a. melanura* between the Rios Aripuanã and Roosevelt and that *C. h. humeralifer* occurred on the right bank of the Rio Aripuaná, east to the Rio Juruena. As pointed out by Rylands (1981), *C. h. intermedius* is alone between the Rios Aripuanã and Roosevelt, taking in the entire basin of the Rio Guariba, and *C. a. melanura*, not *C. h. humeralifer*, occurs on the opposite side (right bank) of the Rio Aripuanã. This means that the geographic distribution of *C. h. intermedius* is entirely surrounded by *C. a. melanura* and, west of the Rio Roosevelt, by what we consider likely to be a new *C. argentata* form (the Rondônia marmoset described as *C. emiliae* by Vivo 1985, 1988, see under *C. a. melanura*).

**Status**

1988 IUCN Red List of Threatened Animals—Vulnerable
1989 List of Brazilian Fauna Threatened with Extinction
Endemic to the Brazilian Amazon
The geographical distribution of *C. h. intermedius* is within one of the principle areas of forest destruction, the north of the state of Mato Grosso, which has been a cause for international concern since the late 1980s. The rate of destruction has increased dramatically since the late 1970s, and is marked especially by the construction of highways connecting such towns as Vilhena, Aripuanã, Fontanilhas, and Humaitá, and by rapid and large-scale colonization (the Polonoroeste Colonization Programme), cattle ranching, and gold mining. By 1988, 208 000 km$^2$ or 23.6 per cent of the forests of the state of Mato Grosso had been cleared (Mahar 1989). 154 700 km$^2$ were cleared just between 1980 and 1988, principally in the north of the state. It competes with *C. a. melanura* as the candidate for the marmoset occurring in the Iquê Ecological Station (200 000 ha), Mato Grosso, in the far south of its range (headwaters of the Rio Aripuanã). This marmoset, with its restricted distribution, should be considered vulnerable.

## ***Callithrix humeralifer chrysoleuca*** (Wagner 1842)

Golden-white tassel-ear marmoset (Engl.) (*v*. Hershkovitz 1977), sagui-claro-de-Santarém (Br.) (*v*. Vivo 1988).

**Type locality** Borba, lower Rio Madeira, Amazonas, Brazil (*v*. Hershkovitz 1977).

**Taxonomy and distribution** As mentioned for *C. h. humeralifer*, Boer (1974) considered this form a valid species on the basis of differing diploid chromosome numbers. This scheme was recognized by Vivo (1988, 1991) and Coimbra-Filho (1990), although Hershkovitz (1977) argued that the '*C. chrysoleucos*' specimen was misidentified.

*C. h. chrysoleuca* occurs south of the Rio Amazonas, between the Rios Madeira and Aripuanã in the west and the Rio Canumã (=Cunumã) in the east (Hershkovitz 1977). The southernmost locality is Prainha, a short distance north of the mouth of the Rio Roosevelt, on the east (right) bank of the Rio Aripuanã. It is probable that Prainha is near to the southern limit to its distribution, the area south of this being occupied by *C. a. melanura* or some other as yet undescribed marmoset.

**Status**
1988 IUCN Red List of Threatened Animals—Vulnerable
1989 List of Brazilian Fauna Threatened with Extinction
Endemic to the Brazilian Amazon
This species has a very small and narrow distribution in an area undergoing considerable deforestation and agricultural development and should be considered vulnerable, if not already endangered. It does not occur in any conservation units.

## *Callithrix jacchus* (Linnaeus 1758)

Common marmoset (Engl.), sagui-do-nordeste (Br.), sagui-comum (*v*. Cabrera 1957; Ávila Pires 1969)

**Type locality** America, restricted to Pernambuco, Brazil, by Thomas (1911) (*v*. Hershkovitz 1977).

**Distribution** This species occurs in the scrub forest (forest patches in dry *caatinga* thorn scrub) and Atlantic forest of the north-east of Brazil, in the states of Alagoas, Pernambuco, Paraíba, Rio Grande do Norte, Ceará, and Piauí, originally extending south as far as the Rios Grande and São Francisco. Hershkovitz (1977) indicated that it also probably extends north-west into the state of Maranhão, to the left bank of the Rio Parnaiba and the Serra do Valentim (Hershkovitz 1977). The western limits are not clearly defined. Hershkovitz (1977) extends the distribution no further west than the middle reaches of the Rio Grande (left bank) and the upper Rio Parnaiba (right bank), with a lacuna between these points and the Rio Tocantins. It has spread into various other regions as a result of introductions, south of the Rio São Francisco, accompanying the destruction and degradation of the natural ecosystems, particularly the Atlantic coastal forest. Introduced and recent populations include those in the state of Sergipe and the north and north-east of Bahia, including the 'Recóncavo da Bahia', as well as in the state of Rio de Janeiro, in south-east Brazil (Coimbra-Filho 1984). Mittermeier (personal observation in 1971) observed this species within the city of Rio de Janeiro, and they are also reported to have established themselves in Buenos Aires. Alonso *et al.* (1987) indicated that the Recóncavo da Bahia shows a relatively narrow zone of mixing between *C. penicillata* and *C. jacchus*. Coimbra-Filho *et al.* (1991), however, showed that this region was originally forested, and argued that the destruction of the natural vegetation over vast areas since the European discovery of Brazil in 1500, along with frequent and repeated introductions, certainly of *C. jacchus* but probably also of *C. penicillata*, has resulted in a confused picture of hybrids between these species and probably between *C. penicillata* and *C. kuhli* (see Coimbra-Filho *et al.* this volume). They argued that pure *C. kuhli* was the orginal form occurring there. Mittermeier *et al.* (1988*b*, p.19) show an as yet undescribed form from the still partially forested area of Valença which has characters intermediate between *C. kuhli* and *C. penicillata* (see under *C. kuhli*).

**Status**

Endemic to Brazil

Although widespread and common in many localities, and even replacing other *Callithrix* species where it has been introduced, *C. jacchus* populations are severely depleted and declining, through habitat destruction, in many parts of their original geographic distribution (Mittermeier *et al.* 1981; Coimbra-Filho 1984). The Tijuca National Park (3200 ha), state of Rio de Janeiro, contains an introduced population of *C. jacchus*. The following conservation units are within the species' geographical range (* indicates possibly introduced and mixed populations of *C. jacchus* and *C. penicillata*):

Brazil
- Sete Cidades National Park (6221 ha) PI
- Serra da Capivara National Park (97 933 ha) PI
- Ubajara National Park (563 ha) CE
- Serra Negra Biological Reserve (1100 ha) PE
- Saltinho Biological Reserve (548 ha) PE
- Pedra Talhada Biological Reserve (4469 ha) AL
- Guariba Biological Reserve (4321 ha) PB
- Mamanguape Ecological Station (9992 ha) PB
- Seridó Ecological Station (1116 ha) RN
- Itabaiana Ecological Station (1100 ha)* SE
- Uruçuí-Una Ecological Station (135 000 ha) PI
- Aiuaba Ecological Station (11 525 ha) CE

Foz do São Francisco Ecological Station (5322 ha) AL
Raso da Catarina Ecological Reserve (99 772 ha)* BA
Ponta do Cabo Branco State Park (379 ha) PB
Guaramiranga State Park (55 ha) CE
Dunas Costeiras State Park (1160 ha) RN
Buraquinho State Biological Reserve (471 ha) PB
Tapacurá State Ecological Station (392 ha) PE

## ***Callithrix penicillata*** (E. Geoffroyi 1812)

Black-tufted-ear marmoset (Engl.) (*v*. Hershkovitz 1977), sagui-de-tufos-pretos (Br.) (*v*. Mittermeier *et al.* 1988*b*), mico-estrela (Br.).

**Type locality** Brazil, restricted to Lamarão, Bahia, by Thomas (1904) (*v*. Ávila-Pires 1969).

**Taxonomy and distribution** Although it is usually credited to E. Geoffroy (1812), Vivo (1988) argued that Humboldt (1812) had prior authorship of this species. *C. penicillata* we recognize here was referred to as *C. p. jordani* Thomas (1904) by Cabrera (1957), Napier and Napier (1967), Ávila-Pires (1969, 1986), Coimbra-Filho (1971) and Coimbra-Filho and Mittermeier (1973*b*). According to Ávila Pires (1969, 1986), the type locality for *C. p. jordani* is Araguarí, Rio Jordão, 800 m altitude, Minas Gerais, Brazil. The marmoset of the type locality given by Hershkovitz (1977) for *C. jacchus penicillata* (above) was regarded by Ávila-Pires (1969) as belonging to *C. p. penicillata* (Humboldt 1812), which is a coastal forest species, the distribution of which, according to Ávila-Pires (1969), extended from the Recóncavo da Bahia, south to Santa Tereza, Espírito Santo. His inclusion of the state of Espírito Santo probably resulted from the mistaken identity of a hybrid of *C. geoffroyi* × *C. flaviceps* (Hershkovitz 1977; Mendes, personal communication), which are the only forms occurring naturally in the state of Espírito Santo. The *C. p. penicillata* of Ávila-Pires (1969, 1986) is considered by us to be a distinct species, *C. kuhli* (Atlantic coastal forest of southern Bahia). Further taxonomic studies are required to clarify the confused situation regarding the black-tufted-ear marmosets, and we predict that a better knowledge of the variation (at least in pelage) of this form over its wide distribution may yet reveal further distinct geographic races, in the north of the state of Goiás and Tocantins, for example.

*C. penicillata* has a very wide distribution, occurring in the *cerrado* region of east central Brazil, in the states of Bahia, Minas Gerais, Goiás, the south-west tip of Piauí, Maranhão and the north of São Paulo, north of the Rios Tieté and Piracicaba (Hershkovitz 1977). In the north, it is restricted to the south of the Rio Grande and Rio São Francisco (however, see *C. jacchus* for a discussion of the situation in northern Bahia, south of the Rio

São Francisco). Vivo (1988) identified two skins in the Museu Nacional, Rio de Janeiro, from the north-east coast of Maranhão, at Miritiba (now called Humberto dos Campos), which may indicate that *C. penicillata* extends its range right through eastern Maranhão, along the left bank of the Rio Parnaiba. These were not located by Hershkovitz (1977) and presumed by him to be *C. jacchus*, following Ávila-Pires (1969). The large gap between the previous northernmost locality (Canabrava, Rio Tocantins, Goiás, locality 275a of Hershkovitz 1977, p.490) and this northern Maranhão locality, indicates that these are probably introduced animals.

Recent surveys in the north of the state of Minas Gerais showed that *C. penicillata* extends its range through the region left blank by Hershkovitz (1977, p.490), between the upper Rio São Francisco and the Rio Jequitinhonha. The species occurs both sides of the Rio Jequitinhonha as far east as the Rio Araçuaí, beyond which it is restricted to the north of the river, with *C. geoffroyi* occurring to the south (Rylands *et al.* 1988), result of a recent introduction (c. 1975) in the vicinity of Belmonte (Coimbra-Filho, unpublished data). *C. penicillata* is typically of the *cerrado* region of Minas Gerais (in the central, south-west, west, and north of the state). Those parts originally covered by Atlantic coastal forest in the east and south-east (the *Zona da Mata*) are the domain of *C. geoffroyi*, *C. flaviceps*, and in part of the Rio Dôce valley, *C. aurita*. However, with the destruction of the forest and also resulting from introductions (misguided release of confiscated animals), *C. penicillata* is taking a hold and probably replacing the other species in numerous localities east and south (states of Espírito Santo, Minas Gerais and São Paulo) of its original range. This is happening in the Rio Dôce State Park (35 973 ha), and is possibly also the case of *C. penicillata* reported from the Itatiaia National Park (30 000 ha, on the border of the states of Rio de Janeiro and Minas Gerais) by Vivo (1991; see also Coimbra-Filho 1984). In both cases *C. aurita* is the species naturally occurring in the area. The sympatry indicated by Vivo (1988) is probably recent, unnatural, and transient (sympatry between any *Callithrix* is considered unlikely, *v.* Coimbra-Filho 1971; Rylands and Faria this volume). Populations in the Recóncavo da Bahia, north of the city of Salvador, were introduced (see *C. jacchus*).

We do not agree with Hershkovitz (1975, 1977) and Vivo (1991) regarding their assertion that the marmosets occurring in the Atlantic forest of southern Bahia are hybrids of *C. penicillata* and *C. geoffroyi* (Hershkovitz) or indistinguishable from *C. penicillata* (Vivo). They are very different animals in terms of their morphology, pelage colour patterns, and ecology (Coimbra-Filho 1984; Mittermeier *et al.* 1988*b*; Santos *et al.* 1987; Stevenson and Rylands 1988; Natori 1990; Coimbra-Filho *et al.* this volume; Rylands and Faria this volume;) and are here considered as a distinct species, *C. kuhli*.

**Status**

Endemic to Brazil

As with *C. jacchus*, although widespread and hardy, and able to survive in extremely degraded habitats, populations of this species have disappeared or are declining in many parts of its range. It has been introduced into part of the Rio Dôce State Park (35 973 ha), the Ibitipoca State Reserve (1448 ha), both in the state of Minas Gerais (Mittermeier and Rylands, personal observations), and the Ilha Grande State Park (56 000 ha), Rio de Janeiro (H.K.M. Corrêa, personal communication). The following conservation units are within its geographical distribution (* indicates possibly introduced and/or mixed populations of *C. jacchus* and *C. penicillata*):

Brazil
- Brasília National Park (28 000 ha) DF
- Emas National Park (131 868 ha) GO
- Chapada dos Veadeiros National Park (60 000 ha) GO
- Serra da Canastra National Park (71 525 ha) MG
- Serra do Cipó National Park (33 800 ha) MG
- Araguaia National Park (?) (562 312 ha) TO
- Grande Sertão Veredas National Park (84 000 ha) MG
- Chapada da Diamantina National Park (152 000 ha) BA
- Pirapitinga Ecological Station (1090 ha) MG
- Raso da Catarina Ecological Reserve (99 772 ha)* BA
- Ibitipoca State Park (1489 ha) MG
- Acauã State Reserve (5000 ha) MG

## ***Callithrix kuhli*** (Wied-Neuwied 1826)

Wied's marmoset (Engl.), sagui-de-Wied (Br.), mico-estrela (Br.) (*v.* Coimbra-Filho 1984, 1985; Santos *et al.* 1987).

**Type locality** Near mouth of Rio Belmonte (= Rio Jequitinhonha) Bahia, Brazil (*v.* Hershkovitz 1977).

**Taxonomy and distribution** Although the specific named was spelt *kuhlii* by Wied-Neuwied, it refers to H. Kuhl and the correct specific name should have only one 'i'. Vivo (1988, 1991) argued that individual and clinal variation in pelage colour of *C. penicillata* invalidate any separation of these coastal forest marmosets of southern Bahia. Hershkovitz (1977) argued that it is a hybrid of *C. penicillata* and *C. geoffroyi*. Experimental hybrids reared at the Rio de Janeiro Primate Centre (CPRJ) failed to produce any forms similar to *C. kuhli* (Coimbra-Filho *et al.* this volume), and have demonstrated that it has a dominant phenotype (Coimbra-Filho 1984; Coimbra-Filho *et al.* this volume). As pointed out by Rosenberger

(1984), the *kuhli* phenotype does not show the intermediate characters of a *C. geoffroyi* × *C. penicillata* hybrid, and is closer in appearance to the former species than the latter. The phenotype is also constant over all parts of its currently known range where it has been observed, extending about 200 km between the Rio de Contas (in the north) and the Rio Jequitinhonha (in the south) in southern Bahia, and is distinct in its pelage colouration from *C. penicillata* in both infants and adults (Coimbra-Filho 1985). Natori (1990) studied the dental morphology (postcanine) of *C. kuhli*, *C. penicillata*, and *C. geoffroyi* and, on these grounds, also argued that *C. kuhli* is not a hybrid but a distinct species.

As mentioned above, *C. kuhli* occurs between the Rio de Contas and Rio Jequitinhonha in southern Bahia, just entering the northeasternmost tip of the state of Minas Gerais (Santos *et al*, 1987; Rylands *et al.* 1988). The western boundary is not well known, but undoubtedly defined by the inland limits of the Atlantic coastal forest (Rylands *et al.* in press). I.B. Santos (in Rylands *et al.* 1988) observed hybrids between *C. penicillata* and *C. kuhli* in the region of Almenara, Minas Gerais, left bank of the Rio Jequitinhonha (16° 41′ S, 40° 51′ W).

Surveys in 1986/1987 by Oliver and Santos (1991) demonstrated the presence of forms intermediate in appearance between *C. kuhli* and *C. penicillata* north from the Rio de Contas, along the coast up to the regions of Valença and Nazaré, just south of the city of Salvador (Mittermeier *et al.* 1988*b*). Individuals observed by Rylands near to Nazaré, just south of the city of Salvador (unpublished data), lacked the white frontal blaze, and, although retaining the pale cheek patches typical of *kuhli*, were uniformly paler in colouration. A photograph of the marmoset from Valença, Bahia, north of the Rio de Contas, is provided in Mittermeier *et al.* (1988*b*, p.19). The variation in pelage colouration of the marmosets in this region is considerable, but Coimbra-Filho *et al.* (1991), showed that true *C. kuhli* extended north through coastal Bahia into the state of Sergipe as far as the Rio São Francisco in the recent past. The present-day confusion has arisen from the widespread forest destruction, most marked and nearly total in Sergipe (Sick and Teixeira 1979), and the introductions and invasions of *C. jacchus* and *C. penicillata.*

Ávila-Pires (1969) included this region in his description of the distribution of *C. p. penicillata*, along with southern Bahia (and further south to Santa Tereza, Espírito Santo, more correctly the domain of *C. geoffroyi* and *C. flaviceps*, see below).

**Status**

Endemic to the Atlantic coastal forest of Brazil

This species is considered the most abundant and adaptable of the southern Bahian primates, but is threatened by widespread forest destruction

(Coimbra-Filho 1984; Santos *et al.* 1987; Coimbra-Filho *et al.* 1991; see also Rylands *et al.* in press). Its abundance in relation to the sympatric golden-headed lion tamarin was demonstrated by the large numbers killed (1829 and just 14 *Leontopithecus chrysomelas*) for sylvan yellow fever research in the region of Itabuna and Buerarema in southern Bahia (Laemmert *et al.* 1946; see also Rylands 1989*b*). It is being bred with success in the Rio de Janeiro Primate Centre (CPRJ) (Coimbra-Filho *et al.* 1989). Mittermeier *et al.* (1981) recommended that it should be considered vulnerable. It occurs in the Una Biological Reserve, Bahia, which although decreed in 1980 with 11 400 ha, has serious problems of squatters and deforestation and only approximately 5500 ha is currently owned by the Brazilian Institute for the Environment and Renewable Natural Resources (Ibama) responsible for its administration. The Una Reserve is receiving attention, however, from Brazilian and international conservation organizations, particularly because of its importance for the protection of *L. chrysomelas*. Oliver and Santos (1991) did not regard it as seriously threatened and recommended that its IUCN status should be 'rare'. It occurs in the following protected areas:

Una Biological Reserve (6250 ha) BA
Lemos Maia Experimental Station (495 ha) BA
Canavieiras Experimental Station (500 ha) BA

## ***Callithrix geoffroyi*** (Humboldt 1812)

Geoffroy's tufted-ear marmoset (Engl.) (*v.* Hershkovitz 1977), sagui-caratinga (Br.), sagui-de-cara-branca (Br.) (*v.* Coimbra-Filho 1986*c*).

**Type locality** Brazil, restricted to near Victoria, 'between the Rios Espírito Santo and Jucu' by Cabrera (1957) who attributes the restriction of the type locality to Wied (1826). Hershkovitz (1977) notes that the names of the two rivers cited are synonyms.

**Distribution** This species occurs in the state of Espírito Santo and the forested eastern and northeastern part of Minas Gerais, north as far as the Rios Jequitinhonha and Araçuaí and south to near the state border of Espírito Santo and Rio de Janeiro (Ávila-Pires 1969; Hershkovitz 1977; Coimbra-Filho 1984; Rylands *et al.* 1988). The populations just south of the Rio Jequitinhonha resulted from animals released near its mouth, at Belmonte, around 1975 (Coimbra-Filho 1986*c*). From there it spread eastward, and today also occurs in gallery forests throughout the region of dry thorn scrub (*caatinga*) of the middle reaches of the river (Rylands *et al.* 1988). Vivo (1988) limits it to the east of the Serra do Espinhaço in Minas Gerais. Hybrid populations of *C. penicillata* and *C. geoffroyi* have been observed in some parts of this mountain range, for example the Serra da Piedade (I.B. Santos and C.M.C. Valle, personal communication) and in the municipality of Santa Barbara (Rylands personal observation), where the Atlantic coastal forest gives way to the *cerrado* (see Coimbra-Filho *et al.* this volume). The range of *C. geoffroyi* overlaps with *C. flaviceps* (see below) in southern Espírito Santo (south of the Rio Dôce) and south-east Minas Gerais. *C. geoffroyi*, however, is generally restricted to lowland areas, below 500–700 m, and *C. flaviceps* to altitudes above 400–500 m (Coimbra-Filho 1971; Coimbra-Filho *et al.* 1981*a*). Hershkovitz (1977) points out that the highest recorded locality for *C. geoffroyi* is Santa Teresa, 659 m above sea level, but Mendes (personal communication) has recorded mixed bands of *C. geoffroyi* and *C. flaviceps* at altitudes of 800 m. Hybrid populations have been recorded for intermediate elevations (Mendes 1989).

**Status**

Endemic to the Atlantic forest of south-east Brazil

Populations of this species are declining, because of the widespread destruction of the Atlantic forest in the states of Minas Gerais (less than 6.8 per cent of the original area of Atlantic forest; Fonseca 1985) and Espírito Santo (approximately 13 per cent of the original forest cover remaining; Mittermeier *et al.* 1981; Mendes 1991*a*). Mittermeier *et al.* (1981) and Coimbra-Filho (1986*c*) recommended that it be considered endangered. Oliver and Santos (1991) found it be patchily distributed but locally abundant, and concluded that this species is not seriously threatened at the present time and should be considered 'rare'. Mendes (1991*a*) reported on the occurrence of this species in the protected areas of Espírito Santo. Its occurrence in the Monte Pascoal National Park is in some doubt, and Oliver and Santos (1991) reported that *C. jacchus* had possibly been introduced there. The following conservation units protect populations of this species:

Brazil

- Serra do Cipó National Park (33 800 ha) MG
- Monte Pascoal National Park (14 000 ha) BA?
- Córrego Grande Biological Reserve (1504 ha) ES
- Córrego do Veado Biological Reserve (2392 ha) ES
- Sooretama Biological Reserve (24 000 ha) ES
- Comboios Biological Reserve (833 ha) ES
- Duas Bocas State Reserve (2910 ha) ES
- Linhares Forest Reserve (21 787 ha) ES
- Porto Seguro Forest Reserve (6069 ha) BA
- Santa Lucia Biological Station (350 ha) ES
- Pau Brasil Experimental Station (900 ha) BA
- Gregório Bondar Experimental Station (710 ha) BA

Goitacazes Reserve Forest (1400 ha) ES
Fazenda Córrego de Areia Reserve (60 ha) (privately owned) MG
Fazenda São Joaquim (Klabin Reserve) (1505 ha) ES

## *Callithrix aurita* (Humboldt 1812)

Buffy-tufted-ear marmoset (Engl.) (Hershkovitz 1977), sagui-da-serra-escuro (Br.) (Coimbra-Filho 1986*b*).

**Type locality** Brazil, restricted to the vicinity of Rio de Janeiro, Guanabara, by Vieira (1944) (*v*. Hershkovitz 1977).

**Taxonomy and distribution** Coimbra-Filho (1986*a*, *b*, 1990, 1991) has argued that the similarity in dental morphology (Natori 1986), behaviour and pelage (infants of the two forms are practically identical in appearance), and vocalizations (Mendes, personal communication) and the recent discovery of wild groups composed entirely of hybrids (Ferrari and Mendes 1991, Mendes in prep.) of the southernmost marmosets, *C. aurita* and *C. flaviceps*, argue for their subspecific status.

*C. aurita* occurs in the montane rain forests of south-east Brazil, in the southern part of the state of Minas Gerais, the state of Rio de Janeiro, and the east and north-east of the state of São Paulo. Vieira (1944) was mistaken in restricting the type locality to the lowland, former state of Guanabara. In the present-day state of Rio de Janeiro it occurs (or occurred) only in mountainous regions such as the Serra dos Orgãos. Hershkovitz (1977) marks the northern limit in Minas Gerais as the Rio Muriaé. However, it has also been recorded for the Rio Dôce State Park in Minas Gerais (Mittermeier *et al.* 1982), and at Carangola in the Serra do Brigadeiro, Minas Gerais (Ferrari and Mendes 1991) to the north of the river. In the Rio Dôce valley it occurs in scattered localities forming a mosaic with *C. flaviceps*. The southeasternmost locality is the Rio Ribeira de Iguapé in São Paulo. From there it extends west between the upper reaches of the Rios Tieté/Piracicaba and Paranapanema. It has also been recorded north of Piracicaba at the following sites: Mogi-Guaçú (Rio Mogi-Guaçú) by Mittermeier (unpublished data) and Muskin (1984*a*); Alfenas, upper Rio Grande in Minas Gerais (Hershkovitz 1977; Muskin 1984*a*); Vargem Grande, São Paulo (Muskin 1984*a*); and Fazenda Monte Alegre, Monte Belo, Minas Gerais (Muskin 1984*a*). The westernmost locality shown by Hershkovitz (1977, p.490) is Boracéia, north-east of Bauru, on the upper Rio Tieté (22° 10′ S, 48° 45′ W).

**Status**
1988 IUCN Red List of Threatened Animals—Endangered
1989 List of Brazilian Fauna Threatened with Extinction

US Endangered Species List
CITES Appendix I
Endemic to the Atlantic forest of south-east Brazil
The apparent rarity of *C. aurita* and the widespread destruction of the forests within its range, especially, for example, along the valley of the Rio Paraiba, are reasons for its endangered status (Coimbra-Filho 1986*b*). It is nowhere common, and populations in the National Parks of Bocaina and Serra dos Orgãos are minimal (Coimbra-Filho 1984). Nine months of surveys in the Carlos Botelho State Park in the Serra do Paranapiacaba, in the south of its range, failed to provide any evidence for populations of this marmoset (Paccagnella 1986; see also Torres de Assumpção 1983*a*). Although recorded as rare in the Itatiaia National Park by Ávila-Pires and Gouveia (1977), Mittermeier *et al.* (1981) reported it extinct there, although it may still occur in very reduced numbers (Coimbra-Filho 1986*b*). It is extremely rare in the private reserve of the Fazenda Barreiro Rico (3259 ha of forest), São Paulo. Milton and Lucca (1984) estimated no more than 8–12 animals in the entire area. Torres de Assumpção (1983*b*) estimated a density of 15 individuals/km$^2$. São Paulo has an extensive system of state parks, reserves, and ecological stations. The following conservation units are within its geographical distribution:

Brazil
Serra da Bocaina National Park (120 000 ha) RJ/SP
Serra dos Orgãos National Park (11 000 ha) RJ?
Itatiaia National Park (309 000 ha) MG/RJ?
Piraí Ecological Station (4000 ha) RJ?
Rio Dôce State Park (35 793 ha) MG
Serra do Mar State Park (30 938 ha) including Núcleo Cunha (6451 ha) SP
Cantareira State Park (7000 ha) SP
Serra do Brigadeiro State Park (32 500 ha) MG
Campos do Jordão State Park (8172 ha) SP?
Desengano State Park (22 400 ha) RJ?
Pedra Branca State Park (12 500 ha) RJ?
Mogi-Guaçu State Biological Reserve (469 ha) SP
Serra de Paranapiacaba State Biological Reserve (336 ha) SP?
Bananal State Ecological Station (884 ha) SP?
Itapeti State Ecological Station (89 ha) SP?
Mogi-Guaçú State Ecological Station (981 ha) SP
Bauru Ecological Station (288 ha) SP?
Juréia-Itatins Ecological Station (8200 ha) SP?
Valinhos Ecological Station (17 ha) SP?
Fazenda Monte Alegre Reserve (*c*.150 ha) (privately owned) MG
Fazenda Barreiro Rico Reserve (3259 ha) (privately owned) SP

## ***Callithrix flaviceps*** (Thomas 1903)

Buffy-headed marmoset (*v.* Hershkovitz 1977), sagui-da-serra (Br.) (*v.* Coimbra-Filho 1982, 1986*a*).

**Type locality** Engenheiro Reeve (now Rive), municipality of Alegre, southwestern Espírito Santo, eastern Brazil, altitude 500 m. (*v.* Hershkovitz 1977).

**Distribution** As discussed for *C. aurita*, Coimbra-Filho (1986*a*,*b*, 1990) argues that *C. flaviceps* is a subspecies of *C. aurita*. The distribution of *C. flaviceps* is described by Hershkovitz (1977), Coimbra-Filho *et al.* (1981*a*), and Coimbra-Filho (1986*a*). It occurs in the highlands of southern Espírito Santo, south of the Rio Dôce, and, although as yet there is no proof, probably—at least in the past when suitable forest was still available—in adjacent parts of the north of the state of Rio de Janeiro, in the municipalities of Natividade, Porciuncula and the north of Bom Jesus do Itabapoãna. It extends west into Minas Gerais in scattered localities as far as Manhuaçú 40° 02′ W, as noted by Coimbra-Filho (1972, 1986*a*) and Coimbra-Filho *et al.* (1981*a*).

**Status**
1988 IUCN Red List of Threatened Animals—Endangered
1989 List of Brazilian Fauna Threatened with Extinction
US Endangered Species List
CITES Appendix I
Endemic to the Atlantic forest of south-east Brazil
*C. flaviceps* has a scattered distribution, being restricted to upland areas, altitudes above 400 m, with *C. geoffroyi* occurring in more lowland areas. Most of the localities known for this animal are disjunct and small (Coimbra-Filho 1972), and probably contain minimal populations. An exception is the Augusto Ruschi (Nova Lombardia) Biological Reserve, where Mittermeier *et al.* (1981) found it to be abundant, although one group observed at the edge of the Reserve was greyish rather than yellowish, and probably a hybrid. Mendes (1989, 1991*a*; Ferrari and Mendes 1991) and Oliver and Santos (1991) reported its occurrence in the Pedra Azul and Forno Grande State Reserves in Espírito Santo, and the Biological Stations of Santa Lucia and São Lourençõ. Although the area is well within its distribution, no evidence has been obtained for its presence in the Caparaó National Park in Minas Gerais and Espírito Santo (Ferrari and Mendes 1991; Oliver and Santos 1991). It is considered to be in danger of extinction because of widespread forest destruction throughout it range (Mittermeier *et al.* 1981; Coimbra-Filho 1984) and also because it is patchily distributed within its range (Oliver and Santos 1991). It occurs in the following conservation units:

Brazil
Augusto Ruschi Biological Reserve (4492 ha) ES
(formerly Nova Lombardia Biological Reserve)
Pedra Azul State Reserve (993 ha) ES
Forno Grande State Reserve (340 ha) ES
São Lourenço Biological Station (? ha) ES
Santa Lucia Biological Station (350 ha) ES
Caratinga Biological Station (800 ha) (privately owned) MG
Mata do Sossego Biological Station (221 ha) (privately owned) MG

## Genus SAGUINUS Hoffmannsegg 1807

### ***Saguinus nigricollis nigricollis*** (Spix 1823)

Spix's black-mantle tamarin (Engl.) (*v.* Hershkovitz 1977), sagui (Br.), bebeleche (Co.) (*v.* Hernandez-Camacho and Defler 1989), pichico (Pe.) (*v.* Burgos 1974).

**Type locality** North bank of the Solimões, near São Paulo de Olivença, Amazonas, Brazil (*v.* Hershkovitz 1977).

**Distribution** *S. n. nigricollis* occurs between the Rios Solimões-Amazonas and Içá-Putumayo, at least as far west as the mouth of the Río Napo (Hershkovitz 1977). In Colombia, its distribution is poorly known, but Hernandez-Camacho and Cooper (1976) reported that it occurs north of the Rio Putumayo to the Rio Caquetá, east to the Brazilian border, indicating its, as yet undocumented, presence between the Rios Japurá and Iça in Brazil (Hershkovitz 1977, 1982). It is sympatric with *Saguinus fuscicollis* throughout its range.

**Status** Regarded as common in Peru by Soini *et al.* (1989), it is not known if it even exists in Brazil. It was heavily exploited for export for biomedical research in Colombia in the 1960/70s but was reported by Hernandez-Camacho and Cooper (1976) to be common. It occurs in the following conservation units:

Brazil
Juamí-Japurá Ecological Station (?) (745 830 ha) AM
Colombia
La Paya National Natural Park (442 000 ha)
Amacayacú National Natural Park (293 500 ha)
Cahuinarí National Natural Park (575 500 ha)
Peru
None

## ***Saguinus nigricollis graellsi*** Jimenez de la Espada 1870

Graell's black-mantle tamarin (Engl.) (*v*. Hershkovitz 1982), bebeleche (Co.) (*v*. Hernandez-Camacho and Cooper 1976), pichico (Pe.) (*v*. Burgos 1974), pichico negro (Pe.).

**Type locality** Banks of Río Napo near Tarapoto, and Destacamento, near confluence with the Marañon, Loreto, Peru, restricted to Tarapoto by Cabrera (1957), further restricted by Hershkovitz (1982) to right bank of Río Napo, opposite Tarapoto and above the mouth of the Río Curaray.

**Distribution** This subspecies occurs in eastern Ecuador and northeastern Peru, south of the Río Putumayo, south to the mouth of the Río Napo, and west to the Río Santiago, on the eastern slopes of the Andes (Hershkovitz 1982). The altitudinal range is between 100 m and 1000 m (Hershkovitz 1982). The southern range limit is in the region of the upper Ríos Pastaza, Tigre, and Marañon, but is not clearly known (Hershkovitz 1982). To the east, Freese *et al.* (1982) reported that it probably does not occur to the north of the lower Río Nanay, indicating that it does not extend to the Rio Amazonas between the Ríos Napo and Marañon.

**Status** Described as common in Peru (Soini *et al.* 1989). The following conservation units are within its known range:

Ecuador
- Yasuni National Park (400 000 ha)
- Sangay National Park (370 000 ha)
- Cayambé-Coca Faunal Ecological Reserve (350 000 ha)
- Cuyabeno Faunal Reserve (30 000 ha)

Peru
- None

## ***Saguinus nigricollis hernandezi*** Hershkovitz 1982

Hernandez' black-mantle tamarin (Engl.) (*v*. Hershkovitz 1982), bebeleche (Co).

**Holotype locality** Río Peneya, a small tributary of the Río Caqueta entering from left (north) about 15 km above the mouth of Río Caguan, and about 50 km in straight line below village La Tagua, Intendencia de Caquetá, Colombia; altitude approximately 150 m above sea level (Hershkovitz 1982).

**Distribution** *S. n. hernandezi* occurs in eastern Colombia between the Ríos Caquetá, Caguan, and Orteguaza and the base of the Cordillera Oriental, Intendencia de Caquetá. It also occurs in the Department of Meta, Angostura, right bank of the Río Guayabero (Hernandez-Camacho and Defler 1989). The altitudinal range is 150–500 m above sea level (Hershkovitz 1982).

**Status** This tamarin was studied by Izawa (1978). It has a very restricted distribution, but no information was provided on its status by Hernandez-Camacho and Defler (1989). There are no conservation units within its known range.

## ***Saguinus fuscicollis fuscicollis*** (Spix 1823)

Spix's saddle-back tamarin (Engl.) (*v.* Hershkovitz 1977), sagui (Br.), pichico bocablanca (*v.* Burgos 1974), pichico comun (Pe.).

**Type locality** 'Near the district of São Paulo de Olivença in the forests between the Solimões and Içá.' Restricted by (Hershkovitz 1977) to the vicinity of São Paulo de Olivença on the south bank of the Rio Solimões.

**Distribution** *S. f. fuscicollis* occurs to the south of the Rio Solimões, between the Rio Javarí in the west and the Rio Jutaí and upper Rio Juruá in the east. It is possible that it also extends further east to the lower Rio Juruá (Hershkovitz 1977). Hodun *et al.* (1981) reported it west of the Río Yavarí as far as the Río Tapiche, an eastern tributary of the Río Ucayali. It occurs north from there only as far as the Río Blanco (left bank), where it meets the range of *S. f. nigrifrons* (right bank of Río Blanco). *S. f. illigeri* occurs west of the Río Tapiche. Distributions of *S. fuscicollis* are not known on the upper Río Ucayali, south of the Río Tapiche and likewise between the Rio Jutaí and the middle and lower Rio Juruá. The altitudinal range is 75–150 m above sea level (Hershkovitz 1977).

**Status** Unknown in Brazil, but Soini *et al.* (1989) reckoned all the Peruvian subspecies of *S. fuscicollis* to be common. The following conservation units are within its known range:

Brazil
  Jutaí-Solimões Ecological Reserve (284 285 ha) AM
Peru
  None

## ***Saguinus fuscicollis fuscus*** (Lesson 1840)

Lesson's saddle-back tamarin (Engl.) (*v.* Hershkovitz 1977), leoncito (Co.), bebeleche (Co.) (*v.* Hernandez-Camacho and Defler 1989), sagui (Br.).

**Type locality** 'Plaines de Mocoa', that is lowlands in the Mocoa district, between the Ríos Putumayo and Caquetá, Putumayo, southeastern Colombia (*v.* Hershkovitz 1977).

**Distribution** *S. f. fuscus* occurs north of the Rio Solimões, ranging northwest between the Rios Japurá-Caquetá and Içá-Putumayo (Hershkovitz 1968). In Colombia, it also extends north of the middle Río Caquetá to the

Río Yari, west to the Andean foothills to an altitude of 500 m above sea level (Hernandez-Camacho and Cooper 1976). On the upper Río Putumayo it follows the left (north) bank of the Río Sucumbíos (Hernandez-Camacho and Defler 1989). In the north-west of its range, it extends to the right bank of the Río Guayabero and possibly east to the headwaters of the Ríos Vaupes and Apaporis in the southern department of Meta (Hernandez-Camacho and Cooper 1976), as was also proposed by Ávila-Pires (1974). Hershkovitz (1977) limits the range in the north to the Ríos Caquetá and Caguan, and is doubtful of the evidence that *S. f. fuscus* occurs beyond the Río Caguan.

**Status** Moynihan (1976*b*) found the distribution of *S. f. fuscus* to be patchy but reported it common in many areas between the Ríos Caquetá and Putumayo. Hernandez-Camacho and Cooper (1976) and Hernandez-Camacho and Defler (1989) concluded that it is not threatened. It status in Brazil is unknown. The following conservation units are within its geographic distribution:

Brazil
- Juamí-Japurá Ecological Station (745 830 ha) AM
- Mamirauá State Ecological Station (1 124 000 ha) AM

Colombia
- Cahuinarí National Natural Park (575 500 ha)
- La Paya National Natural Park (422 000 ha)
- Serrania de Chiribiquete National Natural Park (?) (1 280 000 ha)

## ***Saguinus fuscicollis avilapiresi*** Hershkovitz 1966

Ávila-Pires' saddle-back tamarin (Engl.) (*v.* Hershkovitz 1977), sagui (Br.).

**Type locality** Mouth of the Lago de Tefé, south bank of the Río Solimões (Amazonías), Amazonas, Brazil (*v.* Hershkovitz 1977).

**Distribution** The only evidence regarding the distribution of *S. f. avilapiresi* is its type locality and a specimen from Ayapuá, left bank of the Rio Purus (Napier 1976). Hershkovitz (1977) proposed that it occurs south of the Rio Solimões between the Rios Juruá and Purus. The southern limits are not known but possibly in the region of the north bank of the Rio Tapauá, an affluent of the Rio Purus, but *S. fuscicollis* are not known for this region. Johns (1985, 1986) recorded *S. f. avilapiresi* at his study site on the Lago de Tefé, left bank of the Rio Tefé.

**Status**

Endemic to the Brazilian Amazon

Its status is unknown but there is no evident reason to believe it is vulnerable or threatened. The Lago Ayapuá was decreed a State Environment

Protection Area (610 000 ha) in 1990. It is probable that *S. f. avilapiresi* occurs in that part of the Abufarí Biological Reserve (288 000 ha), Amazonas, which is on the left bank of the Rio Purus.

## *Saguinus fuscicollis cruzlimai* Hershkovitz 1966

Cruz Lima's saddle-back tamarin (Engl.) (*v.* Hershkovitz 1977), sagui (Br.).

**Type locality** Said to be from the upper Rio Purus region, Amazonas, Brazil (*v.* Hershkovitz 1977).

**Distribution** The distribution of this subspecies is not known. Hershkovitz (1968) placed it tentatively in the upper Rio Purus region, south of the Rio Tapauá, in Brazil. However, following the discovery of *S. f. primitivus* in 1970, Hershkovitz (1977) placed it equally tentatively north of the Rio Tapauá, east of the Rio Purus.

**Status** Unknown

## *Saguinus fuscicollis illigeri* (Pucheran 1845)

Illiger's saddle-back tamarin (Engl.) (*v.* Hershkovitz 1977), pichico bocablanca (Pe.) (*v.* Burgos 1974), pichico comun (Pe.)

**Type locality** Left bank of the lower Río Ucayali, near its mouth, Loreto, Peru (*v.* Hershkovitz 1966*b*).

**Distribution** This subspecies occurs in Peru, between the Ríos Huallaga and Ucayali, south of the Río Marañon (Hershkovitz 1977). *S. f. illigeri* also extends east of the Río Ucayali, from the mouth of the Río Tapiche, along its left bank (Hodun *et al.* 1981). How far south it occurs, east of the Río Ucayali, is not known, although Hershkovitz (1977) indicated that it is limited to the Ríos Caxiabatay or Pisqui, west of the Río Ucayali.

**Status**
Endemic to the Peruvian Amazon
Reported by Soini *et al.* (1989) to be common. It occurs in the Pacaya-Samiria National Reserve (1 478 800 ha).

## *Saguinus fuscicollis leucogenys* (Gray 1866)

Andean saddle-back tamarin (Engl.) (*v.* Hershkovitz 1977), pichico bocablanca (*v.* Burgos 1974), pichico comun (Pe.).

**Type locality** Said to be Brazil, but restricted to the Department of Huanaco, Peru by Hershkovitz (1966*b*).

**Distribution** This subspecies is confined to north central Peru, from San Martin, through Huanaco and Pasco to the Río Perene, northern Juno, east to the Río Ucayali in Loreto to as far north as the Río Pisqui (Hershkovitz 1977). To the west it is limited by the Andes, probably not occurring above altitudes of about 900 m (Hershkovitz 1977). Freese *et al.* (1982) observed no primate species at Moyobamba, in the upper Mayo Forest Reserve, at an altitude of 1000 m above sea level.

**Status**
Endemic to the Peruvian Amazon
Freese *et al.* (1982) reported that *S. fuscicollis* (presumably *S. f. leucogenys*) was common in their Pucallpa survey area to the west of the Río Ucayali. Soini *et al.* (1989) reported it as common. It occurs in the Tingo Maria National Park (18 000 ha).

## ***Saguinus fuscicollis nigrifrons*** (I. Geoffroy 1815)

Geoffroy's saddle-back tamarin (*v*. Hershkovitz 1977), pichico bocablanca (*v*. Burgos 1974), pichico comun (Pe.).

**Type locality** Unknown, but restricted by Hershkovitz (1977) to the lower Río Yavarí, Loreto, Peru.

**Distribution** *S. f. nigrifrons* occurs in Peru, between the Ríos Ucayali and Javarí, south of the Rio Solimões (Hershkovitz 1968), as far south as Cerro Azul (Hershkovitz 1977). However, according to Hodun *et al.* (1981), the southern limit to its range is marked by the Río Blanco, an eastern tributary of the Río Tapiche, and *S. f. illigeri* occurs west of the Río Tapiche further north. *S. f. fuscicollis* occurs to the south of the Río Blanco.

**Status**
Endemic to the Peruvian Amazon
This subspecies is common (Soini *et al.* 1989). It is not known to occur in any conservation units.

## ***Saguinus fuscicollis lagonotus*** (Jimenez de la Espada 1870)

Red-mantle saddle-back tamarin (Engl.) (*v*. Hershkovitz 1977), bebeleche (Co.), leoncito (Co.) (*v*. Hernandez-Camacho and Cooper 1976), pichico bocablanca (*v*. Burgos 1974), pichico comun (Pe.).

**Type locality** In Ecuador, on the Río Napo at Coca and Tarapoto (*v*. Hershkovitz 1977). Hershkovitz (1977) argued that the type specimens must have been collected from the right bank of the Río Napo, opposite these sites.

**Taxonomy and distribution** According to Hershkovitz (1968, 1977), *S. f. lagonotus* occurs between the Ríos Napo and Marañon, west to the Andes in Peru and eastern Ecuador. Hodun *et al.* (1981) confirmed its presence on the left bank of the Río Marañon. The altitudinal range is from 100 m to approximately 1200 m (Hershkovitz 1977). Thorington (1988) argued that Hershkovitz's (1977) restriction of *S. f. lagonotus* to the right bank of the Río Napo is not necessarily valid and that it may extend north of the river, where it would be sympatric with (but not necessarily occurring in the same habitats) as *S. tripartitus*. For this reason, Thorington (1988) regards *S. tripartitus* as a full species, whereas Hershkovitz (1977) regards it as subspecific to *S. fuscicollis* (see below).

**Status** Freese et al. (1982) reported that *S. f. lagonotus* was one of the most frequently encountered primates on the Río Nanay (see also Mittermeier *et al.* 1978). Soini *et al.* (1989) regarded it as common. The following conservation units are within its known range:

Ecuador
  Yasuni National Park (400 000 ha)
  Sangay National Park (370 000 ha)
Peru
  None

## *Saguinus fuscicollis weddelli* (Deville 1849)

Weddell's saddleback tamarin (Engl.) (*v.* Hershkovitz 1977), sagui (Br.), pichico bocablanca (*v.* Burgos 1974), pichico comun (Pe.).

**Type locality** Province of Apolobamba (= Caupolican), La Paz, Bolivia (*v.* Hershkovitz 1977).

**Distribution** *S. f. weddelli* has the widest distribution of the *S. fuscicollis* subspecies, the largest part of which is in Bolivia. Heltne *et al.* (1976) and Izawa and Bejarano (1981) believe it to be the only tamarin occurring south of the Río Madre de Dios. It ranges from southern Peru, from the Andes to the Rios Madeira and Mamoré in Bolivia. It extends north in Brazil between the Rios Purus and Madeira in Acre and Amazonas and part of Rondônia, but is known only as far north as the Rio Pixuna (Hershkovitz 1977). It crosses the upper Rio Madeira to its right bank in Rondônia (region of the Rio Jamarí, forming mixed-species groups with *Callithrix* (V. F. Leao and S. G. Paccagnella, personal communication, 1986, see Vivo 1985). The southern limit of its range is in the region of the Ríos Mamoré and Apurimac. Heltne *et al.* (1976) observed it at Ixiamas, Bolivia (13° 46′S) where it was found to be less common than other primates found in the area, and they indicated that this may be near to the southern limit of its range. This species has been studied in the Manu National Park (Terborgh 1983; Terborgh and Goldizen 1985).

**Status** Brown and Rumiz (1986) reported it common in Bolivia. The following conservation units are within its geographical distribution:

Bolivia
  Manuripé National Reserve (1 844 375 ha)
Brazil
  Abufarí Biological Reserve (288 000 ha) AM
  Rio Acre Ecological Station (77 500 ha) AC
  Cuniã Ecological Station (104 000 ha) RO
Peru
  Manu National Park (1 532 806 ha)
  Tambopata Natural Wildlife Reserve (5500 ha)

## *Saguinus fuscicollis primitivus* Hershkovitz 1977

Saddle-back tamarin (Engl.), sagui (Br.).

**Holotype locality** Rio Juruá, Amazonas, Brazil (*v.* Hershkovitz 1977).

**Distribution** *S. f. primitivus* is known only from Pauiní, below the mouth of the Rio Pauiní, and from the region of the upper Rio Purus, presumably restricted to the west (left bank) of the river (Hershkovitz 1977). The range probably extends west from the Rio Purus, between the Rios Pauiní and Tapauá to the Rios Juruá and Tarauacá (Hershkovitz 1977).

**Status**
Endemic to the Brazilian Amazon
Unknown. There are no conservation units within its known range.

## *Saguinus fuscicollis melanoleucus* (Miranda Ribeiro 1912)

White saddle-back tamarin (Engl.), sagui-branco (Br.) (*v.* Hershkovitz 1977).

**Type locality** Amazonas, Brazil; restricted to Santo Antonio, Rio Eirú, upper Rio Juruá, Amazonas, by Carvalho (1957*b*) (*v.* Hershkovitz 1977).

**Taxonomy and distribution** The conspicuous pelage colour pattern differences of this tamarin (along with *S. f. acrensis* and *S. f. crandalli*), compared to the other subspecies of *S. fuscicollis* of Hershkovitz (1977), argue for their separation as a distinct species (Coimbra-Filho 1990, following Miranda Ribeiro 1912 and Carvalho 1957*b*), forming a logical cline as they do in terms of the metachromism hypothesis of Hershkovitz (1968).

It occurs west of the Rio Tarauacá, to the east of the Rio Juruá (Hershkovitz 1977). Colour forms of *S. f. melanoleucus* grade into those of *S. f. acrensis* to the south, and the southern limit to its range is not known exactly. The altitudinal range is up to 100 m (Hershkovitz 1977). The region around the Rio Tarauacá and upper Rios Purus and Juruá is poorly known with regard to callitrichid distributions.

**Status**
Endemic to the Brazilian Amazon
Its status is not known, but its small distribution and the rapidly expanding development of the state of Acre would indicate that it is vulnerable. It is not protected in any conservation units.

## *Saguinus fuscicollos acrensis* (Carvalho 1957)

Acre saddle-back tamarin, sagui-do-Acre (Br.) (*v.* Hershkovitz 1977).

**Type locality** Opposite Pedra Preta, about 15 km below Vila Taumaturgo, right bank upper Rio Juruá, Federal Territory of Acre, Brazil (*v.* Hershkovitz 1977).

**Taxonomy and distribution** Following the original description and classification of this subspecies by Carvalho (1957*b*), we regard this tamarin to be subspecific to *S. melanoleucus* (see above). It is believed by Hershkovitz (1977) to occur to the east of the upper the Rio Juruá, south of Cruzeiro do Sul. Although the eastern limit to the range is not known, it probably does not go further than the Rio Tarauacá. The northern limit is possibly in the region of Cruzeiro do Sul, but Hershkovitz (1977) found that northern populations intergrade in pelage colour with *S. f. melanoleucus*. It is possible that it occurs east of the Rio Tarauacá, or even to the Rio Purus, but tamarin distributions in this area are unknown. Any extension of the known range will evidently be small. There is no record of it occurring in Peru.

**Status**
Endemic to the Brazilian Amazon
Like *S. m. melanoleucus*, its status is not known, but its small distribution and the rapidly expanding development of the state of Acre would indicate that it is vulnerable. It is not protected in any conservation units.

## *Saguinus fuscicollis crandalli* (Hershkovitz 1968)

Crandall's saddle-back tamarin (Engl.) (*v.* Hershkovitz 1977), sagui (Br.).

**Type locality** Unknown.

**Taxonomy and distribution** We regard this species as a subspecies of the form *melanoleucus* (see above; Coimbra-Filho 1990, following Miranda Ribeiro 1912, Carvalho 1957*b*). The affinity of this tamarin to *S. f. melanoleucus* and particularly *S. f. acrensis*, in terms of its pelage colour pattern is the only indication of its distribution. On this basis, Hershkovitz (1977) proposed that it occurs south of the range of *S. f. acrensis*, between the upper Rios Purus and Juruá, on the Peru/Brazil frontier.

**Status** If Hershkovitz (1977) is correct in its distribution, *S. f. crandalli*, like the former two subspecies, is in a region which is undergoing recent, rapid and widespread development, although in a rather remoter situation.

## ***Saguinus tripartitus*** (Milne Edwards 1878) (*v*. Thorington 1988)

Golden-mantle saddle-back tamarin (Engl.) (*v*. Hershkovitz 1977), bebeleche (Co.) (*v*. Hernandez-Camacho and Cooper 1976).

**Type locality** Río Napo, Ecuador (*v*. Hershkovitz 1977).

**Taxonomy and distribution** Hershkovitz (1977) regarded this tamarin as a subspecies of *S. fuscicollis*. According to Hershkovitz (1977), it occurs between the Ríos Putumayo (right bank) and Napo (left bank), west to the Andes in Peru and Ecuador. Ávila-Pires (1974) proposed its occurrence between these rivers right up to the Rio Amazonas, at the mouth of the Rio Içá in Brazil. Whether the range extends so far east is not known. Hernandez-Camacho and Cooper (1976) reported that they had examined specimens in Leticia which were supposed to have come from the Colombian bank of the Río Amazonas. Neville *et al.* (1976) and Freese *et al.* (1982) reported that only *S. nigricollis* occurs at Ampiyacu, north of the Río Amazonas, near to Colombia, in Peru. The altitudinal range is reported by Hershkovitz (1977) to be between 100 and 250–300 m.

Thorington (1988) examined the evidence provided by Hershkovitz (1977) for his descriptions of the distributions of *S. f. tripartitus* (north of the Río Napo) and *S. f. lagonotus* (south of the Río Napo). Thorington (1988) concluded that the evidence was confused, but argued that both *S. tripartitus* and *S. f. lagonotus* occur on the right bank of the Río Napo, in sympatry but not in the same localities (different habitats), and that although *S. tripartitus* occurs on the left bank of the upper Río Napo, there is no evidence that it occurs on the left bank near its mouth. Thorington did not consider the assertion of Hernandez-Camacho and Cooper (1976; see also Hernandez-Camacho and Defler 1989) that captive animals in Leticia, referable to the form *tripartitus*, were obtained from Puerto Narino, on the left bank of the Río Amazonas (downstream from the mouth of the Río Napo), as valid, pending further evidence.

Thorington (1988) argued that *S. tripartitus* is a true species, because of the apparent sympatry between it and *S. f. lagonotus*. Thorington also pointed out that many of Hershkovitz's (1977) *S. fuscicollis* subspecies are more distinct morphologically from one another than are *S. f. lagonotus* and *S. tripartitus*, and that a number of them (without saying which) might be valid species rather than subspecies.

**Status** Reported by Soini *et al.* (1989) as common, but Thorington (1988) recommended that its very small known distribution argues for it to be considered endangered, although we consider it only as possibly vulnerable. If, as Hernandez-Camacho and Cooper (1976) believe, it occurs on the left bank of the Rio Amazonas between the Ríos Napo and Putumayo in Colombia, it is protected in the Amacayacú National Natural Park (293 500 ha).

## *Saguinus mystax mystax* (Spix 1823)

Spix's moustached tamarin (Engl.), sagui-de-bigode (Br.) (*v.* Hershkovitz 1977), pichico barbablanca (*v.* Burgos 1974).

**Type locality** Near Saõ Paulo de Olivença, south bank Rio Solimões, Amazonas, Brazil (*v.* Hershkovitz 1977).

**Distribution** *S. m. mystax* occurs in Peru, south of the Río Amazonas, west to the Ríos Ucayali and Tapiche, south to the junction of the Ríos Urubamba and Ucayali (Hershkovitz 1977). Hershkovitz (1977) restricted it to the west of the Rio Juruá, south of the Rio Solimões, in Brazil, but Johns (1985, 1986) found it occurring on the west bank of the Rio Tefé, with *S. m. pileatus* replacing it on the east bank. Soini *et al.* (1989) reported that its Peruvian distribution is more restricted than indicated by Hershkovitz (1977), occurring only as far west as the east banks of the Río Tapiche and lower Río Ucayali. If Soini *et al.* (1989) are correct, a specimen from the mouth of the Río Samiria (locality 96, Hershkovitz 1977, p.684)—the only locality west of the Río Ucayali—is enigmatic, the Río Samiria being a tributary of the Río Marañon, above the Río Ucayali.

**Status** Considered to be common in Peru by Soini *et al.* (1989). The same is undoubtedly true for the Brazilian populations. Johns (1985, 1986) found it the most abundant of the primates at his study site on the west margin of the Rio Tefé. The following conservation units are within its geographical distribution:

Brazil
- Serra do Divisor National Park (605 000 ha) AC
- Jutaí-Solimões Ecological Reserve (284 285 ha) AM

Peru
- None

## ***Saguinus mystax pileatus*** (I. Geoffroy and Deville 1848)

Red-cap moustached tamarin (Engl.), sagui-de-bigode (Br.) (*v.* Hershkovitz 1977).

**Type locality** Lago de Tefé, near its mouth at Rio Solimões, Amazonas, Brazil (*v.* Hershkovitz 1977).

**Distribution** According to Hershkovitz (1977), *S. m. pileatus* occurs south of the Rio Solimões, west of the Rio Purus, south at least as far as the Rio Pauiní or Rio Mamoria (Hershkovitz 1977). Although Hershkovitz (1977) indicated that it occurs west to the lower Rio Juruá, Johns (1985, 1986) found that the Rio Tefé was its western limit. This argues for the restriction of the type locality of *S. m. pileatus* to the 'east margin of the Lago de Tefé'. Judging from a specimen from 'Juruá (Rio), Amazonas' (locality 174b, Hershkovitz 1977, p. 684), *S. m. pileatus* extends west to the Rio Juruá above the headwaters of the Rio Tefé. Despite Hershkovitz's (1977) conviction that *S. m. pileatus* occurs east as far the Rio Purus on its lower reaches (that is east of the Rio Coarí), we argue that the limited evidence for the distributions of these tamarins suggests that it only reaches the Rio Purus above the Rio Tapauá, with *S. m. pluto* occurring from the Rio Tapauá, along the left bank of the Rio Purus downriver, and as far west as the Rio Coarí (see *S. m. pluto* below).

**Status**
Endemic to the Brazilian Amazon
Nothing is known of the conservation status of *S. m. pileatus*. There are no conservation units within its known range.

## ***Saguinus mystax pluto*** (Lönnberg 1926)

White-rump moustached tamarin (Engl.), sagui-de-bigode (Br.) (*v.* Hershkovitz 1977).

**Type locality** Ayapuá, but probably opposite Ayapuá, on the right bank of the Rio Purus, Amazonas, Brazil (*v.* Hershkovitz 1977).

**Distribution** The distribution of *S. m. pluto* is rather problematic. Hershkovitz (1968, 1977) argued that it occurs south of the Rio Solimões, between the Rios Purus and Madeira, south to at least 7° or 8° latitude, but the southern limits are not known. The distribution as described by Hershkovitz is based on three localities: (1) The type locality, Lago Ayapuá is on the left (west) bank of the Rio Purus, but Hershkovitz (1977) argued that it must have come from the right bank. (2) Lago do Mapixí, a lake on the right bank of the Rio Purus. Although Hershkovitz states that specimens from Lago do Mapixí are labelled 'eastern of Rio Purus' (1977, p.700), this is contradicted in Hershkovitz's gazeteer (p.933, locality 185, Lago do Mapixí), which states under *S. m. pluto* 'C. Lako, June 1931 (left bank of Purus)'. A specimen of *S. labiatus labiatus*, of the same collector with the same date and locality, is ascribed to the right bank of the Purus. (3) Jaburú, on the left bank of the Rio Purus, (locality 186, p.933). As in the Ayapuá locality, Hershkovitz (1977) insists that the specimens were in fact taken from the right bank of the Rio Purus and even places the locality on the right bank in the distribution map for *S. mystax* (p.684, 1977). The problem with accepting that *S. m. pluto* is known only from the left bank of the Rio Purus is that it would have to be sympatric with *S. m. pileatus*. Ávila-Pires (1974) placed *S. m. pluto* on the west of the Rio Purus, and regarded it as a different species for this reason. However, the only specimen of *S. m. pileatus* recorded for the Rio Purus is São Luis da Mamoria, Rio Purus (locality 188, p.933, Hershkovitz 1977), which is far to the south, near the Rio Pauiní. Although different species, it would be very difficult to accept that *S. mystax* is sympatric with *S. labiatus*, because, along with *S. imperator*, they are evidently ecologically very similar, if not identical. We suggest that *S. m. pluto* is in fact restricted to the left bank of the Rio Purus, but is not sympatric with *S. m. pileatus*, and probably extends east as far as the Rio Coarí and south only to the rio Tapauá. The Mamoria locality for *S. m. pileatus* is downstream of the mouth of the Rio Pauini and upstream of the mouth of the Rio Tapauá. The Lago do Ipixuna, another locality (183, p.933, Hershkovitz 1977) for *S. m. pileatus*, is west of the Rio Coarí. There is no evidence of *S. m. pileatus* occurring east of the Rio Coarí, nor north of the Rio Tapauá.

Izawa and Bejarano (1981) reported two isolated populations of *S. mystax* from the south of the Rio Acre, north of the Río Tahuamanú, in Bolivia. One between Cobija and Buenos Aires, right bank of the Rio Acre, and another from 10 km northeast of Porvenir, 11°12′S, on the Brazil/Bolivia frontier beyond the headwaters of the Rio Abunã. Izawa and Bejarano (1981) presumed them to be *S. m. pluto* because Hershkovitz (1977) extended its range south as far as the Abunã headwaters. According to Izawa and Bejarano (1981), these two populations of *S. mystax* are sympatric with *S. labiatus*, but do not form mixed species groups as they do independently with *S. fuscicollis*. There is no ready explanation for these Bolivian populations, more than 3° to the south of the southernmost locality for *S. m. pileatus*. We consider that they are unlikely to be *S. m. pluto*, but may be *S. m. pileatus* (if they cross the upper Rio Purus) or, more likely, a different undescribed subspecies. Izawa and Bejarano (1981) also report a population of *S. imperator* within the distribution of *S. labiatus*, along with a new form of *S. labiatus* (undescribed). It would appear that the distributions of these species are mixed, forming fine-grained mosaics, in the southernmost parts of their ranges, traversing headwaters in a region lacking large rivers as barriers. A similar situation evidently pertains in the upper Rio Dôce, south-east Brazil, with *Callithrix geoffroyi*, *C. aurita*, and *C. flaviceps*. Under any circumstances the *S. mystax* distributions are evidently more complex than portrayed by Hershkovitz (1977) and require further study.

**Status**
Probably endemic to the Brazilian Amazon
The known distribution of *S. m. pluto* is very small. The State Environment Protection Area of Ayapuá (610 000 ha), west bank of the lower Rio Purus, was decreed in 1990. *S. m. pluto* should also occur in the part of the Abufarí Biological Reserve (288 000 ha), Amazonas, which extends to the left bank of the Rio Purus.

## *Saguinus labiatus labiatus* (E. Geoffroyi 1812)

Geoffroy's moustached tamarin (Engl.), red-bellied tamarin (Engl.), sagui-de bigode (Br.) (*v*. Hershkovitz 1977), pichico pecho anaranjado (Pe.).

**Type locality** Restricted to Lago do Janoacan (= Janauacá), Amazonas by Cabrera (1957), but Hershkovitz (1977) argued that the type probably originated from somewhere between the Rios Purus and Madeira, south of the Rio Ipixuna (= Paranapixuna).

**Taxonomy and distribution** *S. l. labiatus* occurs south of the Rio Solimões between the Rios Madeira and Purus (Hershkovitz 1977). It extends along the left bank of the Rio Madeira and Abunã, to Bolivia, crossing the headwaters of the Rio Abunã, as far south as both sides of the Rio Acre, probably limited to the Río Tahuamanú, a tributary of the Río Orton (tributary of the Río Beni) (Izawa and Bejarano 1981). According to Hershkovitz (1977), it extends up the right banks of the Rios Purus and Acre. *S. imperator* occurs on the left bank of the Rio Acre (Hershkovitz 1979). However, Izawa and Bejarano (1981) reported that *S. labiatus* is present on both sides of the Rio Acre in Bolivia and Brazil (Acre), with an isolated lacuna occupied by *S. imperator* on the left bank of the Rio Acre, along the Rio Sao Pedro. Encarnacion and Castro (1978, see also Encarnacion 1990) recorded *S. l. labiatus* between Inaparí on the Rio Acre (right bank) and Iberia, on the left bank of the Río Tahuamanú in Peru, confirming the finding of Izawa and Bejarano (1981), but were not able to indicate the western limits to its range.

Izawa and Bejarano (1981) reported the capture of five monkeys from the left bank of the Rio Acre, opposite Buenos Aires, which were not attributable to *S. l. labiatus*, presenting a 'solid dark grey coloration on the back, were red-bellied, and had a larger moustache which was almost as pronounced as that of *S. mystax*'. They indicate it may have been a hybrid with *S. i. imperator*. As indicated above for *S. m. pluto*, it would seem that the distributions of the three species of moustached tamarins form a mosaic in this part of Bolivia and the Brazil/Bolivia border. We indicate that the ecological similarity of the moustached tamarins argues against sympatry between them (Rylands 1987), especially considering that they are sympatric with the saddle-back tamarins.

**Status** *S. l. labiatus* is widespread and probably common in a large part of its range. Encarnacion (1990) reported that suitable habitat for *S. l. labiatus* was decreasing, drastically and rapidly, especially due to deforestation over the period 1987–1990, in the restricted region occupied by this species in Peru. It possibly occurs in the part of the Abufarí Biological Reserve (288 000 ha), Amazonas, to the east of the Rio Purus.

## *Saguinus labiatus thomasi* (Goeldi 1907)

Thomas' moustached tamarin (Engl.), sagui-de-bigode (Br.) (*v*. Hershkovitz 1977).

**Type locality** Tonantins, Rio Tonantins, north bank Rio Amazonas below mouth of Rio Içá, Amazonas, Brazil.

**Distribution** The distribution of *S. l. thomasi* was recently reviewed by Sousa e Silva (1988). Its distribution was, until recently, known only from its type locality (Hershkovitz 1977), but four specimens (now in the Goeldi Museum, Belém) were collected from Barreirinha, left bank of the Auatí-Paraná (between the Rios Solimões and Japurá), 200 km west of its type locality (Almeida and Deane 1970). Sousa e Silva (1988) suggested that it occurs throughout the region between the left bank of the Tonantins to beyond the Auatí-Paraná. It may even extend west as far as the Rio Içá, but there is no record of it occurring in Colombia (Hernandez-Camacho and Defler 1985, 1989). In the east, it is probably restricted to *terra firme* forest, not inhabiting the extensive inundated forest (*várzea*) near the confluence of the Rios Japurá and Solimões (Ayres 1986).

**Status**
Endemic to the Brazilian Amazon
This subspecies evidently has a restricted distribution. It possibly occurs in the Juamí-Japurá Ecological Station (745 830 ha) and certainly in the western part of the Mamirauá State Ecological Station (1 124 000 ha), Amazonas.

## *Saguinus imperator imperator* (Goeldi 1907)

Black-chinned emperor tamarin (Engl.) (*v*. Hershkovitz 1979), bigodeiro (Br.), sagui-imperador (Br.), pichico emperador (Pe.).

**Type locality** Said to have been collected in the upper Rio Purus and Rio Acre region of southwestern Brazil. Cabrera (1957) restricted the type locality to the Rio Acre. According to Carvalho (1959*b*), the lectotype is from Bom Lugar, or perhaps Monte Verde on the upper Rio Purus, near the mouth of the Rio Acre. Hershkovitz (1979) pointed out that Monte Verde and Bom Lugar are outside its distribution, but indicated that they could have been collected from the left bank of the Rio Acre, opposite Monte Verde.

**Distribution** *S. i. imperator* occurs east of the upper Rio Purus, between the Purus and the Rio Acre (Hershkovitz 1979). Izawa and Bejarano (1981) did not record *S. i. imperator* for Bolivia, but reported an isolated population on the left bank of the Rio Acre, in the basin of the Rio São Pedro in Brazil. These authors also indicated a small incursion across the Rio Acre into Peru. Although Soini *et al.* (1989) do not list this subspecies, Encarnacion and Castro (1978) found populations of *S. i. imperator* (but not *S. l. labiatus*) on the right and left banks of the Rio Acre near the Quebrada Río Branco, approximately 20 km west of Inapari, exactly in the region indicated by Izawa and Bejarano (1981). It is not known how far it extends into Peru along the Rio Acre, nor whether it occurs between the Rio Purus and Pauiní and the Rios Purus and Ituxí (Hershkovitz 1979). The distribution is probably more complex than suggested by Hershkovitz (1979) because Izawa and Bejarano (1981) reported that *S. labiatus* extends to both sides of the Rio Acre (excepting the area of the Rio São Pedro occupied by *S. i. imperator*) and it is improbable that the two species are sympatric (Rylands 1987).

**Status**
1988 IUCN Red List of Threatened Animals—Indeterminate
1989 List of Brazilian Fauna Threatened with Extinction
Peruvian Vulnerable Species List
CITES Appendix II
*S. i. imperator* has a very small range, possibly even smaller than indicated by Hershkovitz (1979). The state of Acre has been undergoing widespread development in recent years, involving extensive destruction of its forests. *S. i. imperator* should be considered vulnerable if not endangered. The Rio Acre Ecological Station (77 500 ha), Acre, is within the distribution described by Hershkovitz (1979).

## ***Saguinus imperator subgrisescens*** (Lonnberg 1940)

Bearded emperor tamarin (Engl.) (*v.* Hershkovitz 1977), bigodeiro (Br.), sagui-imperador (Br.), leoncillo (Pe.) (*v.* Burgos 1974), pichico emperador (Pe.).

**Type locality** Santo Antônio, west bank of Eirú near confluence with Rio Juruá, Amazonas, Brazil (*v*. Hershkovitz 1977).

**Distribution** According to Hershkovitz (1979), *S. i. subgrisescens* occurs in Brazil east of the upper Rio Juruá, east to the Rios Tarauacá and Juruparí, to the Brazil/Peruvian frontier. West it occurs as far as the foothills of the Andes in the upper Río Ucayali (east of the Río Apurimac) and the Urubamba basin and to the north of the Río Madre de Dios in Bolivia (Izawa 1979). Heltne *et al.* (1976) believed it occurred at Cobija, south of the Rio Acre, but Pook and Pook (1979) received reports of it being limited to the south of the Río Tahuamanú. This was confirmed by Izawa and Bejarano (1981) who reported it only from the Rio Muyumanu basin, a south bank tributary of the Río Tahuamanú. Izawa and Bejarano (1981) indicated that the Cobija locality reported by Heltne *et al.* (1976) may have resulted from confusion with *S. mystax*.

**Status**
1988 IUCN Red List of Threatened Animals—Indeterminate
1989 List of Brazilian Fauna Threatened with Extinction
Peruvian Vulnerable Species List
This subspecies is moderately abundant in the Manu National Park (1 532 800 ha), where it has been been studied by (Terborgh 1983; Terborgh and Goldizen 1985) but the conservation status of *S. imperator*, in both Peru and Brazil, is not known.

The following conservation units are within its geographical range:

Brazil
  None
Bolivia
  Manuripí Heath Nature Reserve (1 844 375 ha)
Peru
  Manu National Park (1 532 806 ha).

## *Saguinus inustus* (Schwarz 1951)

Mottled-face tamarin (Engl.) (*v*. Hershkovitz 1977), sagui (Br.), mico diablo (Co.), mico hueviblanco (Co.), chichico negro (Co.) (*v*. Hernandez-Camacho and Defler (1989).

**Type locality** Tabocal, Rio Negro, Amazonas, Brazil (*v*. Hershkovitz 1977).

**Taxonomy and distribution** *S. inustus* occurs between the upper Rios Negro and Japurá, west from opposite the Rio Padauarí (64°), a northern tributary of the Rio Negro, into Colombia between the Ríos Apaporis and Guaviare (possibly also occurring in gallery forest of the Río Ariari to the north) to the base of the Serranía La Macarena (Hernandez-Camacho and Cooper 1976; Hershkovitz 1977). Its distribution in Colombia is poorly known (Hernandez-Camacho and Defler 1989). Hershkovitz (1977) puts the western limit in the region of the upper Ríos Apaporis and Guaviare. Hernandez-Camacho and Defler (1989) reported that it is also known from Cano Yaviya, Río Yarí, Caquetá, and although Hernandez-Camacho and Cooper (1976) supposed the Apaporis (left bank) to be the southern limit to its range in Colombia, Hernandez-Camacho and Defler (1989) were able to report that it occurs as far south as the left bank of the Caquetá. Hershkovitz (1977) places the eastern limit in Colombia along the Ríos Atabapo and Guainia and Rio Negro and certainly Handley (1976; see also Bodini and Perez-Hernandez 1987) gives no hint that it may occur in Venezuela. Ávila-Pires (1974) indicated that it occurs throughout the region between the Rios Negro and Japurá, but this remains to be confirmed. It has been observed near the mouth of the Rio Japurá, on the left bank at the Lago Amanã (Rylands and Clutton-Brock, unpublished data, 1980), which extends its distribution further east between these rivers than was indicated by Hershkovitz (1977).

Both Hernandez-Camacho and Defler (1989) and Hershkovitz (1977) indicated the possibility of subspecies, determined by the extent of white on the cheeks and around the eyes.

**Status** Its conservation status in Colombia (Hernandez-Camacho and Defler 1989) and Brazil is not known. The following conservation units are within its geographical distribution:

Brazil
- Jaú National Park (?) (2 272 000 ha) AM

Colombia
- Serranía de Chiribiquete National Natural Park (1 280 000 ha)
- Nukak National Natural Reserve (855 000 ha)
- Puinawai National Natural Reserve (1 092 500 ha)

## ***Saguinus midas midas*** (Linnaeus 1758)

Golden-handed tamarin (Engl.) (*v.* Hershkovitz 1977), midas tamarin, sagui-de-mãos-amarelas (Br.) (*v.* Hershkovitz 1977)

**Type locality** Suriname (*v.* Hershkovitz 1977).

**Distribution** *S. m. midas* occurs north of the Rio Amazonas and east of the Rio Negro in Brazil, extending north and east to the coast of Amapá and the Guianas (Napier 1976). According to Hershkovitz (1977), the western limit is north of the Rio Negro in the region of the Ríos Cassiquiare and Orinoco, but Handley (1976; see also Bodini and Perez-Hernandez 1987) do not report any callitrichids for Venezuela, and it is probable that the western limit is in the region of the western most locality registered by Hershkovitz (1977) on the Rio Deminí. It is also evidently absent from a large part of central Guyana. It does not occur around the city of Manaus, the domain of *S. bicolor bicolor*, and shows only a narrow zone where it mixes with this species on the periphery of *S. b. bicolor*'s range (Ayres *et al.* 1980). The situation is probably the same with the other two *S. bicolor* subspecies.

**Status** Undoubtedly common throughout the majority of its range, it is considered the most widespread and abundant of all callitrichids. Mittermeier *et al.* (1978) found it to be common and only rarely hunted in French Guiana and Mittermeier and van Roosmalen (1982) and Baal *et al.* (1988) discuss its conservation status and its occurrence in the protected areas of Suriname. The following conservation units are within its geographical distribution:

Brazil
- Pico da Neblina National Park (?) (2 200 000 ha) AM
- Cabo Orange National Park (619 000 ha) AP
- Monte Roraima National Park (116 000 ha) RR
- Rio Trombetas Biological Reserve (385 000 ha) PA
- Lago Piratuba Biological Reserve (357 000 ha) AP
- Uatumã Biological Reserve (560 000 ha) AM
- Anavilhanas Ecological Station (350 012 ha) AM
- Maracá Ecological Station (101 312 ha) RR
- Caracaraí Ecological Station (400 560 ha) RR
- Niquiá Ecological Station (286 600 ha) RR
- Jarí Ecological Station (227 116 ha) PA
- Serra do Aracá State Park (1 818 700 ha) AM

Suriname
- Brownsberg Nature Park (8400 ha)
- Raleighvallen-Voltzberg Nature Reserve (56 000 ha)
- Tafelberg Nature Reserve (140 000 ha)
- Eilerts de Haan Gebergte Nature Reserve (220 000 ha)
- Sipaliwini Savannah Nature Reserve (100 000 ha)
- Herten rits Nature Reserve (100 ha)
- Brinckheuvel Nature Reserve (6000 ha)
- Coppename Mouth Nature Reserve (10 000 ha)
- Wia-Wia Nature Reserve (36 000 ha)
- Galibi Nature Reserve (4000 ha)

Guyana
- Kaieteur National Park (11 600 ha)

## ***Saguinus midas niger*** (E. Geoffroy 1803)

Black-handed tamarin (Engl.) (*v*. Hershkovitz 1977), sagui (Br.).

**Type locality** Said to be Cayenne, but restricted by Hershkovitz (1977) to Belém do Pará.

**Taxonomy and distribution** On the basis of morphometric studies of the postcanine dentition, Natori and Hanihara (1992) found *S. m. midas* to be more similar to *S. bicolor* than to *S. m. niger*, and for this reason, *S. m. niger* should probably be raised to species status.

The black-handed tamarin occurs in the state of Pará, Brazil, south of the Rio Amazonas, east of the Rio Xingú and Rio Fresco, to the lower Rio Araguaia, south as far as the region of Santana de Araguaia (Napier 1976; Hershkovitz 1977). Hershkovitz (1977) limits it to the left bank of the Rio Gurupí, but Johns (1986) censused *S. m. niger* in the Gurupí Forest Reserve on the right bank of the river. It is probable that it extends at least to the Rio Pindaré and possibly as far as the Rio Grajaú in the state of Maranhão. The southernmost locality registered by Hershkovitz (1977) is Gradaús, Rio Fresco. *S. m. niger* also occurs on the Island of Marajó in the Rio Amazonas estuary.

**Status**
Endemic to the Brazilian Amazon
Probably common, although its range covers the most developed and fast developing region in the Brazilian Amazon. The current destruction of the Gurupí Biological Reserve exemplifies the devastation occurring in the region. The following conservation units are within its geographical distribution:

Brazil
- Gurupí Biological Reserve (341 650 ha) PA
- Tapirapé Biological Reserve (103 000 ha) PA

## ***Saguinus bicolor bicolor*** (Spix 1823)

Pied bare-face tamarin (Engl.) (*v*. Hershkovitz 1977), sagui-de-cara-nua (Br.), sagui-de-duas-cores (Br.), sagui-bicolor (Br.), sauim (Br.), sagui (Br.).

**Type locality** Near the village of the Rio Negro (= Manaus), Barra do Rio Negro, Amazonas, Brazil (*v*. Hershkovitz 1977).

**Distribution** *S. b. bicolor* occurs north of the Rio Amazonas, east of the Rio Negro, in the vicinity of Manaus, the capital of the state of Amazonas, Brazil. Although Hershkovitz (1977) and Ávila-Pires (1974) postulated the Rio Branco and the Rios Jauapurí and Alalaú as the northern limit to its range, surveys by Ayres *et al.* (1980, 1982) demonstrated a more restricted range—extending only approximately 30–45 km to the north of Manaus and to the east as far as the town of Itacoatiara, approximately 100 km from the capital. Localities beyond these points indicate only the presence of *S. midas*. Ayres *et al.* (1980, 1982) and Egler (1983) suppose that it might occur as far north as the Rio Jauapurí, right bank tributary of the Rio Negro, but this seems unlikely. *S. b. bicolor* groups are surviving in small, highly degraded forest patches around housing estates and in the suburbs of Manaus. Ayres *et al.* (1980, 1982) discuss the possibility that *S. midas* populations are gradually replacing *S. bicolor*, either naturally, or through alteration and degradation of the forests by human activities.

**Status**
1988 IUCN Red List of Threatened Animals—Indeterminate
1989 List of Brazilian Fauna Threatened with Extinction
CITES Appendix I
US Endangered Species List
Endemic to the Brazilian Amazon
Being restricted to a very small region in the vicinity of the city of Manaus, *S. b. bicolor* is probably the most threatened of the Brazilian Amazonian callitrichids. It occurs in the following conservation units and reserves:

Brazil
Sauim-Castanheiras Ecological Reserve (109 ha) AM
Adolfo Ducke Forest Reserve of the Instituto Nacional de Pesquisas da Amazônia (INPA) (10 000 ha) AM
Walter Egler Forest Reserve of the Instituto Nacional de Pesquisas da Amazônia (INPA) (630 ha) AM

## ***Saguinus bicolor ochraceus*** Hershkovitz 1966

Ochraceus bare-face tamarin (Engl.) (*v.* Hershkovitz 1977), sagui (Br.).

**Type locality** Mouth of the Rio Paratucú, a right bank tributary of the Nhamundá, Amazonas, Brazil (*v.* Hershkovitz 1977).

**Distribution** *S. b. ochraceus* is believed to occur on the west bank of the Rio Nhamundá, possibly extending west to the Rio Uatumã, north of the Rio Amazonas (Hershkovitz 1966*b*). The northern limit to its range might be the left bank of the Rio Alalaú.

**Status**
1988 IUCN Red List of Threatened Animals—Indeterminate
1989 List of Brazilian Fauna Threatened with Extinction
CITES Appendix I

US Endangered Species List
Endemic to the Brazilian Amazon
Ayres *et al.* (1980, 1982) and Coimbra-Filho (1987) review the distribution and status of the *S. bicolor* subspecies. It is probably not under any immediate threat, but its vulnerability will depend on how restricted its range turns out to be. It probably occurs in the Nhamundá State Park (28 370 ha) and the Nhamundá State Environment Protection Area (195 900 ha), Amazonas.

## ***Saguinus bicolor martinsi*** (Thomas 1912)

Martin's bare-face tamarin (*v.* Hershkovitz 1977), sagui (Br.).

**Type locality** Faró, north side of the Rio Amazonas, near mouth of Rio Nhamundá, Pará, Brazil.

**Distribution** According to Hershkovitz (1966*b*), *S. b. martinsi* occurs between the Rio Nhamundá (left bank), east to the Rio Erepecurú, north of the Rio Amazonas. It would appear that the mouth of the Rio Trombetas marks the northern limit, but Ávila-Pires (1974) extends the distribution north-east to the upper Rio Erepecurú.

**Status**
1988 IUCN Red List of Threatened Animals—Indeterminate
1989 List of Brazilian Fauna Threatened with Extinction
CITES Appendix I
US Endangered Species List
Endemic to the Brazilian Amazon
Ayres *et al.* (1980, 1982) review the distribution and status of the *S. bicolor* subspecies. It is probably not under any immediate threat, but its vulnerability will depend on how restricted its range turns out to be. Depending on its real range limit in the north, it may occur in the Rio Trombetas Biological Reserve (385 000 ha), Pará.

## ***Saguinus leucopus*** (Gunther 1877)

Silvery brown bare-face tamarin (*v.* Hershkovitz 1977), white-footed tamarin (*v.* Mack and Mittermeier 1984), titi (Co.), titigris (Co.) (*v.* Hernandez-Camacho and Defler 1989).

**Type locality** Near Medellin, Antioquia, Colombia (*v.* Hershkovitz 1977).

**Taxonomy and distribution** Although Thorington (1976) indicated that *S. leucopus* is more similar to *S. oedipus* than either is to *S. geoffroyi*, Hanihara and Natori (1987), studying the dental morphology, found that *S. oedipus* is more similar to *S. geoffroyi*. *S. leucopus* occurs in northern

Colombia, between the Ríos Magdalena and Cauca from their confluence in the Department of Bolivar (including the Isla de Mompos and the region south-east of the Department between the Ríos Cauca and Magdalena) south into the Department of Antiquoia along the west of Río Cauca basin as far west as the region of Valdivia, the Río Nechi valley, and Porce. It probably extends southward along the tropical forested slopes of the Central Cordillera, west of the Río Magdalena, south into western Caldas and northern Tolima (at least as far south as the vicinity of Mariquita, ranging up to 1500 m (Hernandez-Camacho and Cooper 1976; Hershkovitz 1977; Hernandez-Camacho and Defler 1989). Hernandez-Camacho and Defler (1989) indicated the possibility that the range of *S. leucopus* extends further south-west along the east bank of the Río Cauca in Antiquoia, and further south into the gallery forests of the *llanos* and the forested foothills of the eastern slopes of the Central Cordillera in the Department of Tolima.

**Status**
1988 IUCN Red List of Threatened Animals—Vulnerable
US Endangered Species List
CITES Appendix I
Endemic to northern Colombia
*S. leucopus* has been seriously affected by deforestation in a large part of its original range, and Hernandez-Camacho and Defler (1989) regard it as endangered. There are no conservation units within its geographical distribution.

## *Saguinus oedipus* (Linnaeus 1758)

Cotton-top or white-plumed bare-face tamarin (*v.* Hershkovitz 1977), titi (Co.), titi blanco (Co.) (*v.* Hernandez-Camacho and Defler 1989), titi leoncito (Co.), titi pielroja (Co.) (*v.* Hernandez-Camacho and Cooper 1976).

**Type locality** Restricted to the lower Río Sinu, Cordoba, Colombia by Hershkovitz (1949).

**Taxonomy and distribution** *S. oedipus* occurs in northwestern Colombia between the Río Atrato and the lower Río Cauca (west of the Río Cauca and the Isla de Mompos) and Río Magdalena, in the Departments of Atlantico, Sucre, Cordoba, and western Bolivar, northwestern Antiquoia (from the Uraba region, west of the Río Cauca) and northeastern Choco, east of the Río Atrato, from sea level up to 1500 m (Hernandez-Camacho and Cooper 1976; Hershkovitz 1977; Hernandez-Camacho and Defler 1989).

**Status**
1988 IUCN Red List of Threatened Animals—Endangered
Colombian Endangered Species List

CITES Appendix I
Endemic to northern Colombia
Neyman (1978) estimated that 75 per cent of the original distribution of *S. oedipus* had been cleared for agriculture and pasture, and that the remainder of its range was represented by small isolated forest patches and the Paramillo National Natural Park (460 000 ha). Cerquera (1985) reported on the threats regarding the construction of two hydroelectric dams, Urra I and Urra II, on the Ríos Sinu and San Jorge, in the south of its range. Urra II is sited within the Paramillo National Natural Park and is expected to flood more than 54 000 ha of primary and secondary forest, within what is considered to be the last major stronghold for the species. *S. oedipus* also occurs in the Los Colorados Sanctuary (1000 ha).

## ***Saguinus geoffroyi*** (Pucheran 1845)

Red-crested bare-face tamarin (Engl.) (*v.* Hershkovitz 1977), Geoffroy's tamarin (Engl.), titi (Co.), bichichi (Co.) (*v.* Hernandez-Camacho and Cooper 1976).

**Type locality** Panama, restricted to the Canal Zone by Hershkovitz (1949).

**Taxonomy and distribution** Hershkovitz (1977) considers Geoffroy's tamarin to be subspecific to *S. oedipus*. Mittermeier and Coimbra-Filho (1981) argued that it is a distinctive form with no evidence of intergradation. Their conviction was reinforced by studies of the dental morphology, and their comparison with other *Saguinus* forms which Hershkovitz (1977) considered good species (Hanihara and Natori 1987).

*S. geoffroyi* occurs in the tropical forested zones of Panama, the bordering Coto region of Costa Rica, and northwestern Colombia, west of the Ríos Atrato and San Juan in the Department of Choco, from sea level to nearly 900 m (Hernandez-Camacho and Cooper 1976; Hershkovitz 1977).

**Status**
CITES Appendix I
In Colombia, Hernandez-Camacho and Cooper (1976) report that the status of *S. geoffroyi* is poorly known, but it is found in quite remote areas and is not under threat from exploitation as is *S. oedipus*. Skinner (1985) surveyed *S. geoffroyi* populations in Panama and indicated that they are threatened by capture for sale and by timber extraction (even in National Parks). However, the species is still common in remaining forests and even appears in private gardens and reserved areas in parts of the city of Panama, for example Ancon and Balboa (Mittermeier, personal observation). The following conservation units are within its geographical distribution:

Colombia
  Los Katios National Natural Park (72 000 ha)
Panama
  Darien National Park (597 000 ha)
  Soberania National Park (22 000 ha)
  Portobelo National Park (17 364 ha)
  Altos da Campana National Park (4816 ha)
  Barro Colorado Natural Monument (c.5400 ha)
  Comarca San Blas Anthropological Reserve (141 000 ha)

## Genus LEONTOPITHECUS Lesson 1820

The golden lion tamarin, black lion tamarin, and golden-headed lion tamarin are considered valid species, following Rosenberger and Coimbra-Filho (1984) and Mittermeier *et al.* (1988*b*), as opposed to subspecies of *L. rosalia* (*v.* Coimbra-Filho and Mittermeier 1972; Hershkovitz 1977; Mittermeier and Coimbra-Filho 1981; Forman *et al.* 1986). Their separation as species is based on dental and cranial morphology (Della Serra 1951; Rosenberger and Coimbra-Filho 1984; Natori 1989). The black-faced lion tamarin, *L. caissara*, was first described as a species in 1990, although Coimbra-Filho (1990) regards it as subspecific to *L. chrysopygus*, based on his conviction that the two forms represent extremes of a cline. The conservation status of the four species was recently the subject of an international workshop which resulted in population viability analyses and precise recommendations regarding the direction of future efforts for their protection and the maintenance of captive populations (Seal *et al.* 1990).

### *Leontopithecus rosalia* (Linnaeus 1766)

Golden lion tamarin (Engl.), saui-vermelho (Br.), mico-leão-dourado (Br.), saui-piranga (Br.) (*v.* Coimbra-Filho 1969; Coimbra-Filho and Mittermeier 1972).

**Type locality** Brazil, restricted to the coast between 22° and 23° S, from the Cabo de São Tomé to the municipality of Mangaratiba, by Wied-Neuwied (1826), further restricted to right bank, Rio São João, Rio de Janeiro by Carvalho (1965) (*v.* Coimbra-Filho and Mittermeier 1973*a*; Hershkovitz 1977).

**Distribution** The centre of the distribution of *L. rosalia* is considered to be the basin of the Rio São João, state of Rio de Janeiro (Coimbra-Filho and Mittermeier 1973*a*). In the past this species was distributed throughout the lowland forest (generally below 200 m above sea level) of the central and southern parts of the state of Rio de Janeiro (Coimbra-Filho and Mittermeier 1977). It is now restricted to the municipalities of Silva Jardim, Casimiro de Abreu, Cabo Frio, Araruama, São Pedro da Aldeia, and possibly forest patches in Rio Bonito and Saquarema. Coimbra-Filho and Mittermeier (1973*a*) estimated that the remaining forest habitat totalled no more than 900 km$^2$, but this figure is now undoubtedly considerably less (Coimbra-Filho and Mittermeier 1977).

**Status**
1988 IUCN Red List of Threatened Animals—Endangered
1989 List of Brazilian Fauna Threatened with Extinction
US Endangered Species List
CITES Appendix I
Endemic to the Atlantic coastal forest of Brazil
The endangered status of the golden lion tamarin has been well documented (Coimbra-Filho 1969, 1972, 1984; Coimbra-Filho and Magnanini 1972; Coimbra-Filho and Mittermeier 1973*a*, 1977; Magnanini 1978;), and the situation of three species of lion tamarin recognized up to 1990 has attracted considerable attention and efforts for their conservation (Mallinson 1986; Kleiman 1984, in press). There is now a healthy captive population of *L. rosalia*, which has permitted the establishment of a reintroduction programme at the Poço das Antas Biological Reserve (5200 ha, the only conservation unit within its range), Rio de Janeiro, and surrounding forest patches, initiated in 1984 (Dietz 1985; Kleiman 1984, in press; Kleiman *et al.* 1986). Kierulff and Stallings (1991), surveying the remaining forests in the state of Rio de Janeiro, indicated that the wild population does not exceed 400 individuals. The wild groups that they discovered outside of the Poço das Antas Biological Reserve were mostly in forest patches of less than 1000 ha. They identified deforestation for charcoal, cattle pasture, and agriculture as the principal reason for the continuing decline.

## *Leontopithecus chrysomelas* (Kuhl 1820)

Golden-headed lion tamarin (Engl.), sauim-una (Br.), mico-leão-de-cara-dourada (Br.) (*v.* Coimbra- Filho 1970*a*; Coimbra-Filho and Mittermeier 1972).

**Type locality** Ribeirão das Minhocas, left bank of upper Rio dos Ilhéus, southern Bahia, Brazil (*v.* Hershkovitz 1977).

**Distribution** The distribution and status of *L. chrysomelas* was recently reviewed by Rylands *et al.* (in press). Coimbra-Filho and Mittermeier (1973*a*, 1977) described the range as the south of the state of Bahia, between the Rio de Contas (14° S) and the Rio Pardo (15° 05′S), limited inland by the extent of the evergreen (coastal) and semideciduous (further west) forests. The distribution map provided by Hershkovitz (1977, p.808) shows the Rio Pardo as an affluent of the Rio Jequitinhonha, which it is not. However, recent surveys by I.B. Santos have indicated that it does occur south of the Rio Pardo, to the Rio Jequitinhonha, including a small incursion into the state of Minas Gerais, in the municipalities of Jordânia and Salto de Divisa (Rylands *et al.* 1988, Rylands *et al.* in press). This may be a recent range extension due to forest cutting and trafficking. No evidence has been obtained for it ever having occurred between the lower Rios Pardo and Jequitinhonha and it is today either very rare or extinct from the upper Rio de Contas and west of the Rio Gongogí, west of the BR101 highway (Rylands *et al.* in press). The westernmost locality plotted by Hershkovitz (Poçôes, locality 301, p.808, 1977), is in an area of dry liane forest, and Rylands *et al.* (in press) suggested that it may refer to the eastern part of the municipality of Poçôes, which is semi-deciduous Atlantic forest. This reduces somewhat the western extent of the range as given by Hershkovitz (1977).

**Status**

1988 IUCN Red Data List of Threatened Animals—Endangered
1989 List of Brazilian Fauna Threatened with Extinction
US Endangered Species List
CITES Appendix I

Endemic to the Atlantic coastal forest of Brazil *L. chrysomelas* survives in more localities than any of the other lion tamarin species, and undoubtedly has the largest wild population. However, the remaining forests are being destroyed at an unprecedented rate for the region and the populations surviving are already seriously depleted and isolated (Coimbra-Filho 1970*a*, 1972; Coimbra-Filho and Mittermeier 1973*a*, 1977; Oliver and Santos 1991; Rylands *et al.* in press). Recent efforts to establish a viable captive population are being extremely successful (Mallinson 1987, 1989; Ballou 1989; Mace 1990), and international efforts are now being made to upgrade the status of the Una Biological Reserve, Bahia, decreed in 1980 with an area of 11 400 ha, but in fact being approximately 6250 ha at the present time. Oliver and Santos (1991) provided an important recent review of the status of this species and gave recommendations regarding future efforts for its protection (see also Seal *et al.* 1990).

## *Leontopithecus chrysopygus* (Mikan 1823)

Golden-rumped lion tamarin (Engl.), black lion tamarin (Engl.) (*v.* Mittermeier *et al.* 1985), mico leão-preto (Br.), saui-preto (Br.) (*v.* Coimbra-Filho and Mittermeier 1972).

**Type locality** Ipanema (= Varnhagem or Bacaetava, near Sorocaba), São Paulo, Brazil (*v*. Coimbra-Filho and Mittermeier 1973*a*; Hershkovitz 1977).

**Distribution** This species formerly occurred along the north (right) margin of the Rio Paranapanema, west as far as the Paraná, and between the upper Rios Paranapanema and Tieté in the state of São Paulo. Today it is known only from two widely separated forest patches in protected areas, the Morro do Diabo State Park, municipality of Teodoro Sampaio, and the Caetetús State Reserve, municipality of Gália, (Coimbra-Filho 1970*a*,*c*, 1976*a*,*b*; Coimbra-Filho and Mittermeier 1973*a*, 1977). It may also occur in scattered remnant forest patches in the municipalities of Teodoro Sampaio, Presidente Epitácio, and Presidente Wenceslau in the region of the confluence of the Rios Paranapanema and Paraná.

**Status**
1988 IUCN Red Data List of Threatened Animals—Endangered
1989 List of Brazilian Fauna Threatened with Extinction
CITES Appendix I
US Endangered Species List
Endemic to the Atlantic forest of Brazil
This species is extremely endangered, but, as with *L. rosalia* and *L. chrysomelas*, there are now international efforts for its conservation. The flooding of part of the Morro do Diabo State Park, São Paulo, by the Rosana hydroelectric dam on the Rio Paranapanema, has resulted in an increase in the captive population, formerly limited to the offspring of four individuals from the Morro do Diabo State Park (35 000 ha) at the Rio de Janeiro Primate Centre (CPRJ) (Coimbra-Filho 1970*c*). However, the captive breeding programme for this species has not to date achieved the success of the previous two species due to lack of adequate founders (Simon 1989). Ecological studies and environmental education campaigns are being carried out at the Morro do Diabo State Park (Mittermeier *et al.* 1985). A small population survives in the Caetetús State Reserve of 2178 ha, São Paulo.

## *Leontopithecus caissara* Lorini and Persson 1990

Black-faced lion tamarin (Engl.), carinha-preta (Br.) (*v*. Lorini and Persson 1990), mico-leão-de-cara-preta (Br.) (*v*. Persson and Lorini 1991).

**Holotype locality** Brazil, Paraná, municipality of Guaraquecaba, Ilha de Superagui, Barra do Ararapira (Lorini and Persson 1990).

**Taxonomy and distribution** *L. caissara* has a black face and mane, golden-coloured dorsum and thorax, and black feet, forearms, and tail (Lorini and Persson 1990). Coimbra-Filho (1990) argued that *L. caissara* is only subspecifically distinct from the black lion tamarin, *L. chrysopygus*. He bases his arguments on the pelage colouration which he has observed in a number of captive *L. chrysopygus* in the Rio de Janeiro Primate Centre, sometimes permanent, sometimes transient (during moult), which approximates very closely to that typical of *L. caissara*. The dorsal hairs of the type specimen of *L. caissara* are black at the tip. Also, some *L. chrysopygus* individuals from both the Pontal do Paranapanema and Gália, further east, are intermediate between *L. caissara* and *L. chrysopygus*, and Coimbra-Filho (1990) argued that *L. caissara* and black *L. chrysopygus* represent the two extremes in pelage colouration of the same species.

The black-faced lion tamarin occupies the southernmost limits of the distribution of the callitrichids. The type locality is on the northeastern part the island of Superagui on the coast of the state of Paraná. Other groups have been found in the majority of the island, excepting the extreme north and some higher elevations in the south-west (Persson and Lorini 1991). These authors also reported that it occurs on the mainland, in parts of the valleys of the Rio Sebuí and the Rio dos Patos, limited in the north by the Rio Varadorzinho, and to the west by the Serra da Utinga, Morro do Bico Torto, Morro do Poruquara, and Serra do Gigante. Persson and Lorini (1991) estimated that its entire range is less than 30 000 ha. Four groups have been found to the north, also on the coast, in the municipality of Cananéia in the state of São Paulo (Persson and Lorini in press). The possibility remains that other populations may be found in the lowland forests along the Serra do Mar nearby.

**Status** Endemic to the Atlantic coastal forest of Brazil. The very restricted distribution and the few individuals known to exist make this species extremely endangered (Lorini and Persson 1990), probably the rarest and most endangered of all the callitrichids, despite the fact that part of the island of Superagui, along with the Ilha de Peças, was decreed a National Park (without knowledge of the existence of the lion tamarins), the Superagui National Park of 21 400 ha, Paraná, in 1989. Persson and Lorini (in press) estimate a population not exceeding 250 animals.

# 2

# A vocal taxonomy of the callitrichids

*Charles T. Snowdon*

## Introduction

Vocal characters have been used to clarify taxonomy in several cases with birds and primates. One of the major sources of vocal characters have been calls used in mate attraction and intergroup dispersal. In birds, song is used for mate attraction and dispersal and can differentiate species that are otherwise difficult to discriminate on the basis of morphological characters (see for example Smith 1966). In primates, the long or loud call serves a similar function (Gautier and Gautier 1977). These calls are quite stereotyped in form, have species-specific structure, and may play a role in reproductive isolation.

Marshall and Marshall (1976) have used the song structure of different gibbon species to determine their phylogenetic relationships. Struhsaker (1970) analysed the differences in vocalizations of several species of *Cercopithecus* to determine their phylogenetic affinities. Walek (1978) found similarities between the vocalizations she recorded from *Colobus polykomos* and the vocalizations that Marler (1972) recorded from *Colobus guereza*, and suggested that these two populations might really be the same species. Oates and Trocco (1983) used the structure of the loud calls of black and white colobus (*C. polykomos*) monkeys to develop a tentative alpha taxonomy. Wilson and Wilson (1975) used the vocalizations of banded leaf monkeys (*Presbytis* sp.) to describe possible phylogenetic affinities. The most extensive study using vocal characters to examine phylogenetic relationships is Zimmermann's (1990) work on galagos.

In callitrichids long calls have been very well studied. They appear to be used in several contexts: to defend a group against intruders of the same species, to maintain cohesion within the group, to make contact with an animal separated from the group, and, possibly, to attract a mate (Cleveland and Snowdon 1982; Moynihan 1970). Hodun *et al.* (1981) described subspecific differences in the syllable structures forming the long calls of subspecies of *Saguinus fuscicollis*. Subspecies could be discriminated from each other on the basis of a few acoustic variables. Snowdon *et al.* (1986)

described the long calls of *Leontopithecus rosalia*, *L. chrysomelas*, and *L. chrysopygus*. They found significant differences between the structure of the long calls of each species that paralleled the pattern of differences in craniodental morphology used by Rosenberger and Coimbra-Filho (1984) to argue that the genus *Leontopithecus* consists of three discrete species.

The present paper attempts to use vocal characters as a taxonomic tool regarding several problems in the systematics of callitrichids, involving examples from each of the five genera of the family (*Callithrix*, *Cebuella*, *Callimico*, *Leontopithecus*, and *Saguinus*). I will first summarize the divergent cladistic hypotheses that have been proposed for the Callitrichidae based on a variety of morphological, immunological and other characters. I will then describe the qualitative structure of long calls that I have been able to record from 12 of the approximately 25 species of callitrichids and follow this with a quantitative analysis of their differences. Finally, with the addition of some behavioural data, I will hypothesize relationships between the callitrichids based on vocal and behavioural criteria.

## Relationships of the callitrichids

Figures 2.1 and 2.2 illustrate six different recent schemes describing the affinities within the Callitrichidae. Cronin and Sarich (1978), using immunological data, found a close similarity between marmosets (*Callithrix*) and pygmy marmosets (*Cebuella*), and both of these genera were closely aligned with Goeldi's monkey (*Callimico*). Tamarins (*Saguinus*) were depicted as more distantly related, and the lion tamarins (*Leontopithecus*) were deemed to represent the most distantly related branch in the lineage.

Byrd (1981), using data on dental ontogeny, suggested that marmosets and pygmy marmosets were closely related, but placed the tamarins and lion tamarins together and sharing an ancestry with Goeldi's monkeys. Hershkovitz (1977) using a variety of morphological characters, portrayed all of the marmosets and tamarins as distinct from Goeldi's monkey, but indicated that the marmosets and tamarins were closely related to each other, the pygmy marmosets as the most distantly related and the lion tamarins at an intermediate systematic position.

Boer (1974), using cytotaxonomic criteria, concluded that marmosets and pygmy marmosets were monophyletic, with tamarins and lion tamarins as successively more distant branches. All four of these genera were separate from Goeldi's monkey. Ford (1980*a*), using features of the postcranium, argued that marmosets and pygmy marmosets were cladistically sister-groups, most closely related to tamarins followed by lion tamarins and then Goeldi's monkey. Rosenberger and Coimbra-Filho (1984), using craniodental features, inferred that marmosets and pygmy marmosets were

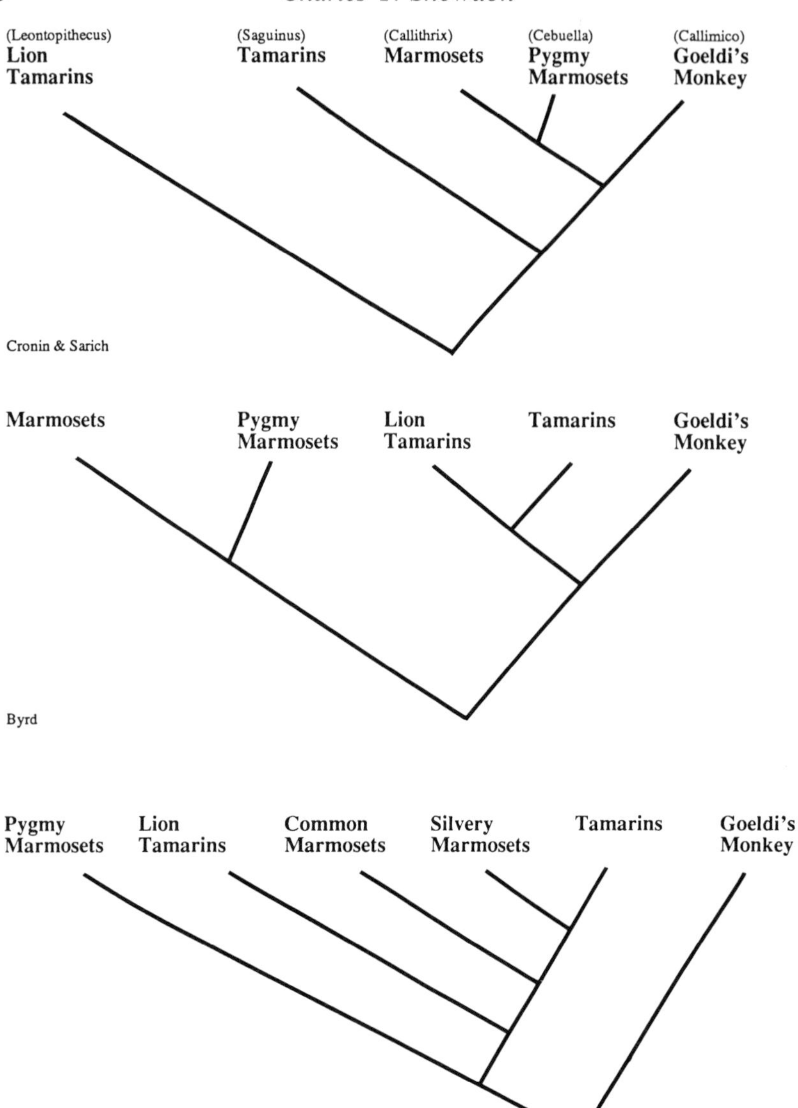


**Fig. 2.1** Depiction of proposed relationships between the Callitrichidae. (Adapted from Fig. 5 of Rosenberger and Coimbra-Filho 1984).

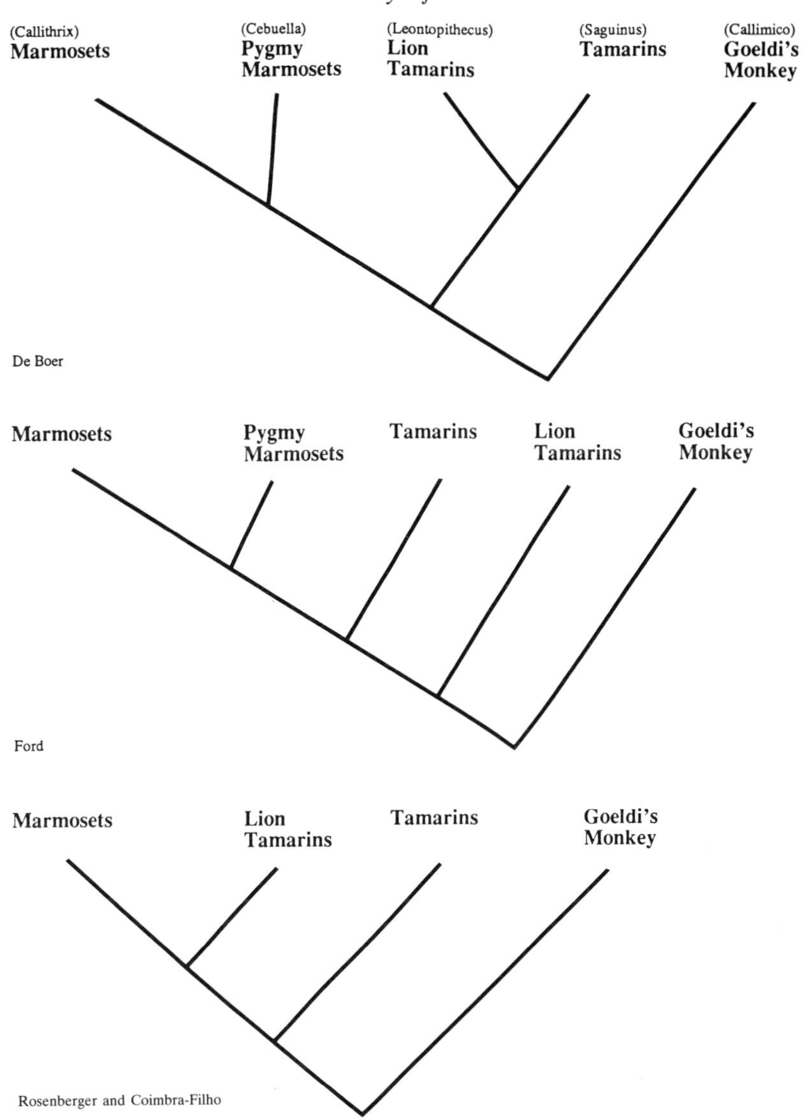


**Fig. 2.2** Depiction of additional proposed relationships between the Callitrichidae. (Adapted from Fig. 5 of Rosenberger and Coimbra-Filho 1984.)

so similar that they should be placed in a single genus, that lion tamarins were the next most closely related genus, and tamarins more distant. All of these genera were monophyletic relative to Goeldi's monkey.

It is clear from this brief review that the variety of morphological and immunological criteria that have been used as evidence of callitrichid genealogy can support a variety of interpretations. There is perhaps no *a priori* reason for choosing one set of characters as being more informative than another, but the greater the number and diversity of systems examined, the greater the likelihood they will converge on the 'proper' genealogical arrangement of the family. What follows next is a description of the vocal characters of the long calls of several different species.

## Features of callitrichid long calls

A recent species-level taxonomic summary of the Callitrichidae (Mittermeier *et al.* 1988*b*) is shown in Table 2.1. Recent modifications for Brazilian species have been made by Coimbra-Filho (1990) including the newly discovered lion tamarin subspecies (*Leontopithecus chrysopygus caissara*). However, for this chapter I will follow the summary of Mittermeier et al. (1988*b*). The species listed in bold type are those for which I have been able to record samples of long calls. Note that all five genera of the callitrichids are represented, although the sample of *Callithrix* is limited to *C. argentata*. Each of the three species of lion tamarins has been recorded as well as one of the two species of the *S. nigricollis* group, two of the three species in the *S. mystax* group, the only species in the *S. bicolor* group, and two of the three species in the *S. oedipus* group.

All of the long calls were recorded spontaneously from animals in the field or housed in normal social groupings in large complex environments in captivity. Captive animals were located at the Centro de Primatologia do Rio de Janeiro in Brazil, Proyecto Primates in Iquitos, Peru, the Jersey Wildlife Preservation Trust in the Channel Islands, Monkey Jungle in Goulds, Florida, and my laboratory in Madison, Wisconsin. Although Cleveland and Snowdon (1982) found that cotton-top tamarins (*Saguinus oedipus*) had different types of long calls used in territory defence and in intragroup cohesion, it was not possible to distinguish such variants in all of the species described here.

Recordings were made with a Sennheiser MD 80 microphone and a Sony TCD-5 professional cassette recorder. All of the vocalizations were analysed using a Kay Model 6061 B Sonograph. Measurements of frequency and duration were made by placing a calibrated graticule over the sonagrams. The number of spontaneously obtained long calls varied greatly among species, as did the number of calls from different individuals. Given

**Table 2.1** The family Callitrichidae (from Mittermeier *et al.* 1988*b*)

*CEBUELLA*
    ***Cebuella pygmaea***

*CALLITHRIX*
  *Callithrix argentata* group
    ***Callithrix argentata***
    *Callithrix emiliae*
    *Callithrix humeralifer*
  *Callithrix jacchus* group
    *Callithrix jacchus*
    *Callithrix aurita*
    *Callithrix flaviceps*
    *Callithrix geoffroyi*
    *Callithrix penicillata*
    *Callithrix kuhli*

*LEONTOPITHECUS*
    ***Leontopithecus rosalia***
    ***Leontopithecus chrysomelas***
    ***Leontopithecus chrysopygus***

*SAGUINUS*
Hairy-faced tamarins
  *Saguinus nigricollis* group
    *Saguinus nigricollis*
    ***Saguinus fuscicollis***
  *Saguinus mystax* group
    ***Saguinus mystax***
    ***Saguinus labiatus***
    *Saguinus imperator*
  *Saguinus midas* group
    *Saguinus midas*

Mottled-faced tamarins
  *Saguinus inustus* group
    *Saguinus inustus*

Bare-faced tamarins
  *Saguinus bicolor* group
    ***Saguinus bicolor***
  *Saguinus oedipus* group
    ***Saguinus oedipus***
    ***Saguinus geoffroyi***
    *Saguinus leucopus*

*CALLIMICO*
  *Callimico goeldii*

the diversity in sampling between species, some caution is in order in interpreting the results.

Figure 2.3 shows examples of long calls and contact trills from silvery marmoset (*Callithrix argentata*) and the pygmy marmoset (*Cebuella pygmaea*). The long calls and trills of both species are remarkably similar. Quantitative measurements (Table 2.2) show that both have similar note

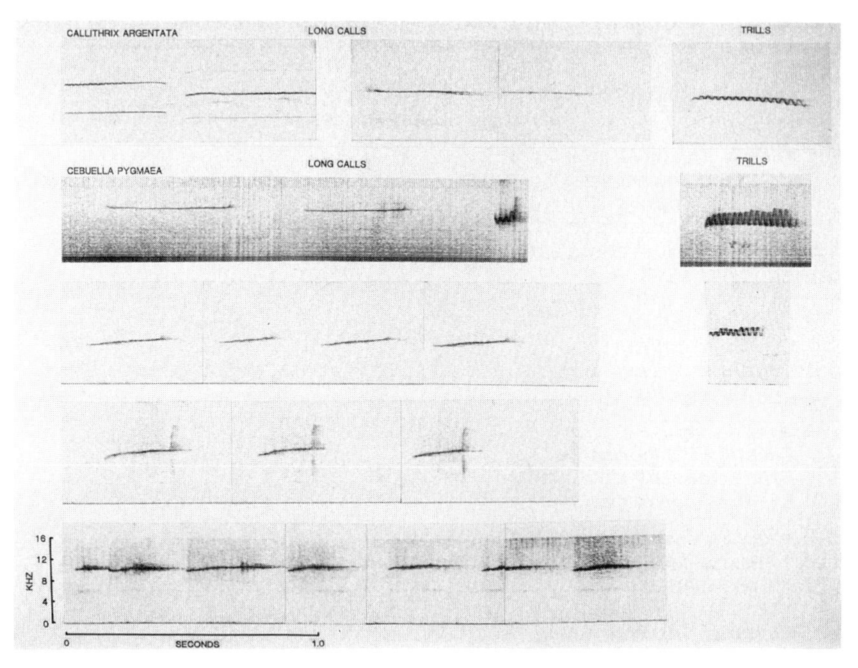


**Fig. 2.3** Spectrograms of long calls and contact trills of the silvery marmoset (*Callithrix argentata*) and the pygmy marmoset (*Cebuella pygmaea*). Note the similarities in structure for both the long calls and the trills.

**Table 2.2** Long call parameters

| Species | Note duration (ms) | Frequency range (kHz) | FM? | Early peak? | Late peak? | FM slope |
|---|---|---|---|---|---|---|
| *Cebuella pygmaea* | 376 | 7.5–10.5 | No | | | |
| *Callithrix argentata* | 508 | 5.5–9.0 | No | | | |
| *Leontopithecus rosalia* | 186 | 5.0–8.0 | Yes | Yes | No | Down |
| *L. chrysomelas* | 134 | 6.0–9.0 | Yes | Yes | No | Down |
| *L. chrysopygus* | 191 | 4.5–7.0 | Yes | Yes | No | Down |
| *Saguinus bicolor* | 813 | 6.5–7.0 | No | | | |
| *Saguinus oedipus* | 600–1000 | 1.1–1.3 | No | | | |
| *Saguinus geoffroyi* | 532 | 1.1–2.2 | No | | | |
| *Saguinus fuscicollis:* | | | | | | |
| *nigrifrons* | 336 | 6.0–10.0 | Yes | Yes | No | Down |
| *illigeri* | 169 | 6.0–10.0 | Yes | No | Yes | Up-down |
| *fuscicollis* | 283 | 6.0–10.0 | Yes | Yes | No | Down |
| *lagonotus* | 211 | 6.0–10.0 | Yes | No | Yes | Up-down |
| *Saguinus mystax* | 88 | 7.5–11.0 | Yes | No | Yes | Up-down |
| *Saguinus labiatus* | 185 | 6.5–10.5 | Yes | Yes | No | Down |
| *Callimico goeldii* | 97 | 7.0–8.5 | No | | | |

duration (370–500 msec) and that the pygmy marmoset calls have a somewhat higher frequency range than those of the silvery marmosets, but with considerable overlap. In both species the main part of the syllable has little or no frequency modulation. The structures of the trills of both species are qualitatively similar as well.

Figure 2.4 shows sonograms of representative long calls of the three species of lion tamarins (Snowdon *et al.* 1986). Although species differences do exist, the calls are all quite similar to one another and quite different from the long calls shown for *C. pygmaea* and *C. argentata* in Fig. 2.3. The durations of individual syllables are quite short, in the range 100–200 msec. There is considerable frequency modulation in each syllable, with the peak frequency occurring early in each syllable and with a downward slope of frequency to the end of the syllable.

Figure 2.5 (top) presents the long call of a pied tamarin (*Saguinus bicolor*). In contrast to the calls shown so far, the pied tamarin long call consists of two or three long syllables with a mean duration of over 800 msec. There is no frequency modulation. In the cotton-up tamarin (*Saguinus oedipus*) and Geoffroy's tamarin (*Saguinus geoffroyi*) (Fig. 2.6), the long calls are remarkably similar in structure to those of the pied tamarin with one major exception. The frequency of the long calls of cotton-top tamarin and Geoffroy's tamarins are in the 1.0–1.5 kHz range, significantly lower in frequency than any of the long calls of other callitrichid species. Cotton-top tamarins and Geoffroy's tamarins are very closely related and have been regarded by some as subspecies rather than separate species (Hershkovitz 1977). The great similarity in long call structure supports the idea that these species are closely related, as well as being related to *S. bicolor*.

Figure 2.7 presents the sonograms of long calls recorded from four different subspecies of the saddleback tamarin (*Saguinus fuscicollis*) (Hodun *et al.* 1981). The calls of each of the subspecies can be differentiated from each other on the basis of a few acoustic parameters, but the main point here is to note the similarities in call structure among the different subspecies and how they differ in turn from the calls shown previously. In the saddle-back tamarin the syllables are shorter than those of the marmosets. The calls are all in the same frequency range, 6.0–10.0 kHz, and show considerable frequency modulation. The patterns of frequency modulation and the mean syllable duration differ significantly between the various subspecies.

Figure 2.8 shows the sonograms of the moustached tamarin (*Saguinus mystax*) and the red-bellied tamarin (*Saguinus labiatus*), both from the same subgroup of tamarins (Hershkovitz 1977). The calls are remarkably similar to one another, having extremely short syllables. Both species have syllables with a great degree of frequency modulation, but the species are

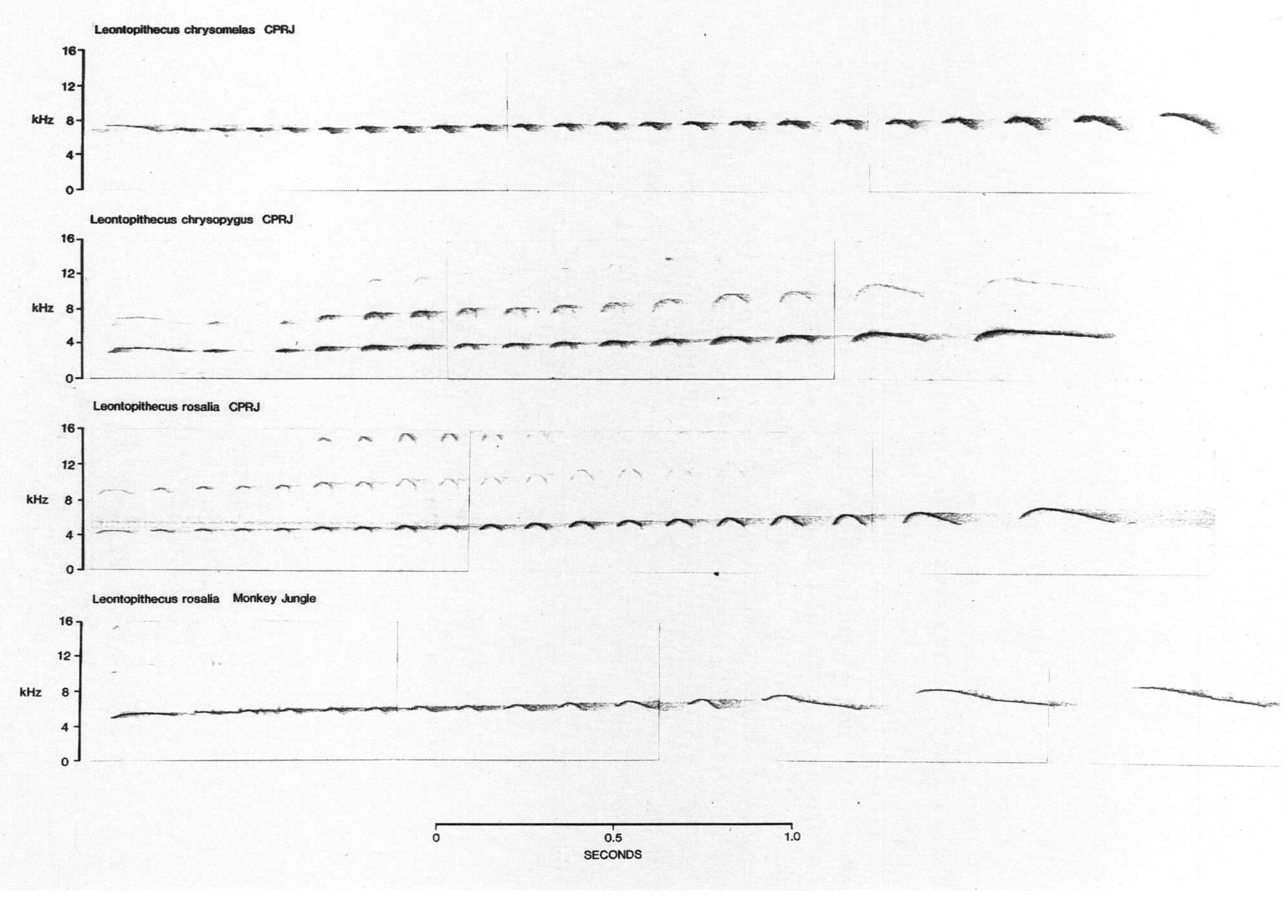


**Fig. 2.4** Spectrograms of long calls of the lion tamarins (*Leontopithecus*). (From Snowdon *et al.* 1986).

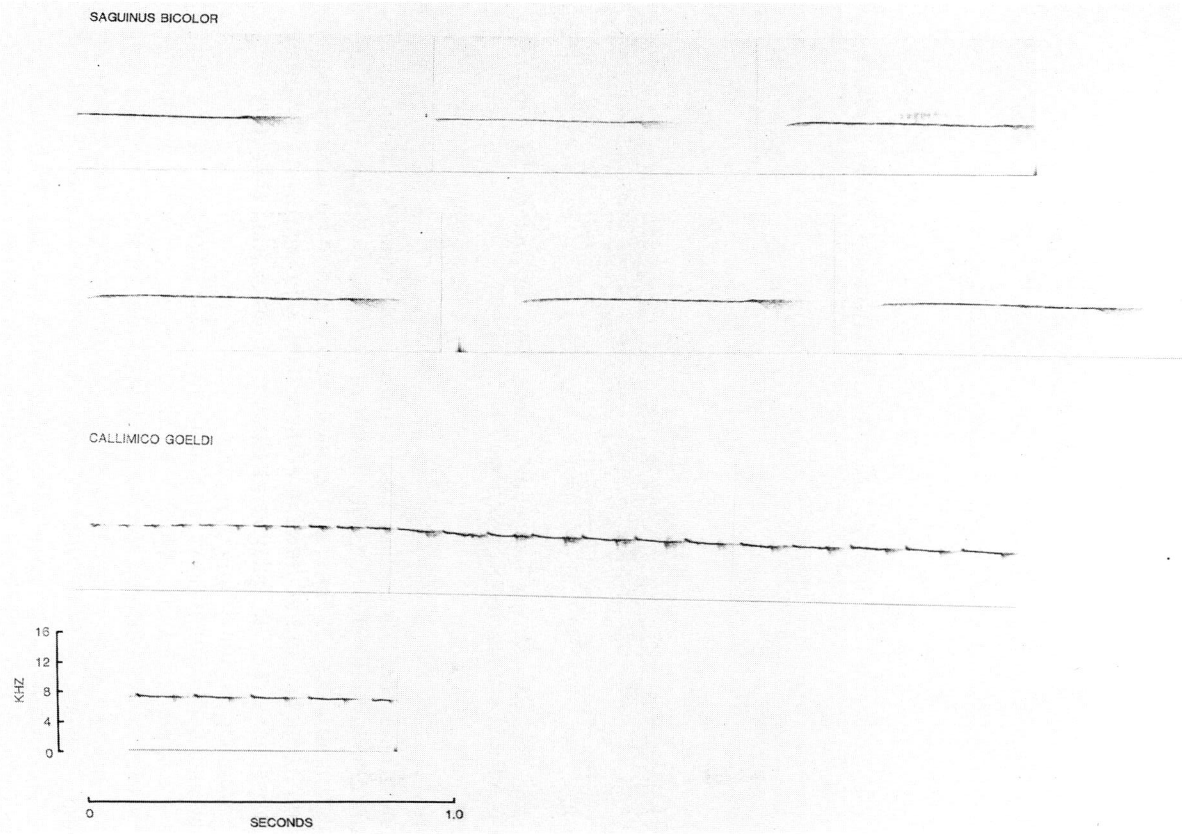

SAGUINUS BICOLOR

CALLIMICO GOELDI

KHZ

16

12

8

4

0

0

SECONDS

1.0

**Fig. 2.5** Spectrograms of the long calls of the bicolor tamarin (*S. bicolor*) and Goeldi's monkey (*Callimico goeldi*).

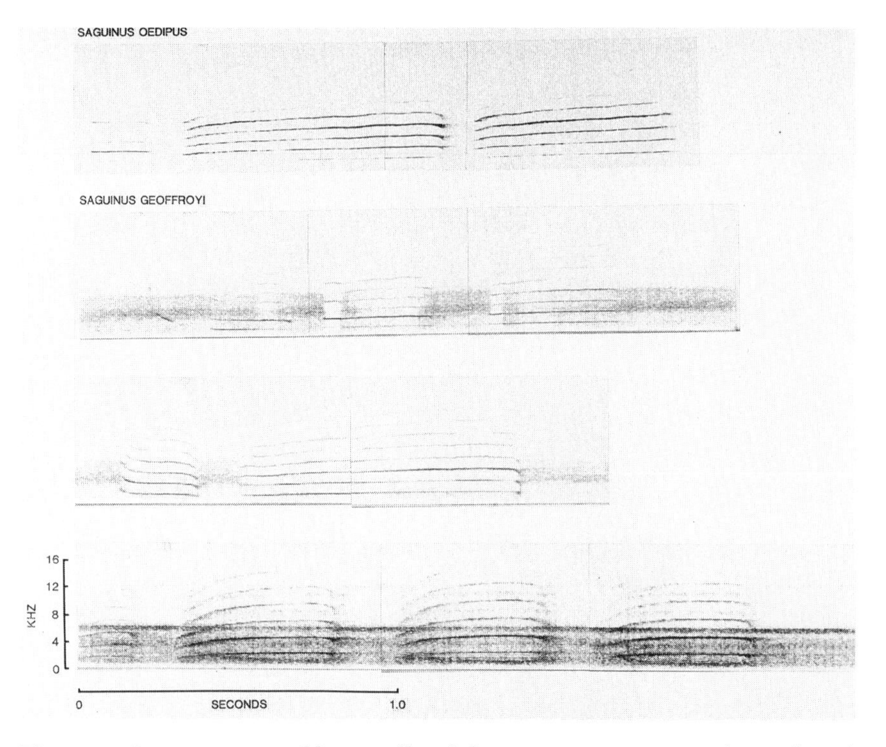


**Fig. 2.6** Spectrograms of long calls of the cotton-top tamarin (*S. oedipus*) and Geoffroyi's tamarin (*S. geoffroyi*).

distinguished from each other by the pattern of frequency modulation. Figure 2.5 (bottom) shows the long calls of Goeldi's monkeys, which are similar in structure to those of the moustached and red-bellied tamarins in both frequency and in duration of the syllables. In contrast to the other monkeys, however, the syllables of Goeldi's monkey long calls have very little frequency modulation.

Several of the quantitative and qualitative characteristics of the long calls are summarized in Table 2.2, which allows us to see the similarities and dissimilarities between the long call structures of the 12 species presented here. First, the long calls of *Cebuella* and *Callithrix* are remarkably similar in structure. Both have 2–3 notes per call with note durations of 300–500 msec, and both span a frequency range of 3 kHz with little frequency modulation. This is consistent with hypotheses of close phylogenetic relationships, as shown in Figs 2.1 and 2.2. Second, the long calls of lion tamarins are also similar to one another but distinct from all other callitrichid species except *S. fuscicollis* which is characterized by having

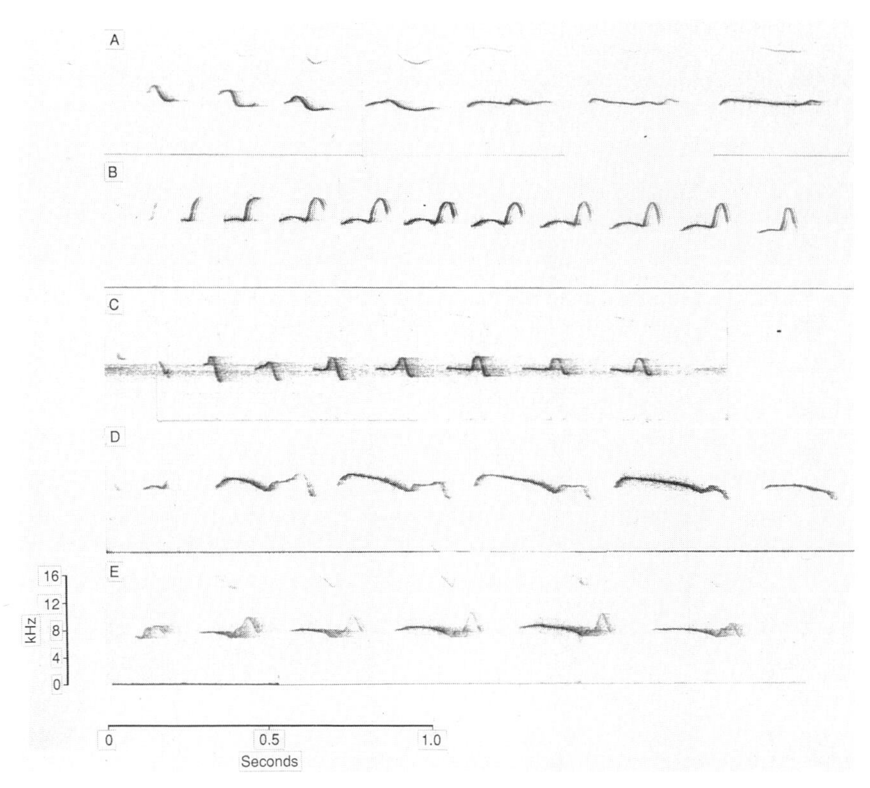


**Fig. 2.7** Spectrograms of the long calls of four subspecies of the saddle-back tamarin (*Saguinus fuscicollis*) A. *S. f. fuscicollis*. B. *S. f. illigeri*. C. *S. f. nigrifrons* (anomaly). D. *S.f. nigrifrons*. E. *S. f. lagontus*. (From Hodun *et al.* 1981).

syllables of moderate duration with extensive frequency modulation. However, there are approximately twice as many notes per call for the *Leontopithecus* species as for the subspecies of *S. fuscicollis*.

The calls of *S. bicolor*, *S. oedipus*, and *S. geoffroyi* are extremely interesting. These are the only long calls among the callitrichids that consist of two or three very long syllables that have little or no frequency modulation. Importantly the calls of *S. oedipus* and *S. geoffroyi* differ from all other species of callitrichids in their low fundamental frequency. Waser and Waser (1977) have shown that there is a 'sound window' in African tropical forest habitats that allows sounds in the frequency range of the long calls of *S. oedipus* and *S. geoffroyi* (1.0–1.5 kHz) to travel for greater distances in the environment than sounds of any other frequency. Thus, it is not surprising to find that the long calls of these species are produced

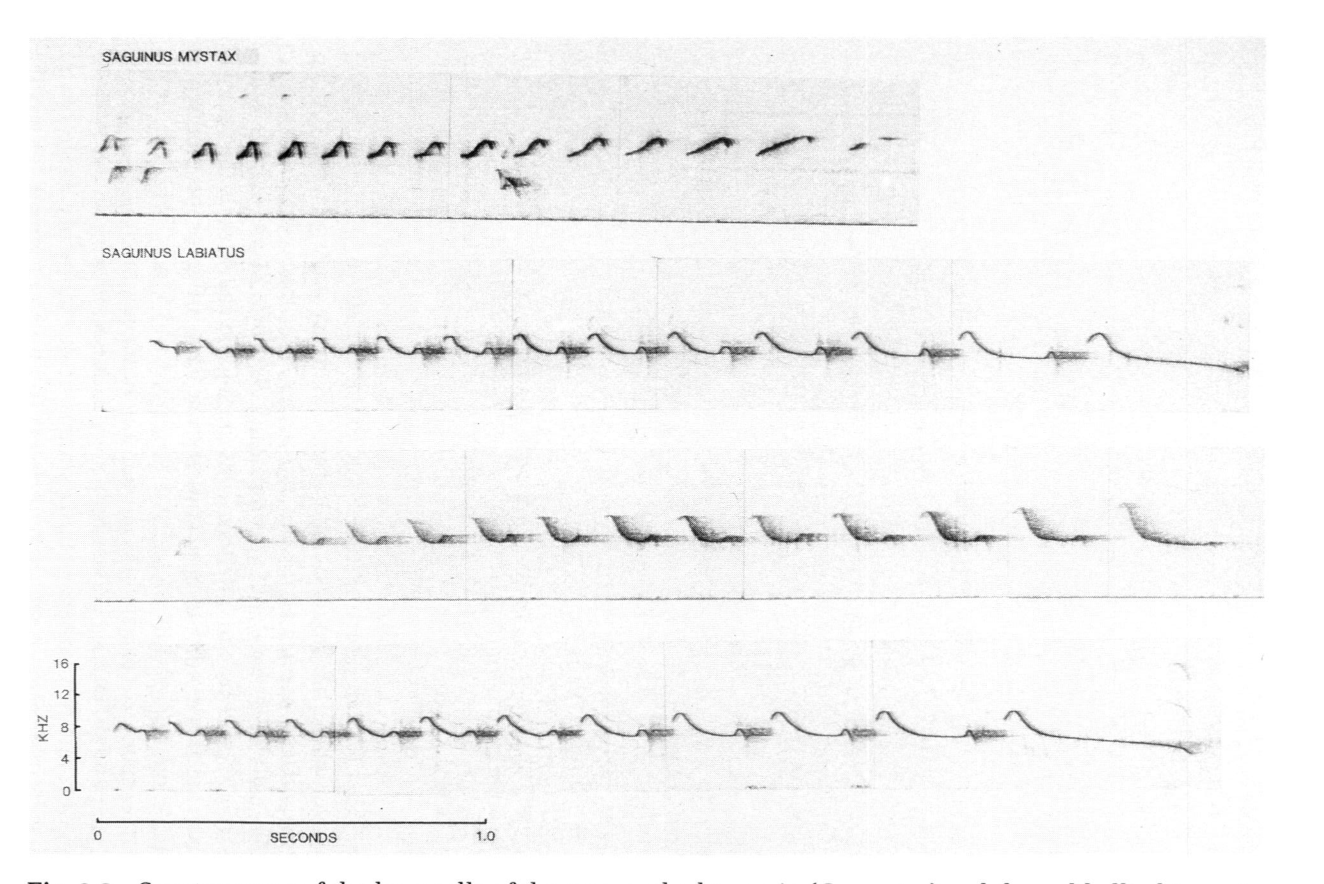


**Fig. 2.8** Spectrograms of the long calls of the moustached tamarin (*S. mystax*) and the red-bellied tamarin (*S. labiatus*).

with a low frequency range. What is surprising is that the long call vocalizations of all of the other callitrichids are so high in frequency. The frequency range of 5.0–10.0 kHz is one where sounds would be relatively rapidly degraded as they passed through the environment.

The long calls of *S. fuscicollis* are quite different from those of the *S. bicolor* group or the *S. oedipus* group. These calls are high pitched with syllables of moderate duration. The syllables have considerable frequency modulation, the pattern of which varies with subspecies. Saddle-back tamarins are sympatric with *S. mystax* and *S. labiatus*, but the long calls of these species are quite different. *S. mystax* and *S. labiatus* have extremely short syllables that are highly modulated, but the location of the peak frequency is at the beginning of the syllable for *S. labiatus* and at the end of the syllable for *S. mystax*, leading to different patterns of frequency modulation. Finally, *C. goeldii*, which in some locations is sympatric with *S. mystax* and *S. labiatus*, also shows similar short syllables, but there is very little frequency modulation.

What could account for the great disparity of the frequency range of *S. oedipus* and *S. geoffroyi* long calls compared with all of the other callitrichids? One possibility is that the home ranges of *S. oedipus* and *S. geoffroyi* are larger than those of other callitrichids, and a long call that travels further is therefore more adaptive for them. However, recent data on home range sizes of different tamarins (Snowdon and Soini 1988) suggests that, if anything, *S. oedipus* and *S. geoffroyi* currently have the smallest home ranges of tamarins.

Snowdon and Hodun (1981) noted that in the Amazon there was a great deal of competing noise from insects, birds, and other sounds at lower frequencies, and they suggested that the use of a higher frequency range for vocal communication by pygmy marmosets might be an adaptation to avoid masking by other sounds. If the acoustic environments were noisier for the Amazonian species and quieter for those in the Atlantic coastal forest, then it is plausible that the high-pitched long calls of most callitrichids would represent an adaptation to environmental noise. It is my impression that the habitat in Colombia and Panama where cotton-top tamarins and Geoffroy's tamarins are found has much less competing noise from insects and birds, so there may not have been pressures for moving to higher-pitched calls, but these impressions need to be documented by careful measurements of environmental noise.

In addition, the cotton-top tamarin and Geoffroy's tamarins have been physically separated by the Andes mountains from the distribution of the rest of the callitrichids, and have probably been isolated for a long period of time. I cannot find any other ecological differences in the habitats of these two species versus the other callitrichids that might explain the differences in long call structure, although future research may indicate some important ecological differences that could account for the differences in frequency ranges of these calls.

## Additional notes on behaviour

Marmosets and tamarins are frequently differentiated from each other on the basis of dentition, with marmosets having dental adaptations for bark gouging in order to obtain access to exudate, an important nutritional resource. The pygmy marmosets and true marmosets share this dentition, and also share the bark gouging and exudate excavation behaviour. In addition, there is a unique threat display that I have seen only in *Cebuella* and *Callithrix* spp. that involves turning the caudal part of the body toward the opponent, lifting the tail, and displaying the genitals while depositing a few drops of urine on the substrate. I have never observed this particular pattern in any of the tamarins, lion tamarins, or Goeldi's monkeys. The affinities of exudate eating and the threat display add to the vocal affinities described above between *Cebuella* and *Callithrix*.

## Phylogenetic relationships based on long call structure

Figure 2.9 presents an attempt to describe the relationships between callitrichids based on the results of the long call structure and behavioural measures reported here. The figure shows a great similarity to the diagram proposed by Rosenberger and Coimbra-Filho (1984), illustrated in Fig. 2.2.

In line with most of the interpretations based on physical features, I group common marmosets and pygmy marmosets very close together. They have long calls that are quite similar, and a trill used in intragroup communication that is also virtually identical. In addition, they share many other common behavioural patterns such as exudate eating and a caudal threat display. I have grouped the lion tamarins (*Leontopithecus*) between the marmosets and the tamarins (*Saguinus*), but this location is admittedly arbitrary based on vocal characters alone. Their long call structures are most similar to those of *S. fuscicollis*, which I think is intermediate among the tamarin groups that I have sampled. Data from additional sources will be needed to place *Leontopithecus* appropriately.

I have divided the tamarins into the three major groupings, with the *S. oedipus* group separating from the other tamarins at a much earlier date. This reflects the clear geographical separation of the *S. oedipus* group from all other tamarins as well as the very different vocal structures of these

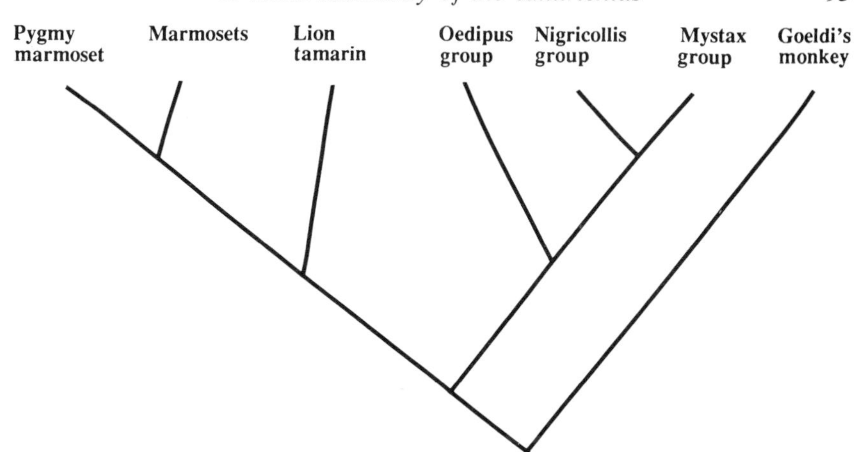


**Fig. 2.9** Hypothesized relationships of the callitrichids based on vocal characters of long calls. Note the similarity between this diagram and the one developed by Rosenberger and Coimbra-Filho (1984) (see Fig. 2.2).

species. *S. bicolor* is not shown in Fig. 2.9, but I see it as an intermediate between the *S. oedipus* group and other tamarin groups. The data on the structure of long calls suggest that there is a clear separation between the *S. nigricollis* group and the *S. mystax* group, although the relative similarity of long call structures in contrast to those of the *S. oedipus* group suggest a more recent divergence of these two groups. Finally, the vocal data of Goeldi's monkey (*Callimico*) suggest very close affinities to the *S. mystax* group of tamarins. However, since *Callimico* produces only one infant at a time while all of the other tamarins and marmosets produce twins, I have kept *Callimico* separate from the tamarins.

This diagram of relationships and the overall results presented here can be used to generate several predictions about species that have not been studied yet. I would predict that the long calls of all remaining species of *Saguinus*, except *S. leucopus*, would have calls in the 5–8 kHz range, while *S. leucopus* should have low frequency long calls with 2–3 long syllables, closer to that of the other *S. oedipus* group tamarins. I would predict that the long calls of *S. nigricollis* would share many similar features to those of *S. fuscicollis* and those of *S. imperator* would share similar features to those of *S. mystax* and *S. labiatus*. *S. inustus* would be an interesting species to study since it is placed as an intermediate between the hairy-faced and bare-faced tamarins (Hershkovitz 1977). Would it also have intermediate long call features?

My data on the genus *Callithrix* are far too limited to attempt any specific predictions. The greatest problem in the analysis presented here is that only one of the nine species of *Callithrix* has been available. Field studies are under-way or in press of many of these species (for example, Mendes 1991*b*) and data on their long call structure will be particularly useful.

In general the use of acoustic variables from the structure of the long calls in callitrichids has been useful in developing hypothesized relationships between species and genera that show parallels to at least a few hypotheses based on morphological criteria. The concatenation of several measures—physical, physiological, and behavioural—into similar patterns across a diversity of taxa provides support for the relationships derived by Rosenberger and Coimbra-Filho (1984) in Fig. 2.2 and those I have developed in Fig. 2.9. Obviously more data from many more species will be needed before a complete picture is available, and I hope that others will seek to fill in the gaps.

## Acknowledgements

The research described here was supported by United States Public Health Service Grant MH 29,775 and a National Institute of Mental Health Research Scientist Award. Travel funds were provided by the Nave fund of Ibero-American Studies Program and by the Graduate School Research Committee of the University of Wisconsin, Madison. I am grateful to Alfred Rosenberger for encouraging me to think about the use of vocal characters in callitrichid taxonomy and for his careful critique of a previous version of this chapter. I am also grateful to the late Alexandra Hodun for her assistance in recording the long calls of various callitrichid species.

# 3

# Experimental multiple hybridism and natural hybrids among *Callithrix* species from eastern Brazil

*Adelmar F. Coimbra-Filho, Alcides Pissinatti, and Anthony B. Rylands*

## Introduction

Although hybridization of primates in captivity has received little attention, it is a subject which can provide important contributions in a number of research fields. Brief comments on this were provided by Wolfe *et al.* (1975) and Deinhardt *et al.* (1976), who judged the hybrid offspring of *Saguinus fuscicollis* × *S. nigricollis* to be a highly satisfactory research model in relation to other callitrichids, being comparable in reproductive performance to the best pure forms. These authors also confirmed the excellent reproductive capacity of *Callithrix jacchus*. Although little used in their laboratories, *C. jacchus* surpassed the fitness shown by a number of *Saguinus* species (*S. fuscicollis*, *S. nigricollis* and *S. oedipus*). For this reason, these authors pointed out that the genus *Callithrix*, although not widely used in research, except for some laboratories in the United Kingdom, would provide an excellent model in biomedical investigation.

In theory it is possible to imagine planning hybrid forms with the aim of obtaining genetic codes for use in specific types of research. Examples include developing hybrids which are susceptible or resistant to certain pathogens, or carriers of genetic anomalies of various origins. The development of planned hybrids could in this way provide significant contributions in immunological and virological investigations, notably where unsatisfactory results have been obtained using pure species or subspecies as models. As an example of this, although *Saguinus mystax* is the only species of this richly varied genus which has to date provided a satisfactory development of hepatitis A virus, the genetic peculiarity which permits susceptibility to this virus is displayed in a less active form among other *Saguinus* forms (Deinhardt *et al.* 1975). It is evident that, despite the similarity of *Saguinus* karyotypes, there are subtle and undetected differences which are sufficient to differentiate these tamarins in their susceptibility to the virus concerned.

Hershkovitz (1977, pp.442–5) listed the reported occurrences of captive hybrids in all callitrichids. Besides our work at the Rio de Janeiro Primate Centre, hybridization amongst the genus *Callithrix* has been recorded only infrequently. Hill (1957) reported on a hybrid *C. jacchus* × *C. penicillata*. Mallinson (1971) and Hampton *et al.* (1971) also reported hybrids between these forms. English (1932, cited in Hershkovitz 1977) reported a hybrid *Callithrix argentata argentata* × *Callithrix jacchus jacchus*, as did Hill (1961). These two forms are considered by Hershkovitz (1977) to be the most distantly related of the *Callithrix* marmosets. Mallinson (1971) reported on a birth (stillborn) of infants to a mated pair of hybrids, each *C. penicillata* and *C. jacchus*.

In this paper we report on the experimental hybridization amongst the six *Callithrix* forms from eastern Brazil, and summarize some recent evidence for localized natural hybrid zones in the wild. The results of the captive hybridization confirm their interspecific, though perhaps somewhat suppressed, fertility, indicating the potential for further experimentation.

It should be emphasized that forms threatened with extinction in the wild were used in these experiments only when there was no short-term possibility of obtaining a mate. It is not our practice to maintain individual marmosets in isolation, and as a result we took advantage of the situation to put different species together, resulting in the hybridization. The experiments were carried out using the descendants of these hybrids.

## *Taxonomy*

Coimbra-Filho (1971), Coimbra-Filho and Mittermeier (1973*b*), Mittermeier and Coimbra-Filho (1981) and Mittermeier *et al.* (1988*b*) recognize six marmoset forms of eastern Brazil as valid species. Coimbra-Filho (1990), however, revised his opinion regarding *C. flaviceps*, considering it a subspecies of *C. aurita* on the basis of the similarity of the phenotypes and their behaviour, besides the hybrid experiments reported below, as well as the extreme similarity of the pelage of the infants, and of the vocalizations of these two forms in comparison to the remaining eastern Brazilian species (Mendes 1991*b*). Wied's marmoset, *C. kuhli*, was previously considered a subspecies of *C. penicillata*, but, following observations of both captive and wild populations, it is here recognized as a good species (see Coimbra-Filho 1985; see also Natori 1990). Hershkovitz (1977) argued that the form *kuhli* is a hybrid *C. penicillata* × *C. geoffroyi*, and placed all eastern Brazilian *Callithrix* as subspecies of *C. jacchus* (see Rylands *et al.*, Chapter 1, this volume).

## Experimental hybridization: methods

The following forms were used during our experiments: *Callithrix a. aurita*, *C. a. flaviceps*, *C. geoffroyi*, *C. kuhli*, *C. penicillata*, and *C. jacchus*. The founder members of these colonies came from numerous localities in the vast region extending from the state of Piauí in the north-east to São Paulo in the south-east of Brazil. The scheme used to follow the hybridization programme is shown in Fig. 3.1 (see also Tables 3.1–3.5).

The symbols used in Tables 3.1–3.5 are as follows:

*, the individual was alive at the time;
**, the individual was exchanged;
***, the animal was donated to another colony;
****, the individual escaped.

Acronyms: CPRJ, Centro de Primatologia do Rio de Janeiro; FEEMA, Fundação Estadual de Engenharia do Meio Ambiente. Abbreviations: Ca, *Callithrix a. aurita*; Cf, *C.a. flaviceps*; Cg, *C. geoffroyi*; Cg s/n°., *C. geoffroyi* without a registration number; Ck, *C. kuhli*, Cp, *C. penicillata*; Cp s/n°; *C. penicillata* without a registration number; Cj s/n°, *C. jacchus* without a registration number: H, hybrid. The terms double hybrid (2H), triple hybrid (3H), quadruple hybrid (4H), quintuple hybrid (5H), and sextuple hybrid (6H) have been adopted in order to differentiate such forms of hybridization from those in which there is a contribution of two, three, four, five, or six factors, referred to in this case as dihybrid, trihybrid, tetrahybrid, pentahybrid, and hexahybrid, respectively. The number of participating genes among the interspecific hybridizations discussed here is high, and as such deserve concepts different from the simple hybridizations between two, three, four, five, or six genes, the terminology of which in no way corresponds to the complexity of the hybridizations in our experiments. Only those hybridizations obtained up to December 1985 are registered (Table 3.1).

The marmosets are fed a regular, controlled diet (see Coimbra-Filho *et al.* 1981*b*) and housed in pairs in cages of approximately 8 m$^2$. Breeding was mostly successful under these conditions, and young and adult hybrids showed similar general behaviour to that of the pure forms.

## Experimental hybridization: results

The experimental hybridizations resulted in several distinct phenotypes. Space does not allow for the necessarily lengthy descriptions of the hybrid forms, but we point out those features which we consider pertinent.

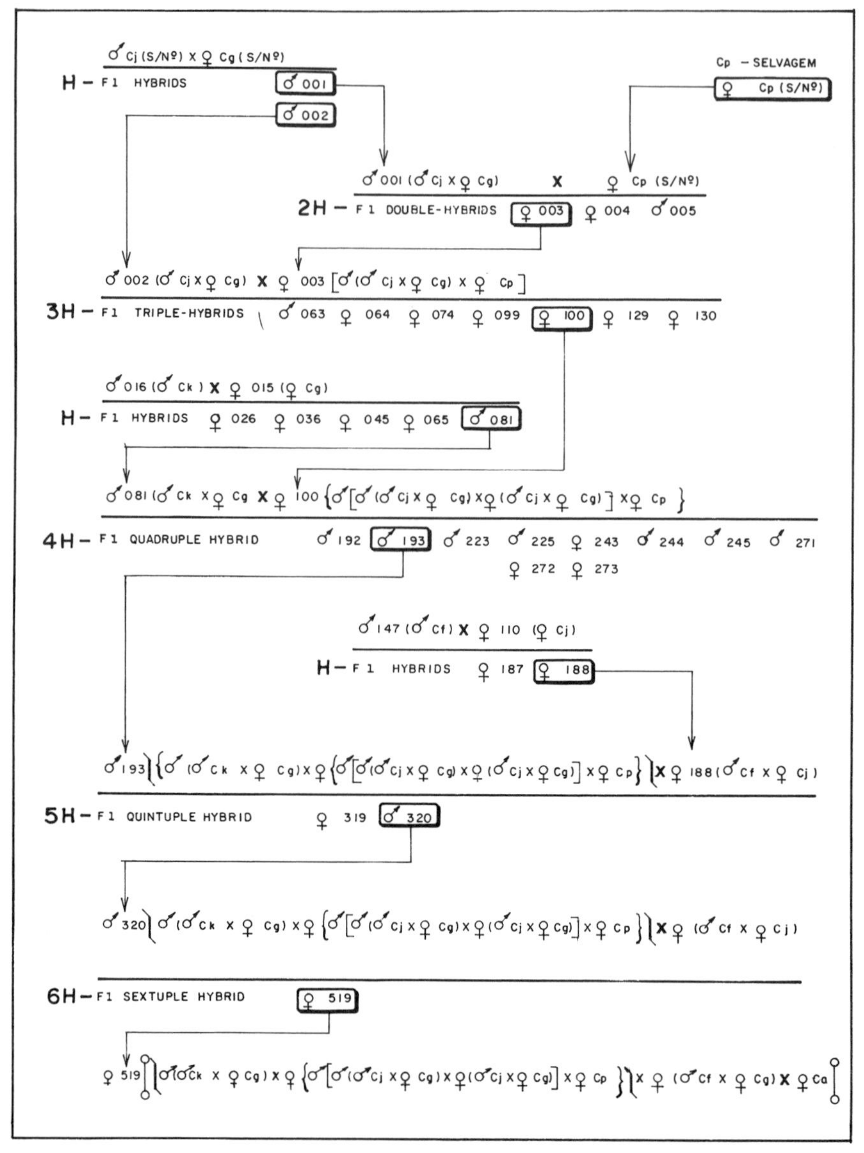


**Fig. 3.1** Genealogy of the sextuple hybrids of the six south-east Brazilian marmosets, *Callithrix*. Abbreviations and symbols are explained in the text.

**Table 3.1** Genealogy, basic information, and biometric data of hybrids produced at CPRJ

| ANCESTORS | | HYBRID (H) | | BIRTH | EXIT | MUSEUM (nº) | COLLECTION MATERIAL | | | | WEIGHT (g) | TOTAL LENGTH (mm) | TAIL LENGTH (mm) |
|---|---|---|---|---|---|---|---|---|---|---|---|---|---|
| ♂ | ♀ | ♂ | ♀ | | | | SKIN | SKULL | FORMALIN. | PARAFFIN. | | | |
| Cj S/nº | Cg S/nº | 001 | | 01-11-68 | 07-04-76 | 028 | X | X | X | X | 408.0 | 550 | 340 |
| | | 002 | | 01-11-68 | 14-09-76 | 049 | X | X | X | X | 430.0 | 510 | 290 |
| Ck 016 | Cg 015 | | 026 | 19-06-73 | 23-11-75 | 016 | X | X | X | X | 475.0 | 565 | 340 |
| | | | 036 | 28-11-73 | 05-05-76 | 033 | | X | X | X | 408.0 | 585 | 360 |
| | | | 045 | 02-05-74 | 20-03-78 | 0121 | X | X | X | | 492.0 | 562 | 337 |
| | | | 065 | 17-01-74 | 22-08-78 | 0127 | X | X | X | | 309.0 | 553 | 337 |
| | | 081 | | 20-05-75 | 20-07-80 | 0221 | X | X | X | | 314.0 | 547 | 344 |
| Ck 016 | Cj 084 | 0131 | | 03-05-75 | 29-06-77 | 075 | X | X | X | | 362.5 | 520 | 320 |
| | | 0155 | | 11-11-76 | 29-03-78 | 0122 | X | X | X | | 560.0 | 547 | 327 |
| | | 0156 | | 11-11-76 | 29-03-78 | 0123 | X | X | X | | 500.5 | 565 | 340 |
| | | | 167 | 22-05-77 | 02-08-80 | 0192 | X | X | X | | 420.0 | 556 | 342 |
| | | | 168 | 22-05-77 | 19-01-79 | 0146 | X | X | X | | 466.5 | 541 | 215 |
| Ck 016 | Cp 199 | | 216 | 09-12-78 | 25-02-80 | 0197 | X | X | X | X | 280.0 | 590 | 212 |
| | | 217 | | 09-12-78 | 07-02-80 | 0191 | | | X | X | 174.0 | 475 | 283 |
| | | 234 | | 18-08-79 | 30-12-82 | 0376 | | | X | X | 283.0 | 500 | 300 |
| | | | 235 | 18-08-79 | 19-10-81 | 0311 | X | X | X | X | 300.0 | 507 | 303 |
| Ck 325 | H 235 | 357 | | 16-10-81 | 20-10-81 | 0313 | | | | X | 34.0 | 193 | 109 |
| | | | 358 | 16-10-81 | 19-10-81 | 0312 | | | | X | 36.5 | 194 | 107 |
| Ck 325 | Cp 295 | 431 | | 25-12-82 | 13-12-83 | ** | | | | | | | |
| | | 432 | | 25-12-82 | 13-12-83 | ** | | | | | | | |
| | | 520 | | 07-11-83 | 13-12-83 | ** | | | | | | | |
| | | | 521 | 07-11-83 | 13-12-83 | ** | | | | | | | |
| Ck 234 | Cp 199 | | 369 | 28-11-81 | 28-11-81 | 0327 | | | X | | 35.0 | 199 | 103 |
| | | | 370 | 28-11-81 | 03-02-82 | 0331 | | | X | | 56.0 | 299 | 182 |
| Cp 072 | Cj 067 | 087 | | 04-06-75 | 17-08-76 | 045 | X | X | X | X | 315.0 | 511 | 316 |
| | | 088 | | 04-06-75 | 17-08-76 | 046 | X | X | X | X | 375.0 | 548 | 328 |
| | | 121 | | 01-76 | 17-05-76 | 036 | X | X | X | | 230.0 | 470 | 285 |
| | | 122 | | 01-76 | 17-05-76 | 035 | X | X | X | | 235.0 | 465 | 280 |
| Cp 068 | Cj 066 | 119 | | 24-12-75 | 18-05-76 | 029 | X | | | | 200.0 | 430 | 250 |
| | | | 120 | 24-12-75 | 18-05-76 | 030 | X | | | | 230.0 | 460 | 280 |
| | | 140 | | 10-09-76 | 04-10-77 | 089 | X | | X | | 374.0 | 503 | 300 |
| | | 141 | | 10-09-76 | 10-10-77 | 091 | X | X | X | X | 415.0 | 535 | 313 |
| | | | 142 | 10-09-76 | 15-09-76 | 050 | | | X | | 30.0 | 194 | 85 |
| Cp 068 | Cj 200 | | 164 | 22-05-77 | 18-12-78 | 0141 | | | X | | 506.0 | 543 | 315 |
| | | 165 | | 22-05-77 | 30-10-78 | ** | | | | | | | |
| | | | 166 | 22-05-77 | 29-05-77 | 074 | | | X | | | | |
| Cg 069 | Cj 066 | 073 | | 02-01-75 | 11-06-76 | 040 | X | X | X | X | 480.0 | 575 | 335 |
| | | | 085 | 03-06-75 | 18-06-76 | 041 | X | X | X | X | 500.0 | 563 | 343 |
| | | 086 | | 03-06-75 | 20-03-78 | 0120 | X | X | X | | 583.0 | 618 | 373 |
| Cg 070 | Cp 135 | 180 | | 16-11-77 | 24-05-80 | 0152 | X | X | X | | 518.0 | 497 | 286 |
| | | 181 | | 16-11-77 | 19-01-79 | 0145 | X | X | X | | 185.5 | 483 | 300 |
| Cg 070 | Cp 232 | | 263 | 14-11-79 | 16-01-80 | 0181 | | | X | | 62.5 | 310 | 190 |
| | | | 264 | 14-11-79 | 11-08-81 | 0277 | X | X | X | X | 316.0 | 491 | 294 |
| Cf 147 | Cj 110 | | 187 | 12-01-78 | | * | | | | | | | |
| | | | 188 | 12-01-78 | 04-12-80 | 0267 | | | X | | 350.5 | 596 | 364 |
| Ca 530 | Ck 468 | | 633 | 14-02-85 | | * | | | | | | | |
| | | | 634 | 14-02-85 | | * | | | | | | | |
| TOTAL | | 24 | 23 | | | | 27 | 25 | 35 | | | | |

SOURCE: CPRJ — FEEMA REGISTER

**Table 3.2** Genealogy, basic information, and biometric data of double-hybrids produced at CPRJ

| ANCESTORS | | DOUBLE HYBRID (2H) | | BIRTH | EXIT | MUSEUM (nº) | COLLECTION MATERIAL | | | | WEIGHT (g) | TOTAL LENGTH (mm) | TAIL LENGTH (mm) |
|---|---|---|---|---|---|---|---|---|---|---|---|---|---|
| ♂ | ♀ | ♂ | ♀ | | | | SKIN | SKULL | FORMALIN. | PARAFFIN. | | | |
| H 001 | Cp S/nº | | 003 | 01-12-70 | 30-04-76 | 0032 | X | X | | | 340.0 | 270 | 100 |
| | | | 004 | 11-08-71 | 28-10-75 | 0013 | X | X | X | X | | 615 | 370 |
| | | 005 | | 11-08-71 | 23-04-76 | 0031 | X | X | X | X | 518.0 | 570 | 340 |
| H 086 | H 065 | 160 | | 05-01-77 | 05-01-77 | 0104 | | | X | | | | |
| | | | 161 | 05-01-77 | 05-01-77 | 0105 | | | | | | | |
| H 140 | Cj 066 | 189 | | 18-01-78 | 24-12-82 | 0374 | | | X | | 259.0 | 476 | 213 |
| | | | 190 | 18-01-78 | 20-07-78 | * * * | | | | | | | |
| | | 191 | | 18-01-78 | 24-01-78 | 0111 | | | X | | 27.0 | 170 | |
| 2H 189 | Cj 110 | 322 | | 05-11-80 | 19-10-81 | 0310 | X | X | X | | 254.0 | 496 | 298 |
| | | 323 | | 05-11-80 | 05-11-80 | 0259 | | | X | | 27.5 | 171 | 99 |
| H 234 | Cp 295 | | 392 | 27-05-82 | 02-05-83 | * * | | | | | | | |
| | | | 393 | 27-05-82 | 02-05-83 | * * | | | | | | | |
| TOTAL | | 6 | 6 | | | | 4 | 4 | 7 | | | | |

SOURCE: CPRJ — FEEMA REGISTER

**Table 3.3** Genealogy, basic information, and biometric data of triple-hybrids produced at CPRJ

| ANCESTORS | | TRIPLE—HYBRID (3H) | | BIRTH | EXIT | MUSEUM (nº) | COLLECTION MATERIAL | | | | WEIGHT (g) | TOTAL LENGTH (mm) | TAIL. LENGTH (mm) |
|---|---|---|---|---|---|---|---|---|---|---|---|---|---|
| ♂ | ♀ | ♂ | ♀ | | | | SKIN | SKULL | FORMALIN. | PARAFFIN. | | | |
| H 002 | 2H 003 | 063 | | 25-10-74 | 18-08-78 | 0126 | X | X | X | | 441.0 | 496 | 295 |
| | | | 064 | 25-10-74 | 30-10-80 | * * | | | | | | | |
| | | | 074 | 03-03-75 | 28-11-77 | 0100 | | | X | X | | | |
| | | | 099 | 30-09-75 | 10-11-75 | 0025 | X | X | | | | 300 | 190 |
| | | | 100 | 30-09-75 | 13-03-80 | 0198 | | | | X | | | |
| | | | 129 | 29-03-76 | 02-05-83 | * * | | | | | | | |
| | | | 130 | 29-03-76 | 12-05-76 | 0034 | X | X | X | X | 48.0 | 250 | 150 |
| H 165 | 3H 129 | 218 | | 21-12-78 | 27-12-78 | 0142 | | | X | | 33.5 | 201 | 114 |
| | | | 219 | 21-12-78 | 19-07-82 | 0357 | | | X | | 439.0 | 520 | 308 |
| | | 220 | | 21-12-78 | 30-10-80 | * * | | | | | | | |
| | | 237 | | 28-08-79 | 02-05-83 | * * | | | | | | | |
| | | | 238 | 28-08-79 | 30-10-80 | * * | | | | | | | |
| | | 268 | | 30-01-80 | 25-06-81 | 0273 | X | X | X | | 397.5 | 515 | 295 |
| | | 269 | | 30-01-80 | 30-06-81 | 0274 | X | X | X | | 348.5 | 493 | 295 |
| | | 270 | | 30-01-80 | 31-01-80 | * * * | | | | | | | |
| 3H 220 | Cp 256 | | 314 | 12-10-80 | 14-10-80 | 0258 | | | X | | | | 114 |
| TOTAL | | 7 | 9 | | | | 5 | 5 | 8 | | | | |

SOURCE: CPRJ — FEEMA REGISTER

**Table 3.4** Genealogy, basic information, and biometric data of quadruple-hybrids produced at CPRJ

| ANCESTORS | | QUADRUPLE HYBRID (4H) | | BIRTH | EXIT | MUSEUM (nº) | COLLECTION MATERIAL | | | | WEIGHT (g) | TOTAL LENGTH (mm) | TAIL LENGTH (mm) |
|---|---|---|---|---|---|---|---|---|---|---|---|---|---|
| ♂ | ♀ | ♂ | ♀ | | | | SKIN | SKULL | FORMALIN. | PARAFFIN | | | |
| Ck 016 | 3H 074 | 182 | | 28-11-77 | 28-11-77 | 0102 | | | X | | 42.0 | 199 | 112 |
| | | | 183 | 28-11-77 | 28-11-77 | 0103 | | | X | | 38.0 | 195 | 108 |
| H 081 | 3H 100 | 192 | | 04-02-78 | 04-10-81 | 0309 | X | X | X | X | 361.0 | 516 | 304 |
| | | 193 | | 04-02-78 | 16-12-83 | ** | | | | | | | |
| | | 223 | | 23-01-79 | 16-12-83 | ** | | | | | | | |
| | | 224 | | 23-01-79 | 27-02-79 | 0151 | | | X | | 36.5 | 232 | 140 |
| | | 225 | | 23-01-79 | 24-01-79 | *** | | | | | 27.0 | 185 | 107 |
| | | | 243 | 13-09-79 | 18-09-79 | 0158 | | | X | | 24.0 | 178 | 100 |
| | | 244 | | 13-09-79 | 24-10-79 | **** | | | | | | | |
| | | 245 | | 13-09-79 | 24-10-79 | **** | | | | | | | |
| | | 271 | | 15-02-80 | 19-02-80 | 0195 | | | X | | 29.5 | 183 | 99 |
| | | | 272 | 15-02-80 | 17-02-80 | 0196 | | | X | | 28.5 | 192 | 110 |
| | | | 273 | 15-02-80 | 04-04-83 | 0386 | X | X | X | | | | |
| 4H 192 | 3H 219 | 309 | | 08-10-80 | 19-10-80 | 0253 | | | X | | 32.0 | 194 | 106 |
| | | 310 | | 08-10-80 | 20-10-80 | 0254 | | | X | | 33.5 | 158 | |
| | | 311 | | 08-10-80 | 02-12-83 | 0425 | X | X | X | | 508.0 | 540 | 320 |
| | | | 344 | 11-08-81 | 16-12-83 | ** | | | | | | | |
| | | | 345 | 11-08-81 | | * | | | | | | | |
| | | | 346 | 11-08-81 | 20-08-81 | 0302 | | | X | | 35.5 | 197 | 105 |
| 4H 223 | 4H 273 | | 399 | 10-09-82 | 10-09-82 | 0359 | | | X | | 52.0 | 216 | 117 |
| 4H 311 | 4H 344 | 481 | | 19-08-83 | 16-12-83 | ** | | | | | | | |
| | | | 482 | 19-08-83 | 16-12-83 | ** | | | | | | | |
| | | 483 | | 19-08-83 | 24-08-83 | 0401 | | | X | | 26.0 | 180 | 100 |
| TOTAL | | 14 | 9 | | | | 3 | 3 | 14 | | | | |

SOURCE: CPRJ – FEEMA REGISTER

Carcasses preserved in formaldehyde, skins, and skulls are maintained in the museum of the Primate Centre (CPRJ).

## *Single hybrids*

Coimbra-Filho (1970*b*, see also Coimbra-Filho and Mittermeier 1973*b*) described the phenotypes of offspring of three births of hybrid *C. jacchus* (male) × *C. geoffroyi* (female) (see Fig. 3.2A, B). Although the first set of twins were healthy and reared successfully, the second set were blind in both eyes, and all of three infants of the third birth were blind in the left eye, two of them dying at an early age. Genetic incompatibility was cited as a possible cause of this blindness. Coimbra-Filho and Maia (1976) reported on the birth of two twins resulting from a *C. geoffroyi* male and a *C. jacchus* female. The first set were abandoned by the inexperienced parents, but the already pregnant female was subsequently paired with a wild male *C. penicillata*, and the offspring were reared successfully. The hybrids showed pelage colouration and pigmentation intermediate between those of the typical pure forms. The white face-mask of *C. geoffroyi* was

**Table 3.5** Genealogy, basic information, and biometric data of quintuple-hybrids produced at CPRJ

| ANCESTORS | | QUINTUPLE-HYBRID (5H) | | BIRTH | EXIT | MUSEUM (nº) | COLLECTION MATERIAL | | | | WEIGHT (g) | TOTAL LENGTH (mm) | TAIL LENGTH (mm) |
|---|---|---|---|---|---|---|---|---|---|---|---|---|---|
| ♂ | ♀ | ♂ | ♀ | | | | SKIN | SKULL | FORMALIN. | PARAFFIN. | | | |
| 4H 193 | H 188 | | 319 | 26-10-80 | 26-10-81 | 0318 | X | X | X | | 243.0 | 515 | 307 |
| | | 320 | | 26-10-80 | | * | | | | | | | |
| 4H 193 | H 187 | 347 | | 31-08-81 | 04-09-81 | 0304 | | | X | | 27.5 | 155 | 76 |
| | | | 348 | 31-08-81 | 04-09-81 | 0305 | | | X | | 21.0 | 142 | 63 |
| | | | 349 | 31-08-81 | 05-09-81 | 0306 | | | X | | 40.5 | 188 | 102 |
| | | 379 | | 07-02-82 | 15-02-83 | 0381 | | | X | | 450.0 | 530 | 320 |
| | | 380 | | 07-02-82 | | * | | | | | | | |
| | | | 381 | 07-02-82 | 13-02-82 | 0336 | | | X | | 31.0 | 206 | 122 |
| | | | 382 | 07-02-82 | 18-02-82 | 0337 | | | X | | 32.5 | 216 | 126 |
| | | | 396 | 29-08-82 | | * | | | | | | | |
| | | 397 | | 29-08-82 | | * | | | | | | | |
| | | 398 | | 29-08-82 | 20-12-83 | 0427 | X | X | X | | 329.0 | 540 | 330 |
| | | 436 | | 01-02-83 | 07-02-83 | 0380 | | | X | | | | |
| | | 437 | | 01-02-83 | | * | | | | | | | |
| | | | 438 | 01-02-83 | | * | | | | | | | |
| 5H 380 | Cf 342 | 676 | | 28-10-85 | 28-10-85 | 535 | | | X | | 52.0 | 227 | 127 |
| TOTAL | | 9 | 7 | | | | 2 | 2 | 9 | | | | |

SOURCE: CPRJ — FEEMA REGISTER

clearly displayed in the phenotypes of the four offspring, which were very similar to the seven infants reported in Coimbra-Filho (1970*b*).

The newborn offspring resulting from the pairing of *C .a. flaviceps* × *C. jacchus* (Fig. 3.3) exhibit a greyish fur on the trunk and limbs, but with a yellowish hue demonstrating the influence of *C. a. flaviceps*. They had a greyish-silvery nape with a black front-spot (Fig. 3.4), which was broader than that observed in a similar area on young *C. geoffroyi*, as well the young hybrids of *C. geoffroyi* with *C. jacchus* and *C. penicillata* (Figs 3.2D, 3.6). The adults are also light greyish, retaining the yellowish hue (Figs 3.5, 3.10). The tail has long fur, with ring-like markings, which are less evident than in *C. jacchus* and *C. penicillata*. They have copious auricular tufts, but the fur is relatively short, with a light yellowish-grey hue. In one of the specimens, the small white spot on the forehead had the form of a half-moon, with a narrow white line extending between the eyes to the nostrils. The colour of the fur on the limbs is similar to the trunk but with a deeper yellowish hue. The eyes are chestnut brown. the circumgenital area is black, similar to the pure individuals of *C. a. flaviceps* and *C. geoffroyi*.

The hybrids of *C. a. aurita* × *C. kuhli* are very like those of *C. a. flaviceps* × *C. jacchus*, and strongly indicate a close affinity between *C. a. aurita* and *C. a. flaviceps*, both providing the expression of very similar and characteristic phenotypes. The marked similarity of hybrids involving either *aurita* or *flaviceps* is one of the aspects which convinces us that they are more

**Fig. 3.2** *Callithrix* hybrids. A and B: adult fraternal twin hybrids. They differ phenotypically due to an accentuated dissociation of characters. The individual shown in A (CPRJ-002), is darker and has a conspicuous cephalic and auricular hypertrichosis. This animal has a triple-hybrid infant (CPRJ-063) on its back (circled) (ancestors *C. jacchus*, *C. geoffroyi*, *C. penicillata*), which displayed the evident influence of *C. geoffroyi*. The hybrid (*C. jacchus* × *C. geoffroyi*) individual shown in B (CPRJ-001) has light, buff-coloured hair. C: quadruple hybrid twins (CPRJ-192 and CPRJ-193, *C. geoffroyi*, *C. kuhli*, *C. jacchus*, *C. penicillata*), which also display an accentuated phenoytypic differentiation. D: a hybrid juvenile (CPRJ-180, *C. geoffroyi* × *C. penicillata*), aged approximately four months.

**Fig. 3.3** A male *C. a. flaviceps* (CPRJ-147), with twins resulting from its having been mated with *C. jacchus* (CPRJ-110). The twins are two and a half months old and the first case of hybridization between these species.

**Fig. 3.4** A five-month old female hybrid, *C. a. flaviceps* × *C. jacchus* (CPRJ-187).

**Fig. 3.5** The same individual as in Figure 3.4 (CPRJ-187), but then an adult, approximately two years old.

closely related to each other than to the remaining south-east Brazilian forms. The cephalic region of the adult hybrids is light greyish, along with the dorsal parts and the limbs. The abdomen is chestnut brownish and the entire circumgenital region black. The spot on the forehead is white and quite large, but with a yellowish hue at its vertex between the eyes. The auricular tufts have short, yellowish-grey fur arising from the anterior part of the pinna. The chin is whitish. The inside of the limbs is more yellowish, and the feet and hands are greyish with a yellow hue. The tail has alternating light and dark grey concentric circles. The eyes are chestnut brown. The remarkable similarity between hybrids of *C. a. aurita* and those of *C. a. flaviceps* is one of the aspects which convinces us that they are subspecies and not species.

The infants of *C. geoffroyi* × *C. penicillata* hybrids maintain the distinct face mask of *C. geoffroyi*, which tends however, to be off-white rather than the brilliant white of pure *C. geoffroyi* (Fig. 3.2 D, Fig. 3.6), a characteristic also found for hybrids observed in the wild (Rylands and Costa

**Fig. 3.6** A male *C. geoffroyi* (CPRJ-070) and a female *C. penicillata* (CPRJ-135), with their infant behind them (CPRJ-180) clearly displaying the chromogenetic influence of its father.

1988). These hybrids are quite variable in pelage colouration patterns and the adult phenotypes do not approximate to that characteristic of *C. kuhli*, as supposed by Hershkovitz (1977).

### *Double, triple, and quadruple hybrids*

Coimbra-Filho (1971) reported on the mating of a double hybrid (male *C. jacchus* × *C. geoffroyi*) with a female *C. penicillata*. At six months the phenotype of the single surviving infant of three born was closest to that of typical *C. penicillata*.

Triple hybridism was reported by Coimbra-Filho (1974), and Coimbra-Filho *et al.* (1976) described a quadruple gestation of a female double hybrid (offspring of a male *C. jacchus* × *C. geoffroyi* mated with a female *C. penicillata*) and a male hybrid (*C. jacchus* × *C. geoffroyi*). Only one infant was born (it showed a deficient grasping reflex and died after 48 hours), the mother expired during the birth, and three further unborn offspring (one a rudimentary fetus) were found during the post-mortem.

The chromogenetic pattern of some of the triple hybrids and adult quadruple hybrids are reminiscent of the *C. penicillata* phenotype, especially regarding the generally greyish colour, the tail, auricular tufts (although rather fuller), the white spot on the forehead (generally triangular but with the contours poorly defined), the blackish crown, black on the gular region, nape, and neck, and whitish sides to the face (see Fig. 3.7, a female triple hybrid—ancestors *C. jacchus*, *C. geoffroyi*, *C. penicillata*, and Fig. 3.10, a male quadruple hybrid—ancestors *C. kuhli*, *C. geoffroyi*, *C. jacchus*, *C. penicillata*). In some individuals the fur of the dorsal region is disorderly, reminiscent of *C. geoffroyi*. The eyes are consistently chestnut brown. The hands and feet are dark, nearly black, similar to the pattern of typical *C. kuhli* and *C. geoffroyi* (the quadruple hybrid has ancestors of these two species). Figure 3.8 shows an infant quadruple hybrid (ancestors *C. kuhli*, *C. geoffroyi*, *C. jacchus*, *C. penicillata*), which shows the phenotypic dominance of *C. geoffroyi*, notably in the white face mask.

### *Quintuple hybrids*

Quintuple hybrids (ancestors *C. a. flaviceps*, *C. jacchus*, *C. kuhli*, *C. geoffroyi*, and *C. penicillata*) resulted on one occasion in a quadruple gestation (Fig. 3.9), with the four offspring all having a general yellowish-grey colouration, demonstrating the persistent and strong influence of the *C. a. flaviceps* phenotype, as with the *C. a. flaviceps* × *C. jacchus* hybrids. The colouring of the inferior and superior limbs of these quintuple hybrids was similar to that observed on their trunks, but in some cases was darker. The tail also had the trunk colouration (yellowish-grey), but becoming darker from the distal half towards the tip. The head was greyish, with a black mask which included the orbital region, following medially in the form of a wedge to the top of the head. The length of this dark area varied, and could even (one individual) be separated into spots, for example on the occipital region and crown. The sides of the face were variably yellowish. They had very dark preauricular tufts.

Figure 3.10 shows an adult male quintuple hybrid (ancestors *C. a. flaviceps*, *C. jacchus*, *C. kuhli*, *C. geoffroyi*, *C. penicillata*), which was greyish, and had a dorsal yellow stripe. The dorsal stripes were not very pronounced. The tail had distinct greyish rings, alternating light and dark,

**Fig. 3.7** A couple comprising a male hybrid (CPRJ-081, *C. kuhli* × *C. geoffroyi*) and female triple hybrid (CPRJ-100, *C. jacchus*, *C. geoffroyi*, *C. penicillata*), which produced the first ever quadruple hybrids.

and the head had a dark greyish hue. The auricular tuft was well developed and fan-shaped (typical of *C. jacchus*). A white, poorly defined spot was present on the lower forehead. There were small white hairs on the rim of the upper lip giving the impression of a narrow white moustache. The lower lip and chin were whitish. There was a light spot on each side of the central axis of the rather dark face. The eyes were light chestnut brown. The dorsal part of the hands and feet had a yellow hue which dominated an otherwise greyish colouration. The hairs of the circumgenital

**Fig. 3.8** A one-day-old male quadruple-hybrid (CPRJ-224, *C. geoffroyi*, *C. kuhli*, *C. jacchus*, *C. penicillata*). Note its head, which displayed the phenotypical dominance of *C. geoffroyi*.

region were dark, nearly black. Overall the face-markings and pelage of the dorsum showed a strong influence of *C. a. flaviceps*.

*Sextuple hybrid*

A single female, born on 6 November 1963, lived only three days. It had a dark stripe extending from the frontal ridge to the dorsum as far as the second sacral vertebra. The same dark colouring surrounded the orbits, extending to the nasal region to give the appearance of a mask. The flanks were brownish grey. The ventrum, neck, and internal arms showed a lighter colour. The fur of the legs, hands, and feet was dark greyish. The tail showed characteristic, if pale and poorly visible, alternating light and dark circles (Fig. 3.11).

## Natural hybridization

The lack of documented hybridization in the wild reported in previous papars (see Coimbra-Filho 1970*b*; Coimbra-Filho and Mittermeier 1973*b*) indicates that it is uncommon and, in the few cases reported recently (see below), they are highly localized.

**Fig. 3.9** Two of the quintuple-hybrids (CPRJ-381 and CPRJ-382, *C. a. flaviceps, C. jacchus, C. kuhli, C. geoffroyi, C. penicillata*) from a quadruple birth. A, front view; B, the occipital region and upper dorsum.

## Callithrix geoffroyi × Callithrix penicillata

Hershkovitz (1977, pp.503–4, 509) argues that the form occurring in southeastern Bahia and northern Espírito Santo is an intergrade between *C. j. penicillata* and *C. j. geoffroyi*. We regard the form in southern Bahia, at least between the Rio de Contas in the north and the Rio Jequitinhonha in the south, to be a valid taxon, *C. kuhli* (see Rylands *et al.* this volume). Coimbra-Filho (1971) reported that the distribution of *C. geoffroyi* was limited far to the south of the Rio Jequitinhonha; in the state of Espírito Santo, between the Rio Jucú in the south (just south of Vitória) and the Rio Itaunas in the north. Our knowledge of the distributions of marmosets in northern Espírito Santo are insufficient and provides a confused picture, complicated further by casual introductions especially of *C. penicillata* in

**Fig. 3.10** Left, a female hybrid (CPRJ-187, *C. a. flaviceps* × *C. jacchus*) with two quintuple hybrid male infants (CPRJ-379 and CPRJ-380, *C. a. flaviceps*, *C. jacchus*, *C. kuhli*, *C. geoffroyi*, *C. penicillata*) from a quadruple gestation (the other two infants are shown in Fig.3.9). Middle, a male quintuple hybrid (CPRJ-320, *C. a. flaviceps*, *C. jacchus*, *C. kuhli*, *C. geoffroyi*, *C. penicillata*) approximately one year old. Right, a male quadruple hybrid (CPRJ-193, *C. kuhli*, *C. geoffroyi*, *C. jacchus*, *C. geoffroyi*).

many parts of Espírito Santo. Hershkovitz (1977, pp.508–9) indicated a northern limit in the eastern part of its range to the Rio Doce valley. The localities for *C. geoffroyi* that he cites to the north (Rio Barra Seca, Rio São Mateus, Conceição da Barra, and Rio Itaunas provided by Ruschi 1964) he discarded as unverified, but suggested that, in all but the last locality, they may be hybrids of the forms *penicillata* and *geoffroyi*. However, Oliver and Santos (1991) confirmed Coimbra-Filho's (1971) assertion that they occur throughout the basin of the Rio São Mateus, north to the Rio Itaunas. At least in this region, therefore, there is no reason to expect hybrid forms, the distribution of *C. penicillata* being far to the west in the *cerrado* region of Minas Gerais. North of the Rio Itaunas, marmoset distributions become somewhat confused. Oliver and Santos (1991) obtained only few scattered reports of unidentified marmosets in the Rio Mucurí basin, and only two definite localities for *C. geoffroyi* at Teixeira de Freitas, Bahia, along the BR-101 highway, just south of the Rio Itanhem. Apart from the outlying Teixeira de Freitas localities, from

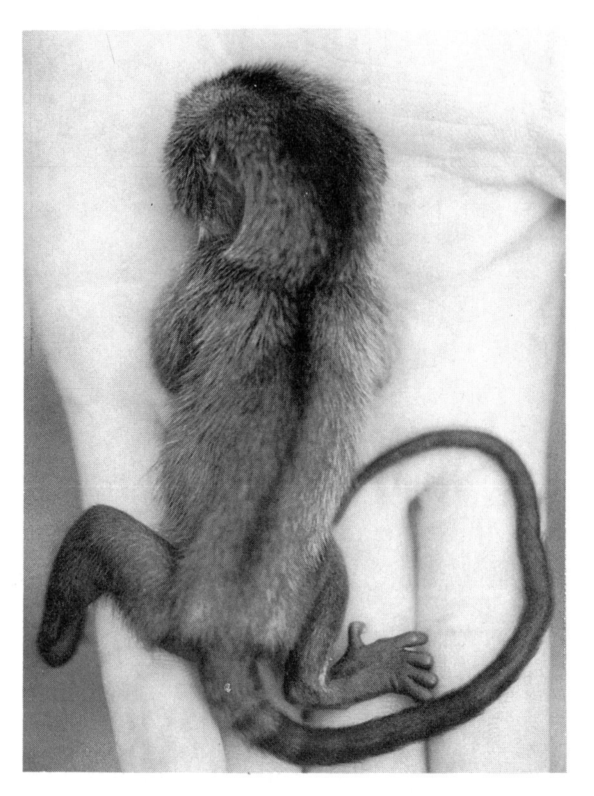

**Fig. 3.11** A three-day-old sextuple hybrid (CPRJ-519), daughter of CPRJ-320 (quintuple hybrid) and *C. a. aurita* (CPRJ-428).

there they observed a hiatus in marmoset distributions between the Rio Mucurí in the southernmost tip of Bahia to the Rio do Frade. Oliver and Santos confirmed the presence of introduced *C. jacchus* in the Monte Pascoal National Park, in southern Bahia in the hiatus area, and also an unidentified marmoset in the vicinity of the National Park, which may have been *C. geoffroyi*. The occurrence of *C. geoffroyi* was again confirmed north of the Rio do Frade along the south bank of the Rio Jequitinhonha, where it extends as far west as the Rio Araçuaí (a right bank tributary of the upper rio Jequitinhonha) as confirmed by Rylands *et al.* (1988), and also by a specimen collected at Araçuai (Hershkovitz 1977, p.490). Interestingly Oliver and Santos (1991) obtained frequent reports and sightings of marmosets in the coastal forests to the north of the Rio de Contas, as far north as the Rio Paraguaçu, which they ascribe to *C. kuhli*, but which is then replaced by *C. jacchus* and/or *C. penicillata* to the north. According to Oliver and Santos (1991) 'the western and northern limits of its (*C. kuhli*) range have not been precisely delineated. It therefore remains unclear whether it gradually intergrades with *C. p. penicillata* in the transitional zone between the Atlantic forests and the caatinga region (*c*.40°) . . .' Santos (in Rylands *et al.* 1988) observed a hybrid group of *C. kuhli* and *C. penicillata* just north of the Rio Jequitinhonha where the Atlantic forest shows a rapid transition to *caatinga* (dry thorn scrub).

In summary, therefore, we would argue that the zone of intergradation from the north of the Rio Doce into southern Bahia, taking in the entire distribution of the form *C. kuhli* north of the Rio Jequitinhonha, postulated by Hershkovitz (1977), is not valid. The primitive distribution of *C. geoffroyi* was probably restricted to between the Rios Jucu in the south and Itaunas in the north (Coimbra-Filho 1971), but due to introductions there are outlying populations on the Rio Itanhem, and also north of the Rio do Frade extending to the Rio Jequitinhonha (notably in the 1970s at Belmonte).

Coimbra-Filho (1971; see also Ávila-Pires 1969) argued that the occurrence of *C. geoffroyi* inland in the state of Minas Gerais may also have resulted from the extensive destruction of the Atlantic forest along with introductions. There still remains, however, the possibility that the primitive distribution of *C. geoffroyi* did include the entire lowland eastern forested region of Minas Gerais extending north to the east of the Serra do Espinhaço and east of the Rio Aracuai south of the upper Rio Jequitinhonha. The middle Rio Jequitinhonha is dominated by an island of *caatinga*, but Rylands *et al.* (1988) observed it along the evergreen riverine forests throughout this region.

Whether its occurrence along the entire south bank of Rio Jequitinhonha west as far as the Rio Araçuaí resulted from recent introductions or not is now difficult to say. However, hybrid or intergradation zones between *C. geoffroyi* and *C. penicillata* may be expected along the distributional limits formed by the transition between *cerrado* and Atlantic forest along at least part of the Serra do Espinhaço in Minas Gerais and in the north in the region of the upper Rio Araçuaí. Regarding the former, hybrid groups, containing animals typical of both species alongside extremely variable mixes, have been observed along the Serra da Piedade (I.B. Santos and C.M.C. Valle, personal communication), and in the municipality of Santa Barbara, near Belo Horizonte, where the Atlantic forest intergrades with *cerrado* (Rylands and Costa 1988). Regarding the upper Rio Araçuaí, Rylands *et al.* (1988) obtained confused reports of mixed groups, but were unable to confirm the exact situation. To the north of the Rio Jequitinhonha, we may expect hybrid zones at the transition between the humid coastal Atlantic forest, the domain of *C. kuhli*, and the dry liane forests and *caatinga* inland (*C. penicillata*). A hybrid group of *C. kuhli* and *C. penicillata* was observed by I.B. Santos (in Rylands *et al.* 1988), at the interface between *caatinga* and Atlantic forest, just north of the middle Rio Jequitinhonha, west of Almenara, Minas Gerais. To the north of the Rio de Contas, to the Rio Paraguaçú the marmosets ascribed to *C. kuhli* by Oliver and Santos may represent a cline gradually changing into the recognized form of *C. penicillata*, although this hypothesis requires further study. To the north of the Rio Paraguaçú extending to central and northern Bahia, the situation is again extremely confused, with widespread forest destruction and repeated recent as well as ancient introductions of *C. jacchus*.

## Callithrix penicillata × Callithrix jacchus

Hershkovitz (1977, p.519) recorded the occurrence of hybrids of *C. jacchus* and *C. penicillata* at Santo Amaro, north of Salvador, state of Bahia. These populations were studied by Alonso *et al.* (1987), who described forms typical of the respective species along with five intermediate pelage patterns. They argued that the narrowness of the hybrid zone is evidence for the specific status of *C. jacchus* and *C. penicillata* and indicated that the reproductive isolation mechanism had partially collapsed due to widespread habitat destruction. We believe that these hybrids are the result of many years of casual introductions, undoubtedly of *C. jacchus* and possibly also of *C. penicillata*, in a region which possibly lacked marmosets in the past or was inhabited by *C. penicillata*, the primitive range of *C. jacchus* being restricted to the north of the Rio São Francisco. In this case, we can expect a mosaic of populations of pure forms and hybrids. Surveys throughout inland Bahia and parts of the states of Sergipe and Piauí are necessary to clarify the true range limits for *C. penicillata* and the occurrence of hybrids in this region.

## Callithrix aurita flaviceps × Callithrix geoffroyi

*C. a. flaviceps* occurs in upland forests in mountainous regions at altitudes above 400 m above sea level, whereas *C. geoffroyi* occupies forests of lower altitudes (Coimbra-Filho 1971). Mendes (1989) found *C. a. flaviceps* occurring above 850 m altitude, whereas *C. geoffroyi* occurs below 500 m. In certain localities, between these altitudes, groups have been observed containing forms typical of both species along with hybrids, but none containing only hybrids (Mendes 1989, in prep.). One of these localities is Santa Teresa, Espírito Santo, and although a specimen from there (kept in the National Museum, Rio de Janeiro) was described by Ávila-Pires (1969) as *C. penicillata*, Hershkovitz (1977, pp.505, 526) indicated that it was in fact a hybrid between the forms *flaviceps* and *geoffroyi*. Ferrari and Mendes (1991) reported on the occurrence of groups containing hybrids at two places in the municipality of Santa Teresa (Santa Lucia and Fazenda dos Irmãos Medanha) and a third at Rio Bonito in the municipality of Santa Leopoldina. These authors concluded that 'while the two species are known to hybridize in these areas, they are well differentiated ecologically and species integrity has been maintained despite the isolation of these populations in relatively small tracts of forest'.

### CALLITHRIX AURITA FLAVICEPS × CALLITHRIX AURITA AURITA

Ferrari and Mendes (1991) reported observing a group of hybrid *C. a. aurita* × *C. a. flaviceps* at their range limits in the municipality of Carangola, eastern Minas Gerais. Interestingly the entire group was composed of hybrid forms in contrast to the situation with *C. geoffroyi* and *C. a. flaviceps* (Mendes personal communication, in prep.). This reinforces our belief that these two forms are only subspecifically distinct.

## Discussion

### *Captive hybrids*

Planned hybridization in non-human primates is a lengthy and costly task because of long gestation periods, late sexual maturity, and the expense of their adequate maintenance and husbandry. The relatively short reproductive cycle and small size of marmosets, however, minimizes these aspects. As indicated above, the hybridization experiments reported here were carried out for the following purposes: (1) to test interspecific fertility between the forms of the eastern Brazilian *Callithrix*; (2) to acquire material for a serial collection with a view to comparative studies; and (3) to examine preliminary zootechnological aspects.

The existence of fertile captive hybrids and the localized and rare occurrence of natural hybrids (in many cases a result of introductions and habitat degradation) is considered by some authors to be sufficient evidence to relegate these forms to subspecies. However, as noted below, infrequent natural hybrids of a number of other primate species have also been recorded without casting doubt on the validity of parent species, and natural hybrids between species are known for numerous other vertebrates (Moore 1977; Barton and Hewitt 1981, 1985). Although they possess a different genetic code from the genome of the ascendant forms, the influence of these hybrids on the population of the recognised species is negligible and this would seem to be the case in the few hybrid localities recorded, excepting perhaps that for *C. a. aurita* and *C. a. flaviceps*. The fact that experimental hybrids of these two forms with the remaining species have almost identical phenotypes reinforces our belief that they have closely similar genotypes and this, combined with their similarity in a number of behavioural aspects, indicates that they should be considered only subspecifically distinct (Coimbra-Filho 1990).

It is not easy to evaluate hybrid fitness in captive conditions. Only a slight reduction, either behavioural or physiological/genetic, would be significant in the wild, but may not be detectable in captivity. These hybridization experiments prove a degree of fertility amongst the six forms, but it would appear to be reduced in at least some cases, most particularly between the more distant species, that is between *C. aurita* and *C. jacchus*. This requires further investigation, but if we are right in surmising that *C. a. flaviceps* and *C. a. aurita* are subspecifically related, we would predict fitter hybrids between these two, than either with the remaining species or between the remaining species.

The fact that we have obtained a sextuple hybrid suggests the enormous potential, through biotechnology and genetic engineering, of providing specific genomes appropriate for specific scientific purposes. The artificial forms of marmosets—true domestic races—obtained in this way, obviously have a genetic constitution different from the natural forms, representing an achievement of great pragmatic potential, although it is necessary to acquire a more thorough understanding of the genomes and their phenotypes in order to obtain an appraisal of the real possibilities regarding scientific interests and demands. However, considering recent biotechnological developments including gene transfer, translocation, gene manipulation, etc., it is easy to foresee important scientific breakthroughs in the near future, exploiting the potential of those forms which are genetically different to the natural species. Due the phenomenon of heterosis, some hybrids are potentially more robust, larger, more productive, and generally more resistant and vigorous than pure individuals, and therefore better adapted to captivity. Although hybrids of *C. a. flaviceps* × *C. jacchus* show in general a high degree of infertility, one of the hybrids we have obtained showed an extraordinary capacity for milk production, raising successfully four and three offspring in successive matings with a male quadruple hybrid (*C. jacchus*, *C. penicillata*, *C. kuhli*, *C. geoffroyi*) (Coimbra-Filho *et al.* 1984, 1989). In addition certain hybrid lineages we have obtained present a loss of hair in various parts of the body (tail, arms, and parts of the head), also of great potential for use in biomedical research and testing.

Programmed hybridization among the callitrichid forms can increase our knowledge of primate cytogenetics. Likewise, zootechnical projects of a utilitarian as well as a conservationist nature can be carried out, since through back-crossing and inbreeding one can obtain forms having genomes similar to those of endangered species. CPRJ has already achieved individuals with genomes which are seven-eighths *C. a. flaviceps* and one-eighth *C. jacchus*, with a phenotype which is practically identical to pure *C. a. flaviceps*, and it is quite possible to obtain 31/32 parts *C. a. flaviceps*, which can be considered basically 'pure'. The genetic patrimony of such individuals will never be the same as the pure natural forms but, although somewhat controversial (see, for example, Fergus 1991), the initiative is a form of avoiding the total extermination of the threatened genome.

### *Natural hybrids*

Although interspecific hybridization in primates is not uncommon (Gray 1954), it is rarely reported for wild populations. At least in some cases, it is due to disturbance, in general originating from human activities and resulting in zones of secondary intergradation or mosaics of hybrid and pure populations. Bernstein (1966) reported a hybrid population of *Macaca fascicularis* (= *M. irus*) and *M. nemestrina* in Kuala Lumpur, Malaysia. He found that the hybridization had resulted from a lack of males in the latter species, which are bigger and more aggressive than in *M. fascicularis*. Locally, *M. nemestrina* were considered dangerous and also harmful to plantations, and were as a result hunted almost to extinction, and a surviving group of females had joined a group of *M. fascicularis*, resulting in hybrids. In a later report, Bernstein (1968) recorded that the group was composed of 18 individuals, including two hybrids (one of each sex) which were perfectly integrated into the social hierarchy, with the male even holding a position of dominance. Unfortunately, the fertility of this male was not ascertained. However, Aldrich-blake (1968) while studying *Cercopithecus mitis stuhlmanni* in the Budongo forest, Uganda, recorded a fertile hybrid with *C. ascanius schmidti*. Interestingly these two species are sympatric, and do not hybridize, over a large part of their ranges. Hybrids have been observed in only four localities. Struhsaker (1984) recorded infrequent fertile hybrids between these species in the Kibale forest in Uganda, at the range limits of the blue monkey (*C. mitis*). The narrow hybrid zone in this case is believed to result from the fact that although redtail females normally reject blue monkey males, the area has a high proportion of solitary blue males. When mating does occur the hybrid female offspring are fully integrated in groups of redtail monkeys, and possibly have advantages (larger size) over redtail females. Hybrid male offspring, which maintain the size and general appearance of blue monkeys, are at a disadvantage, being rejected from the redtail groups with which they are most familiar. Nagel (1971) reported the existence of fertile hybrids between *Papio anubis* and *P. hamadryas* in Ethopia. This is a particularly interesting case of a stable hybrid zone (at least 60 years old), evidently maintained by a tendency for male *P. hamadryas*, which herd females and maintain them in close proximity (a behaviour not characteristic of *P. anubis*), to occasionally take female *P. anubis* from their groups (see Gabow 1975). However, the male offspring of these hybrid matings do not herd as efficiently as their fathers, and are incompetent in this sense in both anubis and hamadryas groups. As the number of incompetent male hybrids increases in hamadryas groups, there is less competition for females, resulting in fewer, or a complete cessation, of the raids for anubis females by hamadryas males, thus providing a brake on the spread of hybrids into the parent populations (Gabow 1975).

Regarding callitrichids, the only cases of natural hybrids are those reported here, except for one possibility of a hybrid form of *Saguinus labiatus* and *S. imperator* in northern Bolivia, sighted by Izawa and Bejarano (1981). The known *Leontopithecus* populations are disjunct, but it is curious that no natural hybrids have been reported for the genus *Saguinus*. The lack of reports of natural hybrid populations for callitrichids is also found for cebids. Kinzey (1981) indicated the possibility of hybrid populations of two subspecies of *Alouatta fusca* (*A. f. clamitans* × *A. f. fusca*) along the right bank of the Rio Doce in south-east Brazil, and two subspecies of *C. apella* (*C. a. libidinosus* × *C. a. nigritus*) along both banks of the Rio Grande in the states of Minas Gerais and São Paulo.

Interspecific and fertile hybridizations in primates do, therefore, occur in the wild, even though they are evidently a rare phenomenon. In the Old World monkeys cited above the hybrid zones are extremely narrow, and at least in two cases (*Cercopithecus* and *Papio*), and probably also the third (*Macaca* ssp.), they are maintained through differences in the behaviour and social organization of the species involved. The cases of natural *Callithrix* hybrids reported here are also extremely localized, and it is of interest to speculate on the mechanisms by which they are maintained. It is necessary to establish whether they are zones of secondary intergradation resulting from such as forest destruction, introductions, or climate changes. In the case of the *C. jacchus* × *C. penicillata* hybrids, studied by Alonso *et al.* (1987) in Bahia, this would appear to be the case. The other hybrid localities, excepting that of *C. a. aurita* × *C. a. flaviceps*, are at ecotones, with populations of the species concerned at their distributional limits. In the case of the *C. a. aurita* × *C. a. flaviceps* hybrid locality, there is no indication of an ecotone. It will also be important to know to what extent the hybrid zones are stable. If ephemeral, they should result in eventual complete speciation or fusion of the two hybridizing taxa (Mayr 1970; Moore 1977). Moore (1977) discusses two mechanisms which would maintain stable hybrid zones. The first is a dynamic equilibrium, maintained by a limited dispersal and hybridization of 'naive' individuals at the distributional limits of each of the species, despite hybrid inferiority (partial reproductive isolation through genetic or behavioural incompatibility). This could apply to all the hybrid localities except perhaps *C. jacchus* × *C. penicillata* in Bahia, since they are all at the distributional limits of the species involved. The second mechanism invokes hybrid superiority at an ecotone. Hybrid zones occur at ecotones in the cases of *C. penicillata* × *C. kuhli* (*caatinga*-Atlantic forest) and *C. penicillata* × *C. geoffroyi* (*cerrado*-Atlantic forest), and *C. geoffroyi* × *C. a. flaviceps* (altitudinal). Although hybrid superiority at the ecotone remains to be established, intuitively we believe that the two gum-feeding specialist marmosets, *C. jacchus* and *C. penicillata* (see Rylands and Faria this volume), would have a competitive advantage over all of the others, as measured by their ability to survive in the 'harsh' environments of the *caatinga* or *cerrado*. In this case, we might suppose that a hybrid zone resulting from secondary contact is ephemeral and expanding, being only a question of time for the species with the competitive disadvantage to disappear; this is possibly the case with *C. penicillata* × *C. geoffroyi*. Hybrid populations, particularly of *C. jacchus* × *C. penicillata* in Bahia, might provide an interesting case of rapid speciation, especially considering the extent of forest destruction and the eventual isolation of hybrid populations, incremented by such processes as the founder effect and genetic drift.

It is difficult to provide any concrete suggestions regarding the causes and processes involved in these hybrid zones, and only long-term studies and monitoring of the hybrid populations, and their comparison with neighbouring populations of the parent species, will provide some answers. Although the social organization of all *Callithrix* marmosets is extremely similar, behavioural studies of the hybrid groups may well provide evidence for fitness differences between hybrids and parent species, and possibly between the sexes of the hybrid offspring, as has been found for the hybrid *Cercopithecus* and *Papio* species mentioned earlier. Any differences found may well provide insights as to the maintenance of the hybrid zones.

To summarize, we may conclude that:

1) *Callithrix* hybrid localities are few, restricted, and of three types, the first being at distributional limits and ecotones, and of ecologically distinct species (*C. penicillata* × *C. kuhli*; *C. geoffroyi* × *C. a. flaviceps*; and *C. penicillata* × *C. geoffroyi*), the second being of ecologically similar (if not identical) subspecies at their distributional limits but not involving an ecotone (*C. a. aurita* × *C. a. flaviceps*), and the third being of ecologically similar species but involving introductions of one or both in areas which may or may not be ecotones (*C. jacchus* × *C. penicillata*);

2) it will be important to examine the stability or otherwise of these hybrid zones;

3) the data on hybridization in captivity has demonstrated that hybrids of all of the south-east Brazilian marmosets are genetically fertile, although their fitness in the wild in terms of survival and reproductive success (determined physiologically and behaviourally) compared to their parent species has yet to be determined;

4) behavioural studies of wild groups will be most important to determine possible differences in the fitnesses of male and female hybrids, possibly providing insights regarding the maintenance of the hybrid localities; and

5) forest destruction, prevalent in the remains of the geographic ranges of these marmosets, may result in isolated hybrid or partially hybrid populations, which along with founder effects and genetic drift may cause rapid 'speciation', or at least the production of localized, uniformly distinct, true-breeding phenotypes.

## Acknowledgements

This chapter is dedicated to Espedito Cordeiro da Silva, formerly of the Laboratory of Comparative Anatomy, Department of Anatomy, Federal University of Rio de Janeiro (UFRJ). We are grateful to Dr Hector N. Seuanez, Head of the Laboratory for Cytogenetics, Department of Genetics, UFRJ, for his critical reading of the original version.

# Part II

## *Behaviour and reproduction*

# 4

# Making sense out of scents: species differences in scent glands, scent-marking behaviour, and scent-mark composition in the Callitrichidae

*G. Epple, A.M. Belcher, I. Küderling, U. Zeller, L. Scolnick, K.L. Greenfield, and A.B. Smith III*

## Introduction

In many mammals, chemical signals are of great importance in socio-sexual communication and in priming of reproductive functions. Such signals are contained in urine, faeces, genital discharge, saliva, and the secretion of specialized skin glands, of which several types may be present on an individual. Often, ingredients from several sources are combined to form complex mixtures. In addition, bacteria may act on substrates such as vaginal discharge or skin secretions to develop volatiles which carry signal function (see recent reviews by Albone 1984; Brown and Macdonald 1985; Duvall *et al.* 1986).

During recent years our knowledge of the chemical composition of scent material employed in communication has increased considerably (Albone 1984). However, decoding chemical signals to the point where chemical structure can be related to biological function continues to be difficult. In a few mammals, for example the golden hamster (Singer *et al.* 1986) and the domestic pig (Melrose *et al.* 1971), specific biological activities rest with a single compound or a small number of compounds. In many cases, however, the biological activity resides in chemically complex mixtures, containing a number of classes of compounds (Crump *et al.* 1984; Belcher *et al.* 1986; Müller-Schwarze *et al.* 1986; Raymer *et al.* 1986). These mixtures may contain compounds of additive or redundant bioactivity, compounds which work in synergy, and compounds which inhibit the biological response (Albone 1984; Novotny *et al.* 1985; Müller-Schwarze *et al.* 1986). Indeed, Albone (1984) and Müller-Schwarze *et al.* (1986) have pointed out that the complexity of many mammalian chemosignals is best described in terms of a scent 'Gestalt'.

The 'Gestalt' concept appears to be well suited to describe chemical signals employed by some primates including the Callitrichidae (Wheeler *et al.* 1977; Crewe *et al.* 1979; Keverne 1978; Schilling and Perret 1987; Epple *et al.* 1989). The communication systems of marmosets and tamarins rely heavily on chemical signals. Among simian primates, callitrichids appear to have the most elaborately developed epidermal scent glands and scent-marking behaviours, suggesting that chemical communication is of great importance in this family.

Our studies on three tamarin species show that scents used in communication are highly complex. The present report reviews some of the observational and experimental studies on scent marking, including our own research on the communicatory function and chemical nature of scent marks, and a discussion of the roles chemical communication plays in socio-sexual behaviour. In this discussion, we will concentrate on our comparative studies on *Saguinus fuscicollis*, *Saguinus o. oedipus*, and *Saguinus leucopus*.

## Sources of chemical signals

All species of marmosets and tamarins studied to date possess specialized scenting organs. These macroscopically differentiated glandular pads are located in two areas, the gular-sternal region of the mid-chest and the area surrounding the genitalia (scrotum, labia pudendi, suprapubic, and perineal regions). In *Callithrix* and *Cebuella*, morphologically differentiated scent glands are located on the external genitalia and the perineal skin (Wislocki 1930; Perkins 1968, 1969*b*; Starck 1969, Sutcliffe and Poole 1978). In the present paper, these will be referred to as circumgenital glands. In many tamarins of the genus *Saguinus* and in *Leontopithecus r. rosalia*, morphologically specialized glandular skin covers the external genitalia (circumgenital gland), and a thick glandular pad, the suprapubic gland, extends rostrally across the symphysis pubis (Wislocki 1930; Perkins 1966, 1869*a*; Epple 1967). The complex of both glands has been referred to as the circumgenital-suprapubic gland. Sternal glands, located above the manubrium sterni, are present in most species studied (Epple and Lorenz 1967). In addition to specialized scent glands, microscopic concentrations of apocrine and/or sebaceous glands occur in other areas of the body such as the suprapubic, perineal and circumanal regions, in the axilla, or on the muzzle. They do not form glandular fields differing in appearance from the surrounding skin (Perkins 1966, 1968, 1969*a*,*b*).

Histological studies of the scent glands were performed in a small number of species. These have shown that glands located in the circumgenital and suprapubic areas generally consist of dense accumulations of very large sebaceous units and of apocrine elements. Apocrine glands, on the other hand, predominate in the sternal area (Perkins 1966, 1968, 1969*a*,*b*; Starck 1969; Sutcliffe and Poole 1978; Wislocki 1930; Zeller *et al.* 1988, 1989).

It must be stressed that there are differences among species in the histological and histochemical characteristics as well as in size and morphology of the glands. In the common marmoset, for example, large sebaceous and apocrine glands are located mainly in the skin of the external genitalia, while the suprapubic area does not show morphologically differentiated glandular fields. There is no sexual dimorphism in the size of the glands (Sutcliffe and Poole 1978). In the saddle-back tamarin, on the other hand, sebaceous and apocrine glands are found in the skin of the external genitalia and extend as a thick, deeply pigmented pad across the suprapubic area.

We have recently studied the anatomy of the circumgenital gland of two species, *Saguinus fuscicollis* and *Saguinus o. oedipus* (Zeller *et al.* 1988, 1989). Our findings show that in saddle-back tamarins a very complex glandular organ, composed of holocrine and apocrine glands, is located beneath the epidermis of the circumgenital skin. In males, specialized holocrine glands, associated with hair follicles, predominate. They have a complex alveolar structure and possess numerous branched excretory ducts. Each group of glands empties into a common duct which enters the hair follicle. The apocrine glands are located predominantly at the periphery of the glandular pad and between the scrotal and perineal areas (Fig. 4.1). The excreting ducts of most apocrine glands empty on to the skin surface in close spatial association with hair follicles. In females, the specialized holocrine glands resemble those of males but are more frequently interspersed with apocrine glands. The apocrine glands are larger and much more numerous than in males, especially in the region of the *labia majora* (Fig. 4.2). In terms of their size and complexity, the holocrine glands of saddle-backs are among the most highly specialized among primates. In males, they form a continuous layer and almost entirely replace the dermis. The apocrine glands of females are exceptionally large and are among the most highly specialized apocrine glands in mammals (cf. Schaffer 1940).

Female *Saguinus o. oedipus* possess a circumgenital glandular organ analogous to that of female *Saguinus fuscicollis* (Perkins 1969*a*; Wislocki 1930; Zeller *et al.* 1989). The glandular organ occupies the suprapubic, labial, and perineal areas and is composed of large holocrine and apocrine glands. Individual apocrine and sebaceous glands are smaller than in

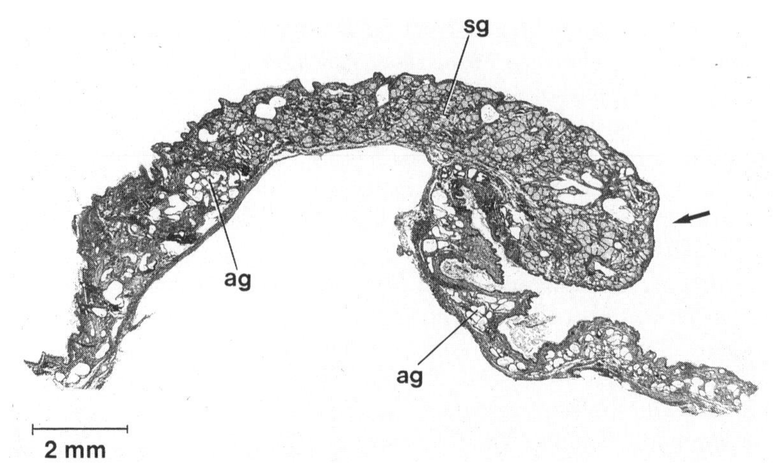


**Fig. 4.1** Sagittal section through the circumgenital skin of an adult male *Saguinus fuscicollis*; peripheral area of scent gland organ. sg, sebaceous glands; ag. apocrine glands. The arrow indicates the caudal edge of the scrotum.

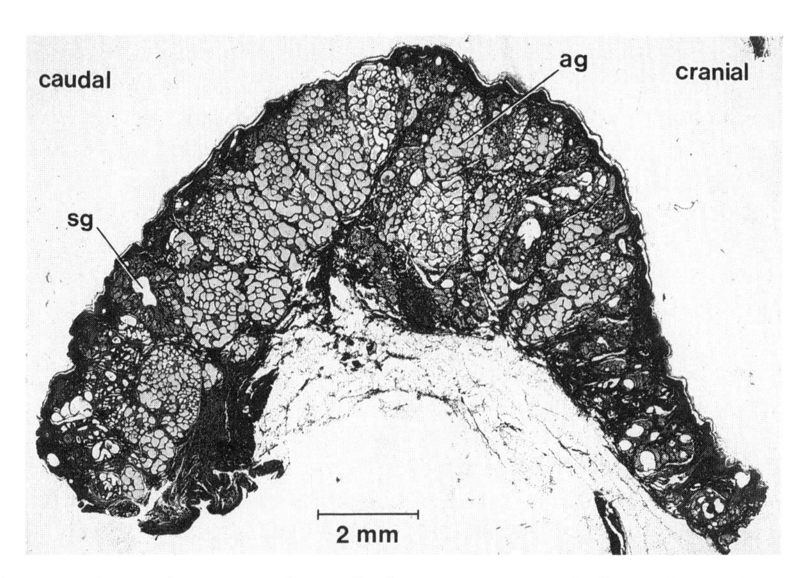


**Fig. 4.2** Sagittal section through the circumgenital skin of an adult female *Saguinus fuscicollis* at the area of the labium pudendi. sg, sebaceous glands; ag, apocrine glands.

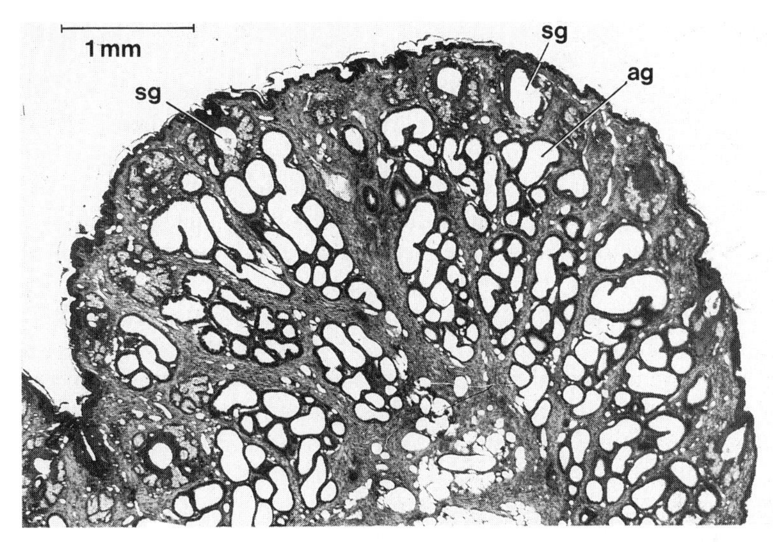


**Fig. 4.3** Sagittal section through labium majus of an adult female *Saguinus oedipus*. The apocrine (ag) and sebaceous (sg) glands are separated from each other by broad septa of connective tissue.

*Saguinus fuscicollis* and are separated from each other by larger amounts of connective tissue (Fig. 4.3). In males of this species, the scent glands of the circumgenital area are much smaller than in females and are nearly absent in the scrotal skin (Perkins 1969*a*; Wislocki 1930; Zeller *et al.* unpublished data). In cotton-top tamarin females, differences have been found between the suprapubic and genital areas of the pad in the size, glycogen content, and phosphorylase activity of the sebaceous glands (Perkins 1969*a*). Such differences suggest different functions for each area, a notion supported by behavioural studies on scent marking (French and Cleveland 1984; French and Snowdon 1981).

To date, little information on the ontogenetic development of the scent glands is available. In infants of *Callithrix jacchus*, no microscopic glandular specializations can be found (Sutcliffe and Poole 1978). The ontogeny of the circumgenital-suprapubic gland of *Saguinus fuscicollis* is currently being studied by Noll. Her findings show that microscopically recognizable circumgenital and suprapubic scent glands begin to develop well before puberty (Noll *et al.* 1989). As the animals mature, their glands increase in size. In *Saguinus fuscicollis*, they reach their full development between the ages of 18–24 months or even later (Epple 1980), in *Callithrix jacchus*, at the age of 12–15 months (Sutcliffe and Poole 1978). The development of the sternal gland has been studied only in *Callithrix jacchus*, where most animals develop a small, diffuse gland at the age of 12–15 months. Fully developed, discrete glands, however, were seen only in established breeding males (Sutcliffe and Poole 1978).

In *Saguinus fuscicollis*, the morphogenetic development of the circumgenital-suprapubic glands is partially controlled by gonadal hormones. Gonadectomy prior to puberty retards their development in males and females but does not inhibit it completely. Moreover, the sexual dimorphism in gland size is not totally dependent on gonadal hormones, since females gonadectomized as juveniles still develop somewhat larger glands than castrated males (Epple 1981*a*, 1982).

In addition to developmental processes, the social environment may influence gland size. French *et al.* (1984) have suggested that in cotton-top tamarins, females develop more active circumgenital-suprapubic glands after being removed from their families and paired with adult males. Very similar observations were made in female *Saguinus fuscicollis* (Epple, unpublished). In *Callithrix jacchus* there is good evidence that the full development of the sternal glands in males depends upon the individual maintaining breeding rank in its group (Epple and Lorenz 1967; Sutcliffe and Poole 1978).

## Scent marking and other chemosignalling behaviours

Marmosets and tamarins possess a rich repertoire of chemosignalling behaviours. These are associated closely with investigatory behaviours. Many observers have reported that callitrichids sniff and lick the bodies, scent glands, and scent marks of conspecifics in a variety of sexual and social situations (cf. Epple *et al.* 1986). In all species, the most conspicuous and most frequently seen chemosignalling behaviours are stereotyped scent-marking patterns involving the circumgenital and suprapubic glands, urine, and the sternal gland. This suggests that the main sources of chemical signals are the secretions of the specialized scent glands and urine. However, vaginal secretion, saliva, nasal secretion, and possibly secretion from microscopic accumulations of non-specialized skin glands in various areas of the body also have been implicated as scent sources.

Most scent marking is performed on items in the environment. In addition, marking the bodies of conspecifics, i.e. partner marking, has also been reported for several species (cf. Epple *et al.* 1986). Three basic marking patterns involve the use of specialized glands in males and females of many species. In addition, several behaviours suspected to have chemosignalling function occur. They do not involve specialized scent glands. The three basic scent-marking patterns have been described for several callitrichid species but have been given different names in the literature. For the sake of clarity, they are briefly described below.

*Circumgenital marking*

Also termed *sit rubbing*, *anogenital marking*, or *anal marking*, this appears to be the lowest-intensity marking pattern in most callitrichids. It is performed by rubbing the circumanal areas against the substrate in a sitting position (Moynihan 1970; Mack and Kleiman 1978; Sutcliffe and Poole 1978; French and Cleveland 1984; Epple *et al.* 1986). During this type of marking, secretions from the circumgenital glands and the circumanal skin and a small amount of urine are applied. In females, vaginal discharge may be added. Our observations on *Saguinus fuscicollis* suggest that the animals do not always add urine and genital discharge to the marks. However, we do not know which factors cause them to vary the amounts they add. Faecal matter adhering to the anogenital area may also be deposited during circumgenital marking, but defecation is not part of the marking behaviour (Epple 1975*a*; Sutcliffe and Poole 1978; French and Snowdon 1981). Circumgenital marking is the most frequent of all marking patterns in *Cebuella*, in some species of *Callithrix*, and in *Saguinus*.

*Suprapubic marking*

Termed *pull rubbing* by Moynihan (1970), this results in the application of secretions from the suprapubic gland. These are deposited when the animal presses the suprapubic pad to the substrate, either pulling itself forward with the hands, legs dangling to either side, or pushing the body with the feet. *Saguinus fuscicollis* also rubs the suprapubic gland against small protuberances while in a sitting position. The motor pattern is adaptable to the characteristics of the substrate. Some species discharge urine during suprapubic marking (Epple *et al.* 1986). Suprapubic marking is common in *Saguinus* and *Leontopithecus*, but has not been observed in captive *Callithrix jacchus* or *Callithrix argentata* (cf. Epple *et al.* 1986). Rylands (1990), however, reports that suprapubic marking is frequent in wild *Callithrix humeralifer*, *Callithrix penicillata*, and *Callithrix kuhli*.

*Sternal marking*

This involves rubbing of the sternal gland against items of the environment (Epple and Lorenz 1967). A monkey may rub the chest against the substrate while lying flat on its stomach or press its sternal area to the item being marked while stomach and rear end are elevated. If the item to be marked is located above the surface supporting the animal it is marked while standing on the hind legs. In most species, sternal marking is shown less frequently than circumgenital or suprapubic marking. In *Leontopithecus r. rosalia*, however, it is the most frequent marking pattern of adults (Epple and Lorenz 1967; Sutcliffe and Poole 1978; Kleiman and Mack, 1980; and others).

### *Ventral rubbing*

This does not involve a single glandular complex. The animal rubs the entire ventral surface over the substrate, probably mixing the secretions from different glandular areas, perhaps with the addition of secretions from unspecialized glands in the abdominal skin. This may result not only in depositing scent on the substrate but may also impregnate the ventral area of the marking animal with the mixture.

### *Muzzle rubbing*

Sometimes accompanied by sneezing, this appears to be a relatively common pattern among callitrichids. It occurs after eating or drinking where its primary function seems to be that of cleaning. However, it is also frequent prior to, during, or after circumgenital and suprapubic scent marking (cf. Epple *et al.* 1986). It may result in the deposition of saliva, nasal discharge, or secretions from sebaceous and apocrine glands located in the skin of the lips (Perkins 1966, 1968, 1969*a*,*b*; Heymann *et al.* 1989), some of which may be of communicatory relevance.

### *Other behaviours*

Other behaviours which may play a role in chemical communication have been observed in several species. These patterns occur more infrequently than the ones described above and some of them appear to be subject to high individual variability. Rubbing of tree branches with the inside of the arms has been reported for wild *Callithrix humeralifer intermedius* (Rylands 1981). No scent glands have been described in this area but Perkins (1966, 1968, 1969*a*) showed that several callitrichids possess ulnar eminences, located on the wrist, in which large apocrine glands are associated with sinus hairs.

Self-marking with glandular secretions and/or urine seems to be quite common. Cotton-top tamarin females mark the heels of their feet by positioning one or both heels underneath the circumgenital gland when marking. This behaviour has been observed in a number of females in our colony, but does not appear to be displayed by all of them. Circumgenital marking of the tail, which is rolled into a tight curl underneath the glandular area, was observed in a *S. o. geoffroyi* female (Epple 1967). It is reminiscent of a similar but more elaborate marking complex described in *Callimico goeldii* (Wojcik and Heltne 1978). Urinating on the tail coiled under the body is reported for wild *Callithrix humeralifer* by Rylands (1990), but it does not seem to be associated with other stereotyped behaviours.

Urinating into the hand, which is held in the stream of urine while the animal urinates in a standing or sitting position, is shown by *Saguinus mystax*. Males and females also rub their cheeks in urine voided by their sexual partner (cf. Epple *et al.* 1986). Urine-washing of both hands and feet in a manner identical to that described for some prosimians and several cebid monkeys (Andrew and Klopman 1974) has been observed in a single adult male *Saguinus leucopus*. However, this individual urine-washed very rarely, and the other individuals in our colony never showed this behaviour (Epple unpublished).

Behaviours presumably stimulating the flow or production of glandular secretions also occur. Scratching or kneading the circumgenital-suprapubic glands and/or the sternal area shortly before scent marking has been observed in several species (cf. Epple *et al.* 1986). In addition to stimulating the scent glands, these manipulations also transfer secretions to the hands, which could then transfer them to the substrate during locomotion.

A species comparison of the motor patterns of scent marking, the social contexts in which it is shown and the frequency with which males and females display this behaviour shows both similarities as well as differences among species (Epple *et al.* 1986). For example, the species we have studied most intensively, *Saguinus fuscicollis*, *Saguinus o. oedipus*, and *Saguinus leucopus* display both circumgenital and suprapubic scent marking. Saddle-back and white-footed tamarins also display sternal marking, but saddle-backs show this pattern less frequently than white-footed tamarins. Cotton-top tamarins, on the other hand, apparently do not display sternal marking (Wolters 1978). In saddle-back tamarins, both sexes mark frequently with their circumgenital and suprapubic glands. Females show slightly but significantly more marking than males (Epple 1980). Cotton-top females on the other hand, display much higher frequencies of scent marking than cotton-top males (French and Snowdon 1981; French and Cleveland 1984; Epple *et al.* 1988). Based on the small number of individuals in our colony, there is no clear-cut sex difference in marking frequency in white-footed tamarins.

*Saguinus fuscicollis* and *Saguinus o. oedipus* differ in the degree of variability in the motor patterns of circumgenital and suprapubic marking and in the behavioural contexts in which both types of markings are performed. In saddle-back tamarins, scent marking is quite variable. The animals show distinct circumgenital and suprapubic marking patterns but also display various combinations of both. On rare occasions, the use of the sternal gland is combined with that of the circumgenital and suprapubic glands as the animal rubs the whole ventral surface against the substrate. At low intensities of arousal, circumgenital marking predominates in male and female saddle-backs. At high intensities, suprapubic marking is added and combined with circumgenital marking in various ways. There is no evidence that circumgenital and suprapubic marking are associated with different behavioural contexts or fulfill different biological roles in this species. *Saguinus leucopus* employs the use of the circumgenital and suprapubic glands in a manner similar to those of *Saguinus fuscicollis*. Golden lion tamarins also appear to show variable combinations of suprapubic and circumgenital marking (Kleiman and Mack 1980).

In cotton-top tamarins, the typical marking pattern at low levels of arousal is circumgenital marking (French and Snowdon 1981). Unlike in saddle-back and white-footed tamarins, suprapubic marking does not seem to be combined with circumgenital or sternal marking (French and Snowdon 1981; French and Cleveland 1984). It is much less frequently performed, and in cotton-top tamarins, appears to be a pattern which is functionally distinct from circumgenital marking. Suprapubic marking is displayed predominantly by females of breeding status during agonistic encounters with conspecifics (French and Snowdon 1981).

## Social contexts of scent marking

Observational and experimental studies on the role of chemical signals in social and sexual behaviours in callitrichids have concentrated on the most frequently displayed patterns: circumgenital, suprapubic, and sternal marking. The frequencies of display of these patterns are influenced by a large variety of physiological, social, and environmental factors. Behavioural studies on scent marking in the Callitrichidae have recently been reviewed by Epple *et al.* (1986). These studies suggest that both scent marking itself and the chemical signals thus deposited are important in many areas of the behavioural biology of marmosets and tamarins.

The functions of chemical signals in callitrichid behaviours have to be considered within the framework of the social system of those primates. It appears that callitrichids have a complex social system. Small, mostly territorial groups, each containing one breeding female, are the rule (Sussman and Kinzey 1984). The offspring of the breeding female are reared communally, with most adult group members participating in their care (Sussman and Kinzey 1984). In captive groups, a pair bond between the breeding female and one of the males is evident in all species (Epple 1975*a*; Kleiman 1977, 1978*a* and others). This is expressed in preferential sexual and social association between the partners and has led to the assumption that all callitrichids are monogamous (Epple 1975*a*; Kleiman, 1977, 1978*b*). Recent field studies have revealed that the social system of wild callitrichids is more flexible than is evident under captive conditions. Many wild groups contain monogamous breeding pairs, but polyandry and, very rarely, polygyny, have also been observed (Sussman and Garber 1987; Sussman and Kinzey 1984; Ferrari and Lopes Ferrari 1989).

Several ecological factors, among them the availability of tree gums as a year-round resource used by the tree gouging marmosets (*Callithrix, Cebuella*), appear to be major determinants of the social system (Ferrari and Lopes Ferrari 1989). Marmosets seem to live in rather stable groups, consisting mainly of family members, and to adhere predominantly to a monogamous mating system (Ferrari and Lopes Ferrari 1989). Tamarins of the genus *Saguinus* and *Leontopithecus*, whose food resources are seasonably much more variable, exhibit more flexible mating systems, including polyandry. They live in relatively unstable groups, containing related as well as some unrelated individuals (Ferrari and Lopes Ferrari 1989). Individuals may leave one group and immigrate into another, possibly changing group affiliations repeatedly. Although details on age/sex classes of dispersing individuals are unknown, it is likely that dispersal increases outbreeding and may give individuals who cannot reproduce within their natal families a chance to find a breeding position (cf. Sussman and Kinzey 1984). Until more field data are available, there is no reason to believe that the strong tendency to establish long-lasting pair bonds exhibited by captive tamarins is lacking in their wild counterparts. If such a pair bond exists, it is likely that most offspring of the breeding female are fathered by her pair mate.

Within this social system, individuals adopt different strategies, depending on their socio-sexual roles in the group. These individuals need to communicate the strategies which are necessary for the establishment and maintenance of their social roles. Considering the many sources of scent, the complexity of marking behaviours and the large number of social situations in which chemosignalling occurs, it seems likely that chemical signals mediate some of the necessary information. Our studies on *Saguinus fuscicollis* (see below) have shown that this species produces chemical messages which identify species, subspecies, individual, gender, and reproductive condition, and may also contain information on social status and the age of the material. It is likely that other callitrichids communicate similar information by means of scent. These, and perhaps additional, still unidentified chemical signals, may be important in a number of behavioural contexts.

Hypotheses concerning the manner in which chemical signals influence various behaviours are mainly based on the fact that scent marking frequencies vary among age/sex classes and may be strongly affected by social and sexual context, albeit differently in different species. Only a few possible interpretations of such observations are discussed here. Among them is the role of chemical signals in the control of reproductive events.

Pair bonding, identification of the fertile phase of the ovarian cycle or of an existing pregnancy, sexual arousal, and fertile mating may all be influenced by signals produced by the female, the male, or both partners. For example, an increase in female scent marking during estrus was reported in *Saguinus o. oedipus* (French and Snowdon 1981). It may be interpreted in terms of sexual attraction to a receptive female. On the other hand, a decrease in scent marking by estrous females in *Leontopithecus r. rosalia* might be a means by which females avoid advertising their condition to unattached males, thereby limiting sexual interaction to their bonded males (Kleiman 1978*a*).

Increases in female scent marking during pregnancy have been reported for *Callithrix jacchus* (Box 1978), *Saguinus fuscicollis* (Epple 1975*a*) and *Saguinus o. oedipus* (Muckenhirn 1967). Scents from pregnant females may promote group cohesion, so that helpers are present at the time infants are born. In *Leontopithecus r. rosalia*, on the other hand, pregnancy and the presence of small infants appear to inhibit scent-marking activity in the breeding female and her mate (Kleiman and Mack 1980).

In *Callithrix jacchus*, *Saguinus fuscicollis* and *Saguinus o. oedipus*, most non-breeding females do not display ovarian cyclicity and it is assumed that this is due to suppression by the breeding females (Abbott 1984; Epple and Katz 1984; French *et al.* 1984). This suppression may be partially mediated by scent. Epple and Katz (1984) have shown that the onset of ovarian cyclicity in a *Saguinus fuscicollis* female was delayed by exposing the female to scent from her family, after she left the family and was paired with a male. In a similar experiment on *Saguinus o. oedipus*, exposure to family scent after pairing delayed the time to first ovulation in newly paired daughters (Savage *et al.* 1988). It remains to be documented that it is scent from the mother rather than from other group members which delayed the onset of cyclicity. In a study using groups of non-related *C. jacchus*, Barrett *et al.* (1990) transferred secretion from the alpha-female scent glands directly to the noses of subdominant females, who had been removed from their groups but housed adjacent to them. These isolated females and their groups also exchanged cages daily. The isolated female showed a significant delay in the onset of cyclicity as compared to control females. More recently, these investigators have shown that female scents appear to play an important role in the initiation of ovarian suppression in submissive females at the time of group formation. When cycling females were rendered anosmic and then grouped the dominant female as well as the subordinate females ovulated (Abbott *et al.* 1991).

Scent also appears to be important in mediating interactions between infants and other group members. Infant common marmosets and saddle-back tamarins raised by humans recognize the odour of their human caregivers, and derive comfort from it in stressful situations (Cebul *et al.*

1978). This suggests that under normal social conditions bonding between infants and caregivers is mediated by scent.

In addition to functions in the area of reproduction, there is evidence for a role of scent marks in aggressive behaviour. In several species, the scent-marking activity of socially dominant males and females in established groups tends to be higher than that of non-dominant group members, particularly when groups contain non-related individuals. Moreover, the scent-marking activity of dominant male and female common marmosets and saddle-back tamarins increases dramatically when strange conspecifics are introduced to established pairs or groups (cf. Epple *et al.* 1986). In cotton-top tamarins tested under comparable conditions, on the other hand, only suprapubic marking increases in females, while males show no increase in marking activity (French and Snowdon 1981). A number of field studies also support the hypothesis that scent marking is important in the context of intragroup and intergroup aggressive behaviour (cf. Rylands 1990). In spite of the differences between species, the results of all of these studies indicate that high social rank and aggressive motivations are communicated by the frequency of marking, by a specific scent quality, or both. In either case, the scents produced in aggressive marking may serve as olfactory threat signals. The high frequency of marking saturates the environment with the scent of an aggressive individual. Conspecifics who have experienced domination by this individual must recognize its scent and are exposed to it throughout much of their range, even in the absence of the dominant animal. In this way, the scent of the aggressor may effectively control the behaviour of a submissive animal just by virtue of communicating individual identity. Should the quality of the scent be specific for an aggressive or a dominant animal, this effect would be further reinforced. Aggressive scent marking may function in a manner similar to other long-distance threat signals. It prevents submissive animals from challenging dominant individuals, and results in avoidance of physical combat and possible injury to both. Aggressive scent marking may aid males and females to establish and maintain a high social rank within their groups and to prevent sexual competitors from challenging the pair bond between mates.

A variety of features in the physical environment of captive and wild callitrichids influence scent-marking activity. This suggests that scent is also important in the animals' interactions with the environment, but the role of scent in this context is poorly understood. Territorial marking, identification of resting and roosting areas, and marking of food resources may all be of importance. Captive marmosets and tamarins may perform much of their marking activities in specific locations in their cages. For example, captive common marmosets show a high amount of wood gouging, a behaviour related to tree exudate consumption in the wild. Scent marking is concentrated at gouge holes and in other specific locations in the cage, such as nest boxes and food shelves, possibly labelling them as resources (Epple 1967; Box 1978; Sutcliffe and Poole 1978). Captive golden lion tamarins perform a high percentage of their circumgenital and particularly sternal marking during long call displays. In a building housing several groups of lion tamarins, scent marking and long call displays were concentrated at air vents through which the animals could hear long calls of other groups. The animals possibly regarded the air vents as territorial borders. Marking of the feeding platform and the area around it in the process of foraging was a second major context in which scent marking occurred (Mack and Kleiman 1978).

In wild callitrichids as well, marking is not randomly distributed over the entire home range of a group. Yoneda (1984*b*) reports that in *Saguinus fuscicollis* it is concentrated in the centre of the territory, where it is often associated with resting or insect foraging. In a group observed by Bartecki and Heymann (1990), on the other hand, most scent marking was performed at the periphery of home range, which was the area most intensively used by the group. In *Saguinus o. geoffroyi*, Dawson (1979) observed marking in areas where the home ranges of groups overlap. Similar observations are reported for suprapubic and sternal marking in *Callithrix humeralifer* by Rylands (1990). Lindsay (1980), however, saw most scent marking performed by wild *Saguinus o. geoffroyi* along the tree branches used as regular trails within the territory. It is possible that under natural conditions, territorial borders as well as particularly important locations within the home range, such as regularly travelled trails or sleeping sites, are scent marked. Trail marks might aid the whole group or individuals which have strayed from the group to orient themselves in a dense arboreal environment.

In *Callithrix* and *Cebuella*, which obtain much of their diet by gouging holes into the bark of trees and consuming exudate, circumgenital marking is concentrated at gouge holes (Coimbra-Filho and Mittermeier 1978; Lacher *et al.* 1981; Rylands 1981, 1985*b*, 1990; Stevenson and Rylands 1988). Scent marking of such resources may communicate information on the nature of the resources. Lacher *et al.* (1981) found that several groups of wild *Callithrix penicillata* utilized the same sap holes for feeding and scent marking. Groups took turns feeding at these locations, but avoided meeting each other. The authors suggest that scent located at sap holes may inform the animals about the time elapsed between uses, and therefore about the availability of sap. It may also help to prevent the meeting of several groups at the same time, an event which would result in aggressive interactions.

Based on field observations of several species of *Callithrix*, Rylands (1985*b*, 1999) and Stevenson and Rylands (1988), on the other hand, suggest that circumgenital scent marking of gouge holes is not related to gum resources. Circumgenital marking may be primarily involved in intragroup communication and is performed preferably at gouge holes because these localities are highly likely to be sniffed by all members of the group and may represent highly absorbent substrates for the scent material. Heymann's (1991) observation that scent marking by wild *Saguinus mystax* is concentrated in feeding trees lends itself to a similar interpretation.

Under some ecological conditions, callitrichids are strictly territorial (Terborgh 1983; Sussman and Kinzey 1984). Therefore, scent may also play a role in territorial defence and intergroup spacing (Dawson 1979; French and Snowdon 1981). However, scents located at territorial borders are not necessarily a means of preventing invasion of an area. They can inform conspecifics about species identity, sex, individual identity, or reproductive condition of the territorial owners. Such information is useful to individuals who have left their own groups, either to immigrate into others or to find partners with whom to establish a new territory.

## Communicatory information contained in scent material

As pointed out above, many callitrichid species produce complex scent marks which combine ingredients from various sources. Such complex mixtures have the potential to encode a large variety of communicatory information by means of both qualitative and quantitative composition patterns. We have studied the communicatory content of scent from saddle-back and cotton-top tamarins in some detail. The ability of saddle-back and cotton-top tamarins to distinguish between scents from various types of donors were studied in a series of behavioural experiments. Results which showed that the tamarins discriminated among the scents from different donor types were interpreted as evidence that the scent material contains communicatory information which identified the donor types.

All behavioural tests consisted of the presentation of either a single scented stimulus object, or of two identical stimulus objects, each carrying a different scent. The stimulus objects were scented by allowing donor monkeys to mark them or by applying a pre-measured amount of stimulus to them. The stimulus objects were then introduced into the home cage of a tamarin for a standard test period during which the subject was allowed to investigate the sample(s) freely. The stimulus objects used for saddle-backs were wooden, aluminium or frosted glass plates, measuring 61 × 5 × 0.6 cm. Since cotton-top tamarins prefer to mark small protuberances, the stimulus objects used for this species were glass rods with closed, rounded ends. They were inserted into holes in wooden shelves, mimicking small protruding twigs.

All test periods were divided into intervals of five seconds each. Three behavioural responses were recorded, and the subjects received a score of one per stimulus in each response category for every interval in which this response was shown. For saddle-back tamarins contacting each stimulus object, sniffing each stimulus object, and scent marking each stimulus object were recorded. Circumgenital marking, suprapubic marking, and combinations of both were recorded as one category. For cotton-top tamarins contacting each wooden shelf containing the stimulus object, sniffing each stimulus object and circumgenital marking of each stimulus object were recorded. Suprapubic marking by cotton-tops was very rare and was not evaluated.

For choice tests, mean responses directed at each of the two samples were compared statistically. For single sample tests, mean responses directed at one type of sample were compared to responses directed at other types of samples presented under identical conditions. Significant differences in the level of responses directed toward different stimuli were interpreted as evidence of the tamarins' ability to discriminate the stimuli from each other.

A number of control tests were performed with saddle-back tamarins in order to document that the behavioural responses to scent marks from conspecifics differed from those given to control stimuli. Unscented stimulus objects and a synthetic musk, exaltolide, were used as controls. The control studies show that scent marks from conspecifics elicit specifically high levels of contacting, sniffing, and scent marking (Epple *et al.* 1980).

The following experiments were conducted to determine among which categories of donors the tamarins can discriminate on the basis of complex scent marks, urine, or glandular secretion. In one series of tests, the ability of both species to discriminate between scents from conspecifics and from other callitrichids was investigated. Both saddle-back and cotton-top tamarins investigate scents from conspecifics more frequently than scents from other callitrichid species. This discrimination is made when a choice between complex, natural scent marks is given. However, samples of voided urine from saddle-backs and secretion collected from the glandular surface of cotton-tops also contain cues on which discrimination between scent from conspecifics and related species can be based (Epple *et al.* 1979, 1987, 1988). In *Saguinus fuscicollis*, scent marks from closely related subspecies offer cues for subspecies discrimination. *Saguinus f. fuscicollis* and *Saguinus f. illigeri* discriminate between scent marks from these subspecies in choice tests (Epple *et al.* 1979, 1987).

Saddle-back and cotton-top tamarins can discriminate between scent marks from conspecifics and from other callitrichids when the scent is offered underneath a screen so that it can only be sniffed, not contacted directly. However, material which can be contacted with the muzzle appears to be more attractive. This result suggests that contact with the scent material is necessary for the perception of its full biological activity (Belcher *et al.* 1988, 1990, and unpublished).

Saddle-back tamarins discriminate between scents from male and female conspecifics under a variety of experimental conditions. Scent marks and urine samples from males are more frequently contacted, sniffed, and marked than corresponding material from females (Epple 1974*b*, 1978*b*). Even mixtures of scent marks from males and females pooled together in methanol are discriminated from equal amounts of marks from females alone (Epple 1978*b*). When the marks are offered underneath a screen, so that they can be sniffed but not contacted, saddle-back tamarins discriminate between scent marks from males and females and between scent marks from castrated and intact males (Epple 1978*b*; Epple *et al.* 1990). In contrast to saddle-backs, cotton-top tamarins do not discriminate between urine samples from males and from females. Moreover, towels on which males had slept during the previous night were not discriminated from towels on which females had slept (Epple, unpublished). Discrimination between scent marks from males and from females was not studied in cotton-tops because the infrequent marking by males makes collection of stimulus material difficult.

Discrimination between scents from two individuals of the same sex are made by both tamarins. Saddle-back tamarins discriminate between scents from two individuals when they are habituated to the scent from one of them by means of a five-minute exposure to this scent. A test offering a choice between this individual's scent and that from a second tamarin of the same sex follows this habituation period. Habituation to the first tamarin's scent results in a preference for that of the second individual (Epple *et al.* 1979). While these tests show the ability of the tamarins to discriminate between the scent from both donors, the following experiment suggests that scents are actually used to identify known conspecifics. Saddle-back tamarins were offered choices between scent from either two familiar males or two familiar females. One of the donors had been given an aggressive encounter with the subjects prior to a choice test. Under these conditions, scent marks, but not urine, from the recent opponent elicit higher responses than corresponding samples from the neutral familiar donor, and also stimulate visual threat displays (Epple 1974*b*, 1978*c*).

In cotton-tops, discrimination between female individuals was tested by presenting single stimulus objects scented with secretion collected from the glandular surface. The animals were habituated to the secretion from one individual by offering them two successive samples of secretion from the same female. When secretion from a novel female was offered as a third sample, the subjects displayed an increase in sniffing and contacting activities (Epple *et al.* 1988).

The scent marks of saddle-backs contain information relating to the social status of the donor. Stimulus objects scent marked by unfamiliar, socially dominant males elicit higher behavioural responses than objects scent marked by unfamiliar, subdominant males. Since dominant males often show more scent marking than subdominant ones, discrimination between donor types might have been based on quantitative cues (Epple 1974*b*).

The scent marks of saddle-back tamarins also contain cues relating to the age of the material. Freshly deposited scent marks are discriminated from marks deposited by the same donor 24 hours or more prior to testing. Fresh scent marks and 24-hour-old scent marks elicit higher levels of investigation than unscented stimulus objects. Older marks, however, are not discriminated from blank stimulus objects. Scent marks from males are discriminated from those of females when both stimuli are fresh, 24 and 48 hours old. However, no discrimination is made on the basis of marks older than 48 hours (Epple *et al.* 1980).

The failure of the saddle-backs to discriminate aged scent marks from blank stimulus objects and to discriminate between scents from males and females when the material is older than 48 hours does not necessarily mean that the animals are no longer perceiving information relevant to species or gender recognition. Some of our analytical studies have shown that aged scent marks contain higher concentrations of volatile compounds than fresh material (Belcher *et al.* 1982). Therefore, it is possible that during the ageing process scent marks develop additional cues on which recognition of the age of the material could be based. Such information may be important in territorial species such as these (Terborgh 1983), in which groups may keep track of the movements of neighbouring groups by means of scent.

## Chemical composition of scent

Our behavioural studies indicate that the scent marks from *Saguinus fuscicollis* and *Saguinus o. oedipus* contain a number of communicatory messages. Similar messages might well be contained in the scent marks deposited by other callitrichid species. To date, the manner in which communicatory messages are chemically encoded within the matrix of complex biological sources such as scent marks is poorly understood. Our chemical analytical studies are designed to shed some light on this question, using a comparative approach which includes *Saguinus fuscicollis*, *Saguinus o. oedipus*, and *Saguinus leucopus*.

In an attempt to isolate compounds or groups of compounds involved in structuring the communicatory messages present in the scent marks of *Saguinus fuscicollis* and *Saguinus oedipus*, a number of fractionation studies have beeen conducted with scent material from both species. These fractionation studies aimed at the isolation and identification of those components of the scent which are involved in structuring species, subspecies and gender codes. They have been described in some detail by Belcher *et al.* (1986, 1990) and Epple *et al.* (1989, 1990), and therefore only a summary will be presented here. When scent marks from individual saddle-back tamarins are dissolved in organic solvent and tested against chemically untreated marks from the same individual, natural scent marks are preferred to material dissolved in hexane or methanol. This preference does not reflect an aversion to the solvent, and is shown even if a larger number of fractionated than natural marks is presented. It shows that the initial step of taking up the material in organic solvent already reduces its attractiveness. However, in spite of some loss in biological activity, information on subspecies and gender is retained in organic fractions of scent material from saddle-backs. Choice tests were conducted both with scent material dissolved both in hexane and in methanol. In both cases, solutions of material from males are preferred over solutions of material from females. Similarly, methanol-soluble material from *Saguinus f. illigeri* is discriminated from identically prepared material from *Saguinus f. fuscicollis*. Gas chromatographic fractionation of hexane-soluble scent material from *Saguinus fuscicollis* further reduces its activity. Given choices between total gas chromatographic eluates of scent marks from males and females, the tamarins tend to prefer material from males. However, further sub-fractionation of scent material from males and females results in complete loss of information on gender in all fractions.

In contrast to *Saguinus fuscicollis*, organic solvent solutions of scent marks from *Saguinus o. oedipus* do not appear to contain communicatory information. Cotton-top tamarins do not discriminate between scent marks from conspecifics and scent marks from saddle-back tamarins when the material is dissolved in methanol. Moreover, response levels to these stimuli are lower than to natural marks (Belcher *et al.* unpublished).

These results and some of the behavioural studies (see also below) suggest that not all biologically important components of the scent material are soluble in organic solvents. Therefore, some of our analytical studies are conducted with aqueous fractions of scent material. Cotton-top tamarins discriminate aqueous fractions of scent from *Saguinus o. oedipus* from identically treated samples of material from *Saguinus fuscicollis*. Saddle-back tamarins discriminate aqueous fractions of scent donated by conspecific males from those of scent donated by females. Identical results are obtained when aqueous fractions are freeze-dried and reconstituted in water before testing. Bioassays on *Saguinus fuscicollis* however, show that natural scent marks are preferred over aqueous fractions of material from the same donor individuals (Belcher *et al.* 1990).

In summary, our fractionation studies show that even the initial steps of taking up natural scent material in organic solvents or in water reduces its biological activity and that activity loss increases gradually as the material is further fractionated. This suggests that the communicatory code in the scent marks is structured by complex mixtures of compounds soluble and insoluble in organic solvent. Because of this complexity, an attempt to isolate the compounds on which the chemical code is based must involve a variety of methodological approaches.

The studies reviewed above show that, in *Saguinus fuscicollis*, cues on which discrimination between two subspecies and between males and females are based are retained in organic fractions of scent marks. Furthermore, tests with screened scent material had suggested that, in *Saguinus fuscicollis*, male–female and intact male–castrated male discrimination are based on airborne compounds. Therefore, our analytical studies initially concentrated on the identification of volatile compounds soluble in organic solvents. For analytical comparison, these studies included material from *Saguinus o. oedipus* and from *Saguinus leucopus*. Gas chromatography/mass spectrometry was employed as the major analytical tool (Yarger *et al.* 1977; Smith *et al.* 1985; Belcher *et al.* 1986). Figures 4.4–6 show gas chromatograms of the volatile fractions of natural scent marks from females of the three tamarin species.

The major volatile constituents of the scent marks from male and female *Saguinus f. fuscicollis* and *S. f. illigeri* are squalene and 15 esters of *n*-butyric acid. Other volatile compounds, including cholesterol and a number of organic acids, are also present (Belcher unpublished; Yarger *et al.* 1977). Cholesterol, squalene and the butyrates are of glandular origin. They are also found in gland secretions collected by manual expression of the labial/scrotal and suprapubic parts of the gland pad.

*Saguinus f. fuscicollis* and *Saguinus f. illigeri* show significant differences in the concentrations of some of the butyrate esters. In addition, scents from males differ from scents from females in the concentration of some of the other butyrates. Thus, information derived solely from the butyrate–squalene concentration profiles of the marks can categorize the donors as members of one of the two subspecies and, regardless of subspecies, as males or females (Smith *et al.* 1985). However, to date we have not been able to demonstrate the communicative significance of these compounds. A number of studies on the communicatory content of fractions of scent material and of semisynthetic and synthetic formulations of squalene and the butyrates demonstrated that these compounds alone are not sufficient to encode the identity of subspecies, gender, or individual, nor are they necessary for gender recognition (Epple *et al.* 1979; Belcher *et al.* 1986 and

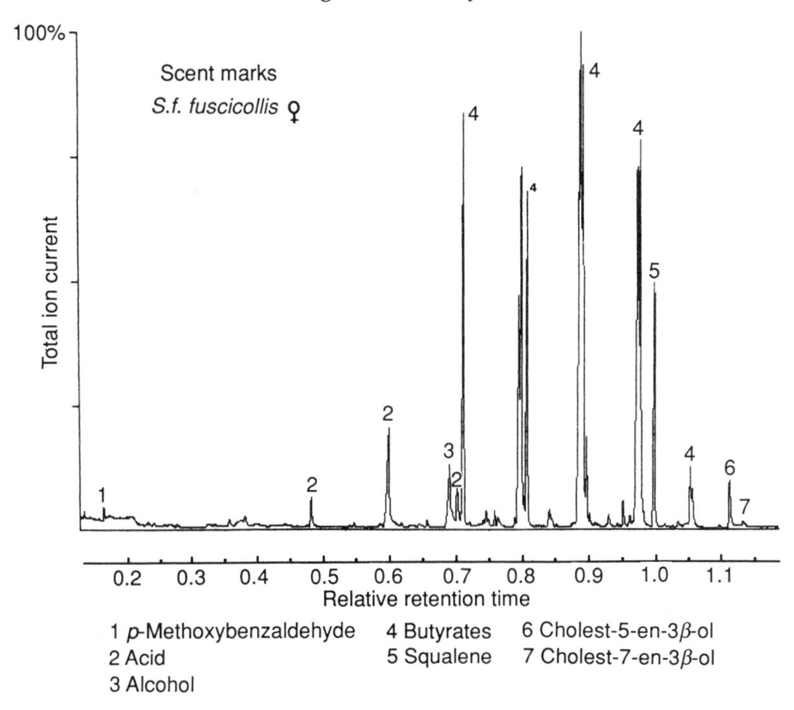


**Fig. 4.4** Mass chromatogram of the volatile components of the scent marks from *Saguinus fuscicollis*. The numbers refer to specifically identified components. Note the relatively high concentrations of the butyrate esters (4). (From Belcher *et al.* 1988).

unpublished). Moreover, acidic fractions containing cues on maleness (see below) did not contain any of these compounds.

These findings leave us at a loss to explain the role of squalene and the butyrates in the scent material from saddle-back tamarins. It appears possible that squalene and the butyrates act as precursors for other, as yet unidentified, constituents, or function as fixatives which control the release of signal compounds. Specific concentrations of the butyrates and squalene may influence release of volatiles in a manner which contributes to identification of donor category.

Figure 4.5 shows that organic extracts of circumgenital scent marks from cotton-top tamarins contain much lower concentrations of volatile compounds than the scent marks from saddle-backs. The major components, squalene, cholesterol, and *p*-methoxybenzaldehyde, were found in all scent marks which were analysed. These compounds account for approximately 36 per cent of the total volatile material in the scent marks of female cotton-tops. In addition, 10 other peaks were found in

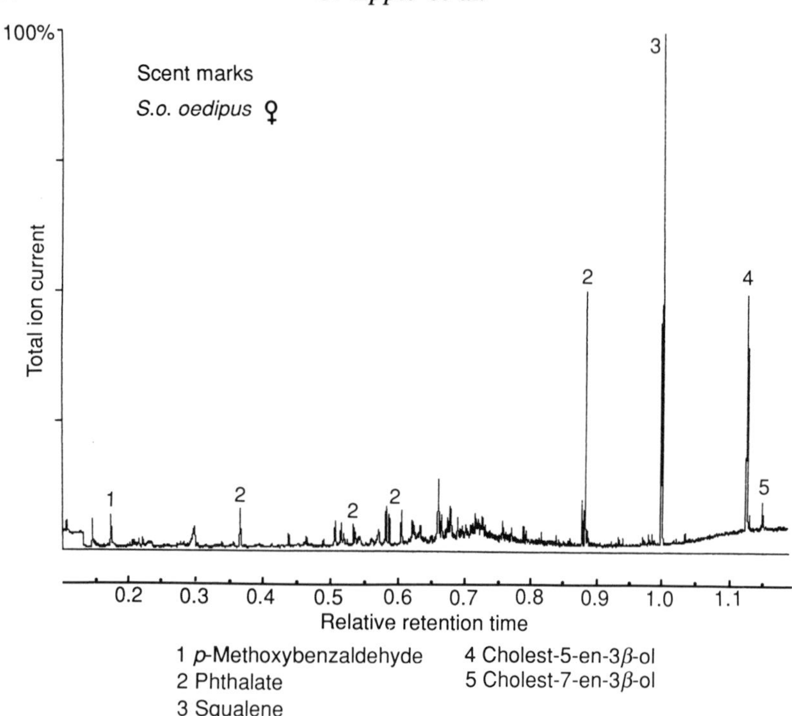


**Fig. 4.5** Mass chromatogram of the volatile components of the scent marks from *Saguinus o. oedipus*. Numbers refer to specifically identified components. (From Belcher *et al.* 1988).

very low concentrations, their presence being quite variable (Belcher *et al.* 1988).

Glandular secretions expressed from the labial-perianal part of the scent glands of female cotton-top tamarins do not contain detectable amounts of the butyrates which constitute a major portion of the marks and glandular secretions of *Saguinus fuscicollis*. Natural circumgenital scent marks from cotton-tops are also devoid of these compounds. However, secretion expressed from the suprapubic part of the scent pad of female cotton-tops contains all the butyrate esters, albeit in much lower concentrations than in *Saguinus fuscicollis*. We were not able to collect suprapubic marks for the purpose of analysis and consequently do not know whether the butyrates produced by the suprapubic pad of *Saguinus o. oedipus* are actually deposited in scent marks. The difference in the constituents of the secretions of the circumgenital and suprapubic parts of the scent gland of female cotton-tops may reflect differences in the fine structure of the gland (Perkins 1969*a*) and are consistent with the notion that in this species circumgenital

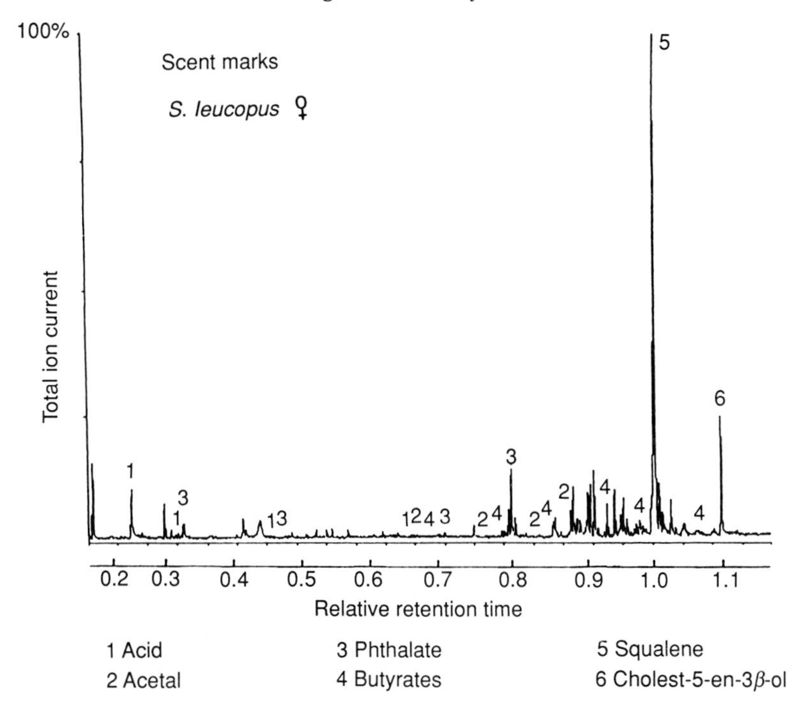


**Fig. 4.6** Mass chromatogram of the volatile components of the scent marks from *Saguinus leucopus*. Numbers refer to the specifically identified components.

and suprapubic marking are functionally distinct (French and Snowdon 1981; French and Cleveland 1984).

Analytical studies on natural scent marks from *Saguinus leucopus* show that the butyrate esters, cholesterol, and squalene are found in the scent from both males and females. In addition, expressed gland secretion resembles the scent mark in the composition of the volatile material. However, the overall composition profile of the scent material from *Saguinus leucopus* resembles that from *Saguinus o. oedipus* in that cholesterol and squalene predominate, with the concentrations of other components, including the butyrate esters, being very low.

Studies employing methods alternative to those previously used have resulted in additional fractions of scent marks from *Saguinus fuscicollis* which contain compounds involved in structuring gender codes. Aqueous pools of scent, retaining cues relating to gender (see above), were separated into basic/neutral and acidic fractions. This was accomplished by making the solution basic with sodium bicarbonate and extracting the basic and neutral components into chloroform. Acidification of the water residue, followed by extraction with chloroform, yielded the acidic components. In choice tests, the basic/neutral fractions of material from males are not discriminated from corresponding fractions of female or castrated-male material. However, acidic extracts of material from intact males are preferred over corresponding material from females and from castrated males. These findings indicate that testosterone-dependent acidic components in the scent mark provide cues relating to maleness.

The following unpublished studies show that these cues are derived from scent gland secretions, not from the urinary constituents of the marks. Bioassay choice tests demonstrated that natural scent marks are preferred over urine from the same donor individual (Epple 1978*b*). Moreover, acidic extracts of scent marks are preferred over acidic extracts of urine. It therefore appears that the active components in the acidic fractions derive from scent gland secretions. Chemical analysis by gas chromatography–mass spectrometry of acidic extracts from scent marks and urine showed the presence of a series of aromatic acids and aldehydes. These are of urinary origin and are higher in concentration in scent from males than from females and from castrated males. Synthetic preparations of these compounds, added to scent from castrated males in an attempt to impart maleness to the scent, did not increase the attractiveness of the material. These preparations were not preferred over female or castrated male scent.

The results of our studies with acidic fractions and of some additional bioassays indicate that constituents, which have not yet been identified with the techniques employed in the studies described above, are important in constituting the full biological activity of the scent material in saddle-back tamarins. It is reasonable to assume that this is the case for the other two species as well. In *Saguinus fuscicollis*, some of the unknown active constituents may be aromatic acids. Others seem to be of high molecular weight.

Saddle-back tamarins discriminate between scents from males and females and between scents from intact and castrated males on the basis of volatile cues alone (see above). However, after dissolving the scent marks in organic solvents, a residue remains. A combination of this residue and the fraction soluble in organic solvent is more attractive to the tamarins than the soluble fraction alone. This preference, however, is only displayed when the stimulus can be contacted with the muzzle, not when it only can be sniffed through a screen (Belcher *et al.* 1990).

Comparable observations were made on *Saguinus o. oedipus*. Cotton-top tamarins are capable of discriminating species identity on the basis of volatile scent components alone. However, most individuals seem not to be motivated to do so. A group of 12 tamarins did not discriminate between scent marks from cotton-tops and scent marks from saddle-backs when they were allowed to investigate marks presented under a screen (Belcher *et al.* 1988). However, four of these subjects appeared to be more interested in the screened marks than were the other animals. Subsequently, more extensive testing of these four tamarins showed that each individual was able to discriminate between the scent marks from the two donor species when tested with screened marks (Belcher *et al.* 1988). Thus, it appears that the cotton-tops can derive some information on species identity from airborne cues alone, but that additional compounds in the scent are necessary to elicit behavioural responses similar to those shown when the scent material is fully accessible to the monkeys (Belcher *et al.* 1988).

These results suggest that in addition to airborne compounds, compounds of lower volatility, some of them not soluble in organic solvents, are important in structuring scent images. Because of the importance of proteins in the scent of some mammals (Vandenbergh *et al.* 1975, 1976; Marchlewska-Koj 1977; Novotny *et al.* 1980; Singer *et al.* 1986) we directed our attention to proteinaceous material in the scent mark of the tamarins. Natural scent marks from male and female saddle-back tamarins contain a number of water-soluble proteins. Gel electrophoresis indicates that the major protein (molecular weight 66 kD) is found in urine, which is one of the constituents of the complex scent mark. A second protein (molecular weight 18 kD) comes from the gland secretion. It is present in natural scent marks and in pure secretion from the scent gland, but not in urine. Under reducing conditions, this protein is not detected. However, two additional bands appear at 4 and 14 kD. This suggests that the 18 kD protein is composed of two non-identical subunits linked by a disulphide bond. Scent marks from *Saguinus o. oedipus* and *Saguinus leucopus* show similar protein patterns. For these species it is not known which of the proteins originate from the urine and which originate from gland secretion.

As mentioned above, extraction of scent material from saddle-backs with organic solvent results in protein-free fractions of scent. Gel electrophoresis shows that the proteins are retained in the water-soluble residue which remains after organic solvent extraction (Belcher *et al.* 1990). Although the organic extracts contain cues on which subspecies and gender discrimination can be based, their attractiveness is increased when they are recombined with the residue (see above). This observation, and the fact that an increase in attractiveness is perceived only when the animals can contact the material with the muzzle, suggest that the proteins in the residue contribute to the sensory quality of the material.

In order to gain some insight into the manner in which proteins influence the sensory characteristics of scent marks, a number of experiments were performed. They consisted of choice tests during which the tamarins were offered scent samples in which the proteins had been enzymatically degraded (Belcher *et al.* 1990). Two experiments assessed the effect of protein degradation on species and gender cues. Previous studies with organic extracts had suggested that the proteins in the scent marks are not necessary for discrimination of species and gender. The following experiments confirmed this. When presented with choices between aqueous pools of marks from *Saguinus fuscicollis* and *Saguinus leucopus* in which the proteins are enzymatically degraded, saddle-backs prefer scent from conspecifics over scent from the related species, a response identical to that shown to natural scent marks. Similarly, digested scent from male conspecifics is preferred over digested scent from female conspecifics (Belcher *et al.* 1990).

Another set of studies assessed the effects of protein degradation on the general attractiveness of the scent. The subjects received choices between aqueous pools of digested scent marks or control marks, and water. Control pools were incubated under the same conditions which were used for digestion, but did not contain the digesting enzyme. The animals did not discriminate between digested scent marks from males and water samples, but did prefer the non-digested control samples over water (Belcher *et al.* 1990).

Another experiment offered a choice between two aliquots from the same pool of male scent marks. The proteins in one of the aliquots were digested while the second aliquot was incubated under the same conditions without enzymes. The animals discriminated between the non-digested and the digested aliquots, marking preferentially over the non-digested sample. In contrast, when digested scent material from females was tested against an undigested aliquot from the same pool of female marks, the tamarins did not discriminate between the samples (Belcher *et al.* 1990).

The results of the behavioural studies with digested scent material indicate that removal of the proteins results in a reduction of attractiveness of the scent from males without causing pronounced changes in its information content. This change in attractiveness appears to be a minor one. Moreover, the tamarins' reaction to it seems to depend on the context in which the scent is encountered. When digested scent material is presented as an alternative to a water blank or to an aliquot of non-digested scent from the same donors, the change in attractiveness caused by protein degradation may be strong enough to reduce the behavioural response to the digested scent. However, when choices between two socially relevant scents from different types of donors are presented, the tamarins seem to ignore the changes in attractiveness caused by degradation of the proteins.

Proteins may not influence the sensory characteristics of scent from females to the same degree as those of male scent. On the other hand, failure to discriminate between digested and non-digested scent from females may be due to the fact that, in the stimulus context in which they were tested, female scents were not attractive enough to cause the tamarins to direct much attention to minor changes in their qualities.

It seems possible that proteins contribute directly to the attractiveness of the scent marks and that the sensory cues they provide are mediated via the vomeronasal organ, a chemosensory organ well developed in callitrichids (Maier 1982). On the other hand, some of the proteins may function as carriers of volatile ligands. These roles are not mutually exclusive. A carrier function may involve the transport of ligands from the secreting skin gland to the surface, rendering ligands soluble in the medium of the scent mark, or protection of the ligands from rapid dissipation. Monkeys discriminate between scent from males and females and from intact and castrated males on the basis of airborne cues alone (see above). However, these cues survive freeze-drying (Belcher *et al.* 1990), a process likely to remove volatile material which is not bound to a carrier. Moreover, the fact that scent marks from males are preferred over marks from females for up to two days after they have been deposited (Epple *et al.* 1980) also suggests that the relevant volatile compounds are protected from rapid dissipation.

Proteins fulfill carrier function in the chemocommunication systems of other species. In mice, the urinary priming pheromones produced by males and females appear to be associated with the major urinary proteins. Urinary protein is associated with the pheromone which accelerates the onset of puberty in females (Vandenbergh *et al.* 1975, 1976; Novotny *et al.* 1980) and with the pregnancy blocking pheromone in male mouse urine (Marchlewska-Koj 1977). The induction of a surge of luteinizing hormone in male mice by exposure to female urine is due to a pheromone weakly bound to urinary protein (Singer *et al.* 1988).

Thin layer chromatography of scent marks from saddle-backs and cotton-tops has suggested that, in addition to proteins, wax esters, steryl esters and triglycerides are present. In human skin lipid, free fatty acids are derived from triglycerides by the action of skin microorganisms, while many other species of mammals produce other classes of esters which are less readily hydrolysed by bacteria (Albone 1984). Although we know nothing about the biological role of the triglycerides in the scent marks of our tamarins, it is possible that they are the sources of the fatty acids found in the material, and that bacteria present on the surface of the scent glands are involved in their production. This notion is supported by the fact that in saddle-back tamarins the scent gland region supports a resident population of bacteria (Nordstrom *et al.* 1989).

Additional compounds found in the scent marks of saddle-backs and cotton-tops are metabolites of gonadal hormones. These have also been documented in the urine of males and females of both species by means of radioimmunoassay (Epple and Katz 1984; French *et al.* 1984; Küderling and Epple, unpublished data), and are probably constituents of urine deposited during scent marking rather than products of the scent glands.

## Conclusions

Our behavioural studies indicate that scent marks from saddle-backs and probably also from cotton-tops communicate a large body of information about the individual that produces them. This informational complexity is paralleled by chemical complexity. Much of our chemical and behavioural work involving fractionations, work with synthetic and semi-synthetic mixtures, and pattern recognition studies has shown that the signal content of the scent material is very likely based on overlapping qualitative and quantitative patterns of a variety of components.

In both species, the volatile constituents of the scent marks communicate some of its behaviourally significant information. Furthermore, in saddle-backs, gender and subspecies categories are mathematically characterized by specific concentration patterns of the major volatile constituents of the marks. However, these patterns alone do not communicate the identity of subspecies, gender, or individual. We have evidence that non-volatile as well as volatile components are necessary to fully constitute the chemical messages. These components may be derived from several sources including glandular secretion, urine, genital discharge, as well as products of bacterial action. It appears that all of these components together present the total 'Gestalt' or 'image' which contains encoded messages of a wide variety. This total image reflects a quite complex system which has evolved to contain many communicatory messages.

Our comparative studies have concentrated on *Saguinus fuscicollis* and *Saguinus o. oedipus*. Additional data on the chemical nature of material from *Saguinus leucopus* have illustrated the complexity of scent in all three species. They also illustrate the fact that there are similarities and differences in scent composition among species which may reflect taxonomical relationships. Our analytical data on only three species have to be interpreted with great caution. However, considering qualitative and quantitative differences between species, the scent marks from *Saguinus o. oedipus* and *Saguinus leucopus* resemble each other more closely than either of these resembles the scent marks from *Saguinus fuscicollis*. This parallels the taxonomical grouping of the three species by Hershkovitz (1977). As our knowledge of scent chemistry increases and extends to other species, the possibility that the chemical composition of the scent marks of different Callitrichidae reflects taxonomic relationships among species can be explored.

## Acknowledgements

The early support of our research by the National Science Foundation is gratefully acknowledged. Our more recent studies were supported by the National Institutes of Health (R01 NS-21790) and the Deutsche Forschungsgemeinschaft.

# 5

# Comparative aspects of the social suppression of reproduction in female marmosets and tamarins

*D.H. Abbott, J. Barrett, and L.M. George*

## Introduction

Marmosets and tamarins certainly have in common a particular adaptation to reproduction: only one female breeds in any single social group (for reviews see Mittermeier *et al.* 1988*a*). This extreme reproductive specialization has become one of the hallmarks of the Callitrichidae where, as far as females are concerned, the winner in social dominance interactions 'takes all'. The dominant female in each callitrichid group is therefore the only breeding female, and she actively attains and maintains her dominant breeding status (Abbott and Hearn 1978; Epple 1967; French *et al.* 1984; Rothe 1975*b*).

This state of reproductive affairs is rather unusual for an anthropoid primate. Most primate species are polygamous, and where reproductive suppression is found amongst subordinate female primates, it is usually manifest as poorer birth rates or poorer rates of infant survival in comparison with dominant females (Abbott 1991; Harcourt 1987). This trait of social control of reproduction is apparently taken to its logical conclusion in callitrichid monkeys, where there is a complete social block to reproduction in subordinate females.

Since Abbott and Hearn (1978) first discovered that dominant female common marmosets (*Callithrix jacchus*) inhibited ovulation in their female subordinates, it has been commonly assumed that an inhibitory physiological mechanism underpins socially-induced reproductive suppression in all female callitrichid monkeys. Recent research, however, suggests that social suppression of ovulation may not operate in all callitrichid species and that other mechanisms may be operating to maintain the one-female breeding system (e.g. in the golden lion tamarin, *Leontopithecus rosalia*: French and Stribley 1987). This paper therefore reviews our current understanding of the social control of female reproduction in callitrichid monkeys and will briefly discuss the possible reasons for this typically callitrichid specialization. It concentrates on the four callitrichid species for which both behavioural and physiological information is available regarding the nature of reproductive suppression among females.

## Common marmoset (*Callithrix jacchus*)

Epple (1967) was the first to describe the behavioural monopoly of reproduction by dominant females in callitrichid monkeys from her studies of social groups of common marmosets. Eleven years later, Abbott and Hearn (1978) showed that dominant female common marmosets achieved this reproductive monopoly by inhibiting the sexual behaviour and reproductive physiology of their female subordinates. This was the first time, in a primate species, that physiological methods had been employed to show that social dominance behaviour suppressed ovulation in subordinates. Since then, the common marmoset has provided one of the best examples, among socially-living mammals, of clear-cut social suppression of reproduction.

All (Evans and Hodges 1984) or most (Abbott 1984) post-pubertal daughters in captive family marmoset groups do not ovulate and no daughters breed. Subordinate females (ranks 2 and below) in captive social groups of unrelated adults, or peer groups, either do not ovulate (75 per cent) or undergo occasional inadequate ovarian cycles with deficient luteal phases (25 per cent: Abbott and George 1991). Sexual interactions with males are also rare for daughters or subordinate female adults (Abbott 1984, 1986). Consequently, whilst there is suppression of sexual behaviour which might prevent fertile matings from taking place involving non-breeding female marmosets, the suppressed reproductive physiology and the resulting total block to fertility is the more relevant reproduction impairment.

In subordinate female marmosets, suppressed ovulation was due to suppressed gonadotrophin secretion from the anterior pituitary gland. Basal and GnRH-stimulated plasma concentrations of luteinizing hormone (LH) were reduced in comparison to dominant females (Abbott *et al.* 1988). Follicle stimulating hormone (FSH) secretion was also apparently reduced because follicles removed from the ovaries of anovulatory subordinate females and exposed to FSH *in vitro* were highly responsive, in terms of their oestrogen secretion (Harlow *et al.* 1986). The indirect nature of the evidence for suppressed FSH secretion arises because of the current lack of an FSH assay for the marmoset. Normally, plasma oestrogen concentrations were also suppressed in subordinate females (Abbott *et al.* 1988). Repressed pituitary and ovarian function were both apparently due to suppressed gonadotrophin releasing hormone (GnRH) secretion from the hypothalamus in the brain. Long-term treatment of subordinate females with exogenous GnRH-stimulated plasma LH concentrations into the range of the ovulatory dominant females, increased ovarian activity and led to one pregnancy which miscarried (Abbott 1987, 1989).

The neuroendocrine source of the hypothalamic suppression of GnRH secretion did not involve weight loss, hyperprolactinaemia, hypercortisolaemia, or alterations in the pattern of melatonin secretion, all of which had previously been associated with suppression of GnRH secretion in other species (Abbott *et al.* 1981; Abbott 1986; Webley *et al.* 1989). However, an increased sensitivity to the negative feedback action of oestradiol upon LH (and probably GnRH) secretion has been identified in subordinate female marmosets. This suggests that the neuroendocrine mechanisms involved in mediating the negative feedback effects of ovarian oestradiol have been intensified to such an extent by the neural effects of social subordination that the ovaries of subordinate females are effectively clamped at a very low level of oestradiol secretion: hence the anovulatory state (Abbott 1988). A second inhibitory neuroendocrine mechanism has also been demonstrated, involving the endogenous opioid peptides (Abbott *et al.* 1989). It is not yet clear whether the inhibitory neuroendocrine mechanisms identified to date suggest that two separate inhibitory pathways suppress GnRH secretion, or whether they form two components of one integral pathway inhibiting GnRH.

Pheromones from dominant female marmosets were implicated as one possible factor triggering this neuroendocrine inhibition of hypothalamic GnRH secretion in female subordinates. As illustrated by a typical example in Fig. 5.1, subordinate females removed from their groups and housed singly rapidly showed increases in plasma LH concentrations and ovulated, on average, 10–11 days following separation (Barrett *et al.* 1990). However, if isolated subordinates are kept in scent contact with their dominant females, the onset of their first ovulation is delayed, on average, to 31–32 days (e.g. Fig. 5.1). Dominant female pheromones may not be the only factor involved in neuroendocrine suppression of GnRH secretion in subordinate marmosets, because subordinate females rendered anosmic at both the vomeronasal organ and the main olfactory epithelium and left in their social groups all remained anovulatory (Barrett and Abbott 1989). Seemingly, despite the olfactory blockade, an effective inhibitory signal from dominant females, probably behavioural intimidation, still maintained reproductive suppression in female subordinates. Auditory cues were not important in this respect.

## Cotton-top tamarin (*Saguinus oedipus*)

One year after the first description of female reproductive endocrinology in this species (French *et al.* 1983), French and his colleagues (1984)

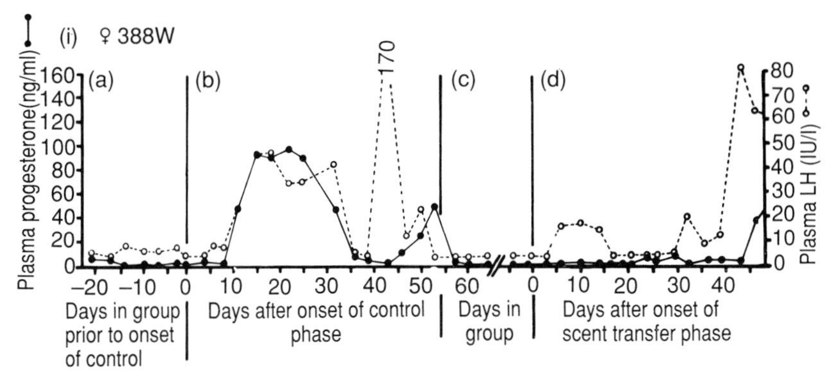


**Fig. 5.1** Plasma progesterone (—) and LH (---) concentrations in subordinate female common marmoset monkey 388W while (a) subordinate in a social group, (b) removed from her social group and housed singly whilst undergoing control saline swabbing procedures for the scent contact phase, (c) returned to her social group and subordinate status for 28 days, and (d) removed from her social group, housed singly, but maintained in scent contact with her dominant female through thrice daily exchange of cages with the dominant female along with concomitant swabbing of the isolated subordinate's external nares and naso-palatine canal openings with material from her dominant female's circumgenital area. Ovulation, as determined by sustained elevations of plasma progesterone concentrations above 10 ng/ml, occurred after 9 days in the control phase and after 46 days during scent contact. Reprinted with kind permission of the Zoological Society of London from Abbott *et al.* (1989).

demonstrated the suppression of ovarian cycles in subordinate daughters living in captive family groups. As shown in Fig. 5.2, the suppression of ovarian function was due only to subordinate social status. In this typical example, ovulation commenced shortly after removal of the subordinate from her family and following pairing with a male. The female conceived on her second cycle. In this tamarin, female reproductive endocrinology is commonly monitored by hormonal determinations of urine rather than of blood. Elevations of urinary concentrations of conjugated forms of oestrone appear to reflect excreted products from the corpus luteum in the ovary, following presumed ovulation and excreted products from the placenta during pregnancy (French *et al.* 1983, 1984; Ziegler *et al.* 1987*b*). Progestagenic secretions into the bloodstream are excreted mostly in faeces in this tamarin (Ziegler *et al.* 1989). However, as in the common marmoset, suppressed ovarian cyclicity may not occur in all subordinate female cotton-top tamarins all of the time. Tardif (1984), in a study of progesterone concentrations in the blood, found evidence of possible ovulatory cycles in non-breeding daughters in approximately 50 per cent

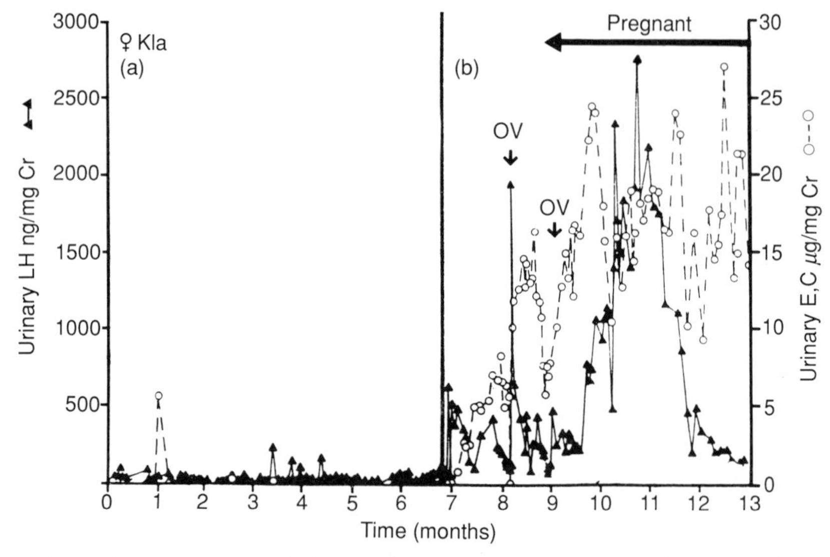


**Fig. 5.2** Urinary concentrations of oestrone conjugates (0– –0) and LH (<--->), expressed per mg creatinine (mg Cr), in subordinate female cotton-top tamarin 'Kla' while (a) subordinate in a family group and (b) removed from her family group and paired with a male. Ovulation, determined as the day after the appearance of an LH peak prior to the sustained rise of urinary concentrations of oestrone conjugates, occurred 38 days after pairing with a male and conception occurred at the next ovulation, 64 days following pairing. Note the prolonged elevation of urinary concentrations of oestrone conjugates and the transient elevation of urinary concentrations of presumed chorionic gonadotrophin (cross-reacting with the LH antisera) following conception. OV, ovulatory LH peak. Reprinted in modified form with kind permission of the Society for the Study of Reproduction from Ziegler *et al.* (1987*b*).

of the families examined. Nonetheless, it was not clear whether the cycles were of normal duration or whether they occurred regularly.

The physiological component of reproductive suppression in subordinate female cotton-top tamarins seems similar to that in common marmosets. Ovulation was inhibited and urinary concentrations of LH were suppressed in subordinate females until the subordinates were removed from their dominant female (Ziegler *et al.* 1987*b*; Fig. 5.2). The evidence suggests that social suppression of hypothalamic GnRH secretion may also occur in subordinate cotton-top females in an analogous fashion to that in common marmosets. There is certainly a similarity to the marmoset in the pheromonal contribution of the dominant female in the suppression of ovulation in subordinates (Savage *et al.* 1988).

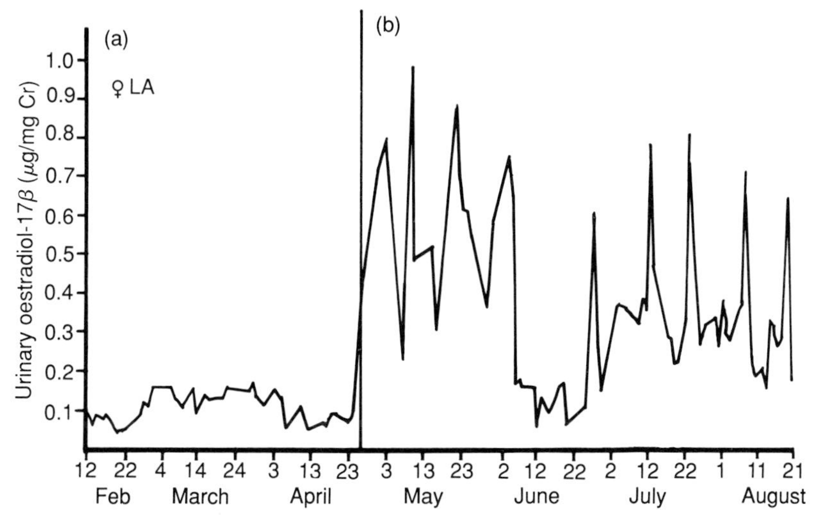


**Fig. 5.3** Urinary concentrations of total oestradiol-17β (oestradiol following β-glucuronidase/aryl sulphatase enzyme hydrolysis) in subordinate female saddle-back tamarin 'LA' while (a) subordinate in her family group and (b) paired with a male. Urinary concentrations are expressed per mg creatinine (mg Cr). Ovulation, estimated as two days following the presumed pre-ovulatory oestradiol peak, occurred approximately 8 days after pairing with a male. Reprinted in modified form with kind permission of Alan R. Liss, Inc. from Epple and Katz (1984).

## Saddle-back tamarin (*Saguinus fuscicollis*)

Suppression of ovarian cycles was also apparent in subordinate daughter saddle-back tamarin monkeys living in captive family groups (Epple and Katz 1984). Figure 5.3 illustrates the low and acyclic pattern of urinary total oestradiol in a subordinate female prior to removal from her family and the immediate onset of ovarian cyclicity on pairing with a male. In this species, cyclical elevations in urinary 'total' oestradiol concentrations may reflect the pre-ovulatory increase in ovarian hormone secretion normally associated with oestrogen secretion from pre-ovulatory follicles immediately prior to the ovulatory LH peak (Hodges *et al.* 1981). In the latter study, urinary concentrations of total immunoreactive oestrogen in a mature female saddle-back tamarin peak the day before each urinary LH peak. The urinary hormonal profile in saddle-back tamarins may therefore be different to the pattern of post-ovulatory increases in urinary oestrogen concentrations from cotton-top tamarins (e.g. Fig. 5.2).

While nothing further is known of the physiological causes of the ovarian suppression in subordinate female saddle-back tamarins, pheromones from dominant females may also play a role in suppressing ovulation in subordinates (Epple and Katz 1984).

## Golden lion tamarin (*Leontopithecus rosalia*)

In contrast to the three other callitrichid species studied to date, there was no evidence for the suppression of ovulation in subordinate daughter golden lion tamarins remaining in their family groups (French *et al.* 1989). This point is illustrated in Fig. 5.4, where three subordinate females showed regular elevations in urinary 'total' oestrogen concentrations (mostly indicative of post-ovulatory corpus luteum function or of pregnancy in dominant females of this species: French and Stribley 1985, 1987). The apparently ovulatory oestrogen profiles of all three were similar to that of breeding females. However, no subordinate daughters have been observed copulating with males in their family groups (French *et al.* 1989). Consequently, in this callitrichid species, the main cause of female reproductive suppression is behavioural rather than physiological and involves the inhibition of sexual interaction with males in otherwise apparently fertile subordinate females.

Nevertheless, golden lion tamarins do exert some form of social control over the ovarian cycles of females living in groups, at least in captivity. French and Stribley (1987) have shown that the ovarian cycles of females living in groups appear to be synchronized: this included both breeding (dominant) and non-breeding (subordinate) females. This tamarin may have developed a physiological specialization which is the next best thing to reproductive suppression. Synchronization of ovarian cycles would mean that all females in a group would ovulate at approximately the same time and therefore provide perhaps the best opportunity for the dominant female to monopolize the breeding male or males during this fertile period for all the females. It is possible, though not yet established, that pheromones from the dominant female may be involved in the synchronization process.

## Discussion

There are two questions which need to be addressed briefly here: (1) What are the possible reasons for this extreme suppression of reproduction among subordinate female callitrichid monkeys, and (2) What reasons lie behind the species differences in achieving this objective?

The first question has been addressed in general by several workers (see Sussman and Garber (1987) for a recent review). The answer appears to

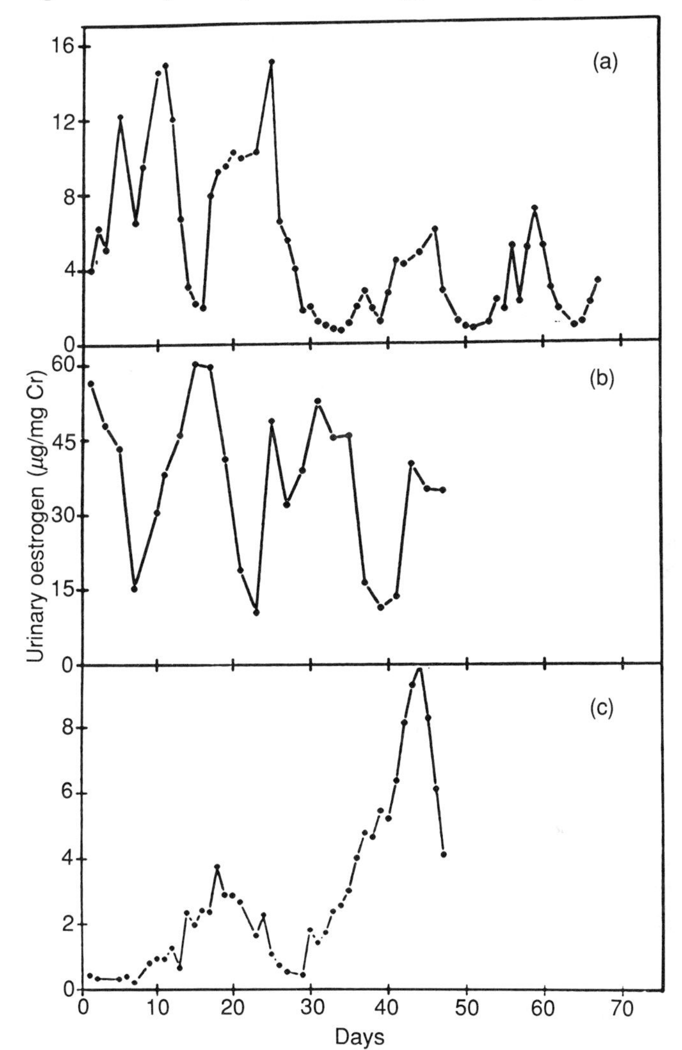


**Fig. 5.4** Urinary concentrations of 'total' oestrogen (oestrone and oestradiol combined following β-glucuronidase enzyme hydrolysis) in three subordinate female golden lion tamarins as either daughters in family groups (a: 'Charl', 24 months of age; b: 'Gerry', 39 months of age) or as a subordinate in a pair of unrelated females (c: 'Lucy', 120 months of age). Urinary concentrations are expressed per mg creatinine (mg Cr). Note the regular ovarian cyclicity exhibited by all three subordinates. Figure reprinted with kind permission of Alan R. Liss, Inc. from French *et al.* (1989).

come in at least three parts. Non-reproductive or subordinate females may benefit from putting off their departure from their social group because of difficulties in either establishing a separate breeding group with a separate territory (Emlen 1984; Macdonald and Carr 1989; Wasser and Barash 1983) or establishing themselves as the breeding female within an already formed social group, whether it be natal or separate (French and Inglett 1991; French *et al.* 1989; Sussman and Garber 1987; Scanlon *et al.* 1988). These difficulties may arise from having to belong to a group so as to be able to defend an area containing sufficient resources necessary for survival against the encroachment of other animals or groups (e.g. Macdonald and Carr 1989) or from aggression received whilst attempting to immigrate into a different group (e.g. French and Inglett 1991) or from an increased risk of predation (mostly from birds of prey: Sussman and Kinzey 1984; see also Caine, this volume), as found in free-living dwarf mongooses (Rasa 1989) with a similar one-female breeding system to callitrichids.

The second reason may well involve the communal rearing system employed by callitrichid monkeys to raise the offspring of the breeding female. Certainly, in free-living moustached tamarins, *S. mystax*, larger groups are more successful at rearing the young of the breeding female than smaller groups (Garber *et al.* 1984; Sussman and Garber 1987): similar findings have been found in other mammals employing a one-female breeding system (Macdonald and Carr 1989). Apparently, the dominant female in a callitrichid group requires help from non-breeding animals to rear her offspring so as to maximize the chance of her young surviving. 'Helper' females cannot then breed themselves as they will probably not be able to compete successfully against the dominant female for the necessary 'helpers' to raise their own offspring. Furthermore, in such a situation, reproducing subordinate females would not be available to help rear the offspring of the dominant female and thus would risk the survival of any infants born in the group. The requirement for 'helpers' by the dominant female may in itself have arisen as exploitation by the dominant female of female subordinates prolonging their stay in a group because of the difficulties (mentioned above) in setting up separate groups of their own (Macdonald and Carr 1989). The dominant female, in such a way, could therefore expand her reproductive potential and take advantage of the ecological circumstances constraining callitrichids to live in groups (Sussman and Kinzey 1984).

A third reason may arise from the requirement of young, non-breeding females to gain essential experience in infant care-taking which will prove invaluable for the rearing of their own offspring (Sussman and Garber 1987). This point has acquired specific attention in captive management of callitrichid monkeys, where prior experience of infant care-taking is a prerequisite for captive-born breeding females to rear their own offspring successfully (Epple 1975*a*; Hearn 1983; Tardif *et al.* 1984; Stevenson and Sutcliffe 1978).

All these possible contributory factors to the development of female reproductive suppression in callitrichid monkeys would be enhanced if there were a high degree of genetic similarity among group members. This may prove to be the case among the marmosets (*Callithrix* species and *Cebuella pygmaea*, the pygmy marmoset) more so than among the tamarins (*Leontopithecus* and *Saguinus*) because the former appear to form more stable nuclear or extended family units of 5–15 animals whereas the latter appear to form smaller, less stable groups, possibly involving unrelated animals (Ferrari and Lopes Ferrari 1989). However, not all field studies fit easily into this dichotomy (e.g. Scanlon *et al.* 1988) and this speculation will only be clarified by genetic fingerprinting studies of free-living callitrichid groups.

Having established the possible socio-ecological factors behind the development of extreme social suppression of female reproduction in callitrichid monkeys, we can then ask why species differences should exist in the suppression mechanisms.

The answer to this question may, in part, lie in the differing degrees in which the dominant breeding females in each species rely on non-breeding 'helpers' to help rear the dominant's own offspring. In the common marmoset, the non-breeding group members participate in co-operatively rearing the young of the breeding female from the day of the infant's birth (Box 1977; Ingram 1977; Stevenson 1976*b*). Similar infant commitments by non-breeding animals have been observed in pygmy marmosets (Soini 1988) and *Saguinus* species (Epple 1975*b*; Cleveland and Snowdon 1984; Snowdon and Soini 1988). However, such infant care-taking by non-breeding animals in golden lion tamarin groups does not generally commence until the infants are approximately 1–3 weeks of age (Hoage 1978; Kleiman *et al.* 1988). Apparently, there is a delay in the onset of 'helping' behaviour in *Leontopithecus*, in comparison to *Callithrix* and *Saguinus* species. The dominant breeding female in golden lion tamarin groups may therefore be less dependent on other group members to help rear her offspring than is the case for the common marmoset and the cotton-top and saddle-back tamarins. Thus, there may be less of an essential requirement that only one female breeds per group. A golden lion tamarin social group in the wild may therefore be able to support the rearing of more than one set of offspring at a time. Some evidence for this has already been found (A.J. Baker, cited in French *et al.*, 1989; Dietz and Baker 1991).

Another possible answer to this question is that the ecological specialization of *Callithrix* species, *Cebuella pygmaea*, *Saguinus* species, and *Leontopithecus* species may have developed quite separately and may not be as unified and as closely-linked as currently thought (Ferrari and Lopes Ferrari 1989). Given this scenario, golden lion tamarins may have developed an effective behavioural control mechanism of female reproduction without the need for physiological suppression. This would certainly be consistent with the intense aggression shown by female golden lion tamarins to unfamiliar conspecifics in comparison to the weak aggressive responses shown by female cotton-top tamarins (French and Inglett 1989; French and Snowdon 1981). It would also fit with observations of free-living groups where non-breeding daughters were aggressively evicted from their groups by their mothers, with fatalities arising amongst the daughters in some cases (Baker 1987; A.J. Baker, personal communication). Such severe female–female aggression within groups has yet to be reported in free-living *Callithrix* and *Saguinus* species (Terborgh and Goldizen 1985; Scanlon *et al.* 1988; Stevenson and Rylands 1988; Sussman and Garber 1987). However, aggressive peripheralization or eviction of sub-adults by dominant males or females does occur in free-living groups of pygmy marmosets (Soini 1988).

## Summary

Reproduction is suppressed in all but one female, the dominant female, in social groups of callitrichid monkeys. This unified specialization may, however, be achieved in different ways in the different species. In the common marmoset and cotton-top and saddle-back tamarin, ovulation is inhibited in non-breeding subordinate females. In golden lion tamarins, ovulation is not inhibited in non-breeding subordinates and some form of suppression of subordinate sexual behaviour appears to operate to prevent ovulatory subordinates from becoming pregnant.

A requirement for group living for survival may have generated the extreme suppression of female reproduction practised by callitrichid monkeys. Either differences in the degree of co-operation required or differences in the development of the suppression mechanisms may have led to the current species differences in the nature of female reproductive suppression in callitrichid monkeys.

## Acknowledgements

The authors would like to thank Dr. H.D.M. Moore for criticism of the manuscript, M.J. Llovet and the animal technical staff for their care and maintenance of the common marmosets, T.J. Dennett and M.J. Walton for preparation of the figures, and R. Gray for typing the manuscript. This work was supported by grants from the Wellcome Trust, the Nuffield Foundation, the Association for the Study of Animal Behaviour, the University of London, and MRC/AFRC Programme Grant and travel grants from the Royal Society and the British Council to DHA.

# 6

# Callitrichid mating systems: laboratory and field approaches to studies of monogamy and polyandry

*A.F. Dixson*

## Introduction

Traditionally, the marmosets and tamarins have been regarded as monogamous primates which live in extended family units (e.g. Kleiman 1977). However, with increasing knowledge of callitrichid field biology it has become apparent that a single reproductive female may mate with more than one partner (e.g. *Callithrix humeralifer*: Rylands 1982, 1986*a*; *Saguinus fuscicollis*: Goldizen 1987*a*). This has led to suggestions that 'facultative polyandry' may occur; i.e. that two or more males copulate with a single female and co-operate in rearing her twin offspring (Goldizen 1987*a*; Sussman and Garber 1987). Paternity of offspring and genetic relatedness of adult group members has not been determined for free-ranging callitrichids. Nor is it known whether facultative polyandry occurs throughout this primate family, or if it is restricted to particular species or populations. Answers to these questions may be achieved by a combined field and laboratory approach. Three avenues of research are described in this review. Firstly, a detailed knowledge of sexual behaviour and its degree of dependence upon hormonal mechanisms is valuable when interpreting the limited field data on mating activity. Secondly, morphological features of the reproductive tract can be useful in studies of mating systems (e.g. relative testis size: Harcourt *et al.* 1981; Dixson 1987*a*). Thirdly, the invention of DNA finger-printing techniques has had far-reaching consequences for studies of biological relatedness and paternity evaluation (Jeffreys *et al.* 1985*a*,*b*; Wetton *et al.* 1987), so that this approach may also prove valuable in studies of callitrichid mating systems.

## Hormones and sexual behaviour

To understand the nature of interrelationships between hormones and sexual behaviour in callitrichids it is necessary to consider several factors.

Sexual interactions are infrequent in established groups (Kleiman 1977 and may continue during pregnancy (*Callithrix jacchus*: Evans and Pool 1984; *Leontopithecus rosalia*: Kleiman and Mack 1977; *Saguinus fuscicollis*: Goldizen 1987*a*). If animals are 'pair-tested' in captivity then copulation is possible at any stage of the female's ovarian cycle (*C. jacchus*: Kendrick and Dixson 1983; *Saguinus œdipus*: Brand and Martin 1983) or after removal of the ovaries and adrenal glands (*C. jacchus*: Dixson 1987*b*). Clearly then, female sexual receptivity is not rigidly dependent upon sex steroids and periods of heightened sexual activity, referred to as 'post-partum oestrus' by some authors (e.g. in free-ranging *Cebuella pygmaea*: 1987*a*), probably also involve hormonal effects upon female attractiveness or proceptivity (Beach 1976). Studies of endocrine changes and sexual activity are therefore required, but these necessitate measurements of circulating hormones or of urinary metabolites since this is the only reliable way of monitoring ovarian cycles in female callitrichids.

In order to obtain comprehensive measurements of sexual behaviour in groups of common marmosets, we employed videotaping techniques to record all interactions which occurred during 24–30 days after parturition in eight family groups (Dixson and Lunn 1987). Blood samples were obtained regularly from females in order to measure plasma luteinizing hormone (LH) and progesterone. In six groups a pre-ovulatory LH peak was measured between days 10–18 post-partum (mean ± s.e.m. 13.8 ± 1.3 days). Plasma LH levels rose to 45.7 ± 245.3 ng/ml (mean 124.4 ± 33.5 ng/ml) at this time followed by increases in plasma progesterone. In three cases females failed to conceive (groups 1, 3 and 6: Fig. 6.1), pregnancies occurred in a further three cases (groups 2, 4 and 5: Fig. 6.2) and in groups 7 and 8 ovulation did not occur during the 30 days after parturition and sexual activity was minimal.

Only the adult pair and twin offspring were present in these marmoset groups. Therefore we cannot comment on possible polyandric interactions. However, the findings on sexual behaviour and hormonal patterns in groups 1–6 are interesting and represent the most complete studies available for a callitrichid species. Mount attempts first occurred between days 1–10 after parturition (mean 4.8 ± 1.5 days) and in all groups males initiated the first mounts (Table 6.1, Sign test $P = 0.06$) and used tongue-flicks as a pre-copulatory invitation before females had exhibited invitational (proceptive) patterns. However, before the peri-ovulatory phase was reached (defined here as ±3 days from the day of the LH peak), ejaculatory mounts were less frequent and this finding applied particularly to the first 5–7 days after parturition. During this 'pre-ovulatory phase', females were more aggressive to males and refused or terminated more mounts (Table 6.2). Trauma to the female's reproductive tract and discomfort after giving birth, her subsequent maternal behaviour, and endocrine changes after parturition could all influence her behavioural responses to the male. The effects

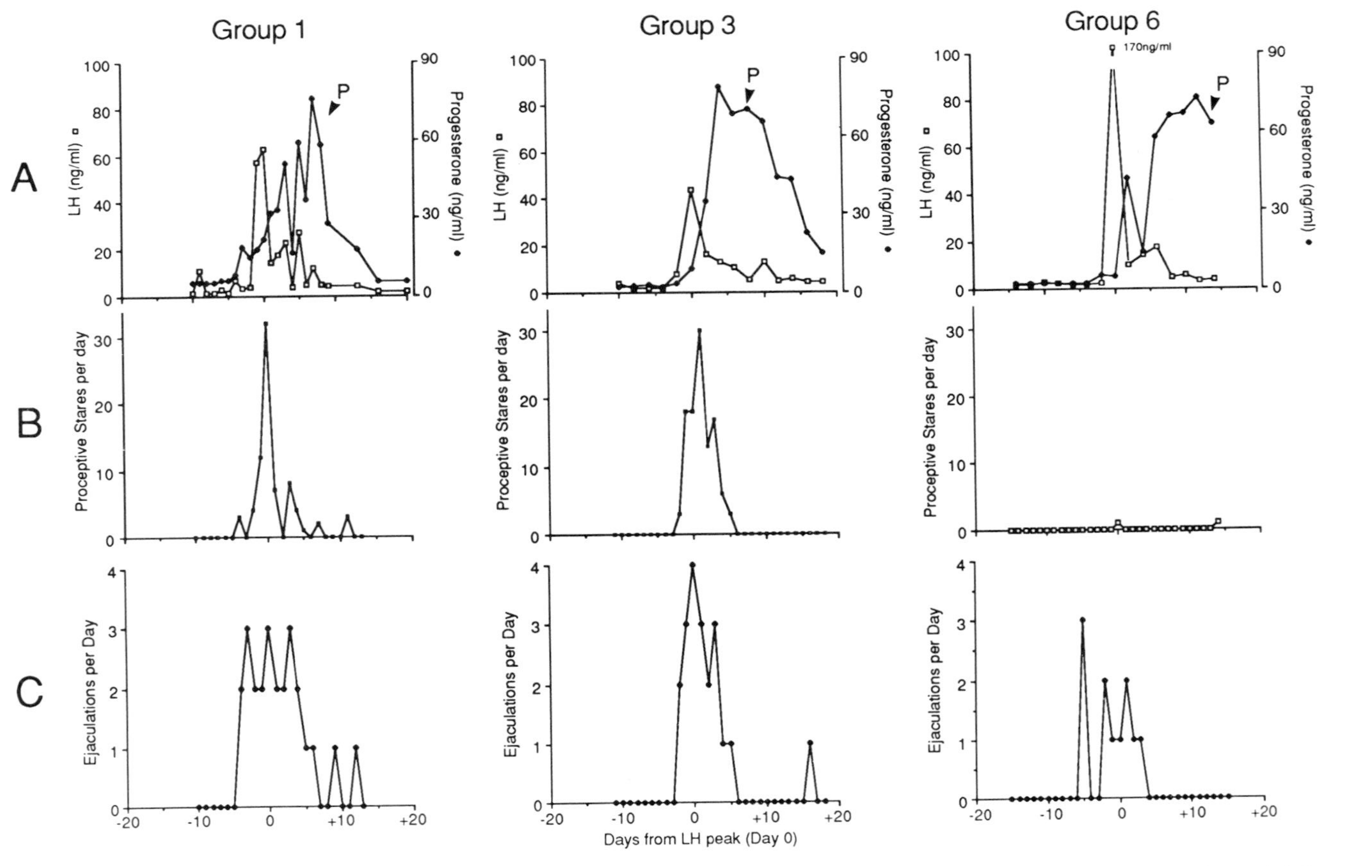


**Fig. 6.1** Hormonal and behavioural changes during the post-partum period in marmoset groups 1, 3, and 6 in which females did not conceive (A). Plasma levels of luteinizing hormone (LH) and progesterone (P). (B) Females : proceptive stares. (C) males : ejaculations. Measurements of hormone levels and daily frequencies of sexual behaviour are plotted with respect to the occurrence of the LH peak (Day 0).

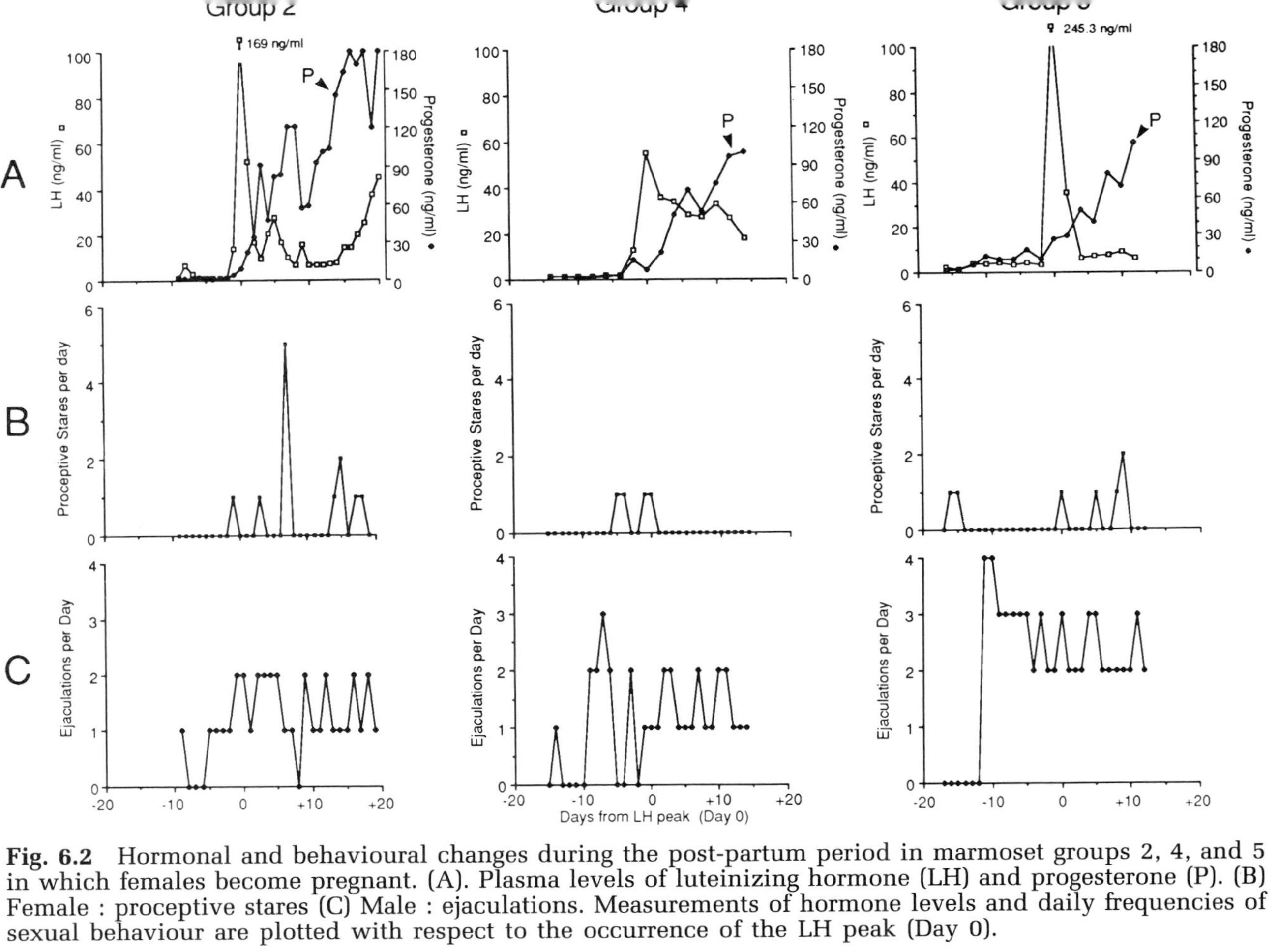


**Fig. 6.2** Hormonal and behavioural changes during the post-partum period in marmoset groups 2, 4, and 5 in which females become pregnant. (A). Plasma levels of luteinizing hormone (LH) and progesterone (P). (B) Female : proceptive stares (C) Male : ejaculations. Measurements of hormone levels and daily frequencies of sexual behaviour are plotted with respect to the occurrence of the LH peak (Day 0).

**Table 6.1** Day of first occurrence of selected features of sexual behaviour during the post-partum period in family groups of common marmosets (After Dixson and Lunn 1987). Day 1 is the day of parturition.

| Behaviour pattern | Group number | | | | | | mean ± s.e.m. |
|---|---|---|---|---|---|---|---|
| | 1 | 2 | 3 | 4 | 5 | 6 | |
| Male: precopulatory display | 1 | 6 | 8 | 5 | 2 | 10 | 5.33 ± 1.40 |
| Female: proceptivity | 6 | 9 | 10 | 9 | 2 | 10 | 7.66 ± 1.28 |
| Mount or mount attempt | 6 | 1 | 8 | 2 | 2 | 10 | 4.83 ± 1.51 |
| Male initiates mount | 6 | 1 | 8 | 2 | 2 | 10 | 4.83 ± 1.51 |
| Female initiates mount | 7 | 9 | 10 | 10 | 18 | 11 | 10.83 ± 1.53 |
| Ejaculation | 7 | 1 | 10 | 2 | 7 | 11 | 6.33 ± 1.66 |
| Day of L. H. peak | 11 | 10 | 12 | 16 | 18 | 16 | 13.83 ± 1.32 |

measured were quite variable, and in some groups ejaculation was possible as early as day 1 or 2 after parturition (Tables 6.1 and 6.2).

If we next consider the peri-ovulatory phase, this was associated in all groups with increased frequencies of mounts and ejaculations. Heightened proceptivity during this phase, as well as decreases in female aggressiveness and subtle changes in receptivity, may have combined to enhance the males' sexual activity (Table 6.2).

The males' copulatory behaviour with females subsequent to the periovulatory phase differed, depending on whether conception had occurred. With females which failed to conceive (Fig. 6.1), ejaculatory mounts decreased in frequency in the post-ovulatory period. By contrast, mounts and ejaculations continued at higher frequencies in those groups where females had become pregnant (Fig. 6.2). Examination of data from individual family groups failed to reveal any consistent differences in proceptivity, receptivity or aggression between these two sets of females during the post-ovulatory phase. It is possible therefore, that non-behavioural, sexually attractive cues continue to operate in females after they have conceived, resulting in maintenance of copulatory initiation by males. Equally, such a mechanism could be adaptive in allowing the female to maintain sexual relationships with the male (or males) in her group which will subsequently assist in rearing her offspring.

Little has been said so far about sexual attractiveness in female callitrichids, but it is necessary to consider this aspect of sexuality. Since callitrichids make extensive use of scent marking displays and olfactory communication (Epple 1974*a*), it is possible that chemical cues influence female sexual attractiveness. Frequencies of genital scent-marking behaviour by both sexes showed no consistent changes during the post-partum period and, although males' olfactory inspections of females' genitalia or scent-marks increased during the peri-ovulatory phase, these effects were not

**Table 6.2** Daily frequencies of sexual, aggressive, and scent-marking behaviour in common marmoset groups during the females' post-partum period. Data are daily means ± s.e.m. for six groups (Dixson and Lunn 1987).

| Behavioural measurement | Post-partum phases | | |
|---|---|---|---|
| | A. Pre-ovulatory | B. Peri-ovulatory | C. Post-ovulatory |
| **Male** | | | |
| Mounts | 1.50 ± 0.71 | 5.17 ± 1.64* | 2.58 ± 0.92 |
| Ejaculations | 0.61 ± 0.26 | 1.86 ± 0.24* | 0.96 ± 0.34 |
| Aggression | 0.32 ± 0.07 | 0.07 ± 0.03* | 0.20 ± 0.06 |
| Scent marks | 124.0 ± 16.0 | 124.0 ± 23.0 | 149.0 ± 24.0 |
| Inspects females' genitalia | 0.88 ± 0.20 | 1.0 ± 0.28 | 0.51 ± 0.16 |
| Inspects females' scent-marks | 8.02 ± 2.2 | 11.05 ± 2.16 | 11.42 ± 3.67 |
| **Female** | | | |
| Proceptive stare | 0.12 ± 0.06 | 4.0 ± 2.48* | 0.46 ± 0.15 |
| Immobile display | 0.45 ± 0.19 | 4.24 ± 2.39* | 0.69 ± 0.17 |
| Initiates mounts (%) | 4.8 ± 2.4 | 24.0 ± 10.8 | 20.0 ± 8.1 |
| Mount refusals or Terminations (%) | 25.5 ± 8.3 | 7.0 ± 4.7 | 1.3 ± 0.7 |
| Aggression | 5.84 ± 1.07 | 1.96 ± 0.42* | 1.95 ± 0.65* |
| Scent-marks | 125.0 ± 31.0 | 136.0 ± 25.0 | 157.0 ± 30.0 |

* $P = 0.05$, Wilcoxon test. A *vs.* B or A *vs.* C.
Peri-ovulatory phase = ±3.0 days from the day of the LH peak: all preceding days constitute the pre-ovulatory phase: all days following peri-ovulatory period are the post-ovulatory phase.

statistically significant (Table 6.2). However, frequency measurements of this kind are of limited value under captive conditions since the cage is presumably permeated with the odour of the animals and repeated close contact with olfactory cues may be rendered unnecessary. More informative are experiments conducted after olfactory bulb transections in male marmosets. We have conducted a number of studies and the results of one experiment are shown in Fig. 6.3. The olfactory bulbs were transected in this male two weeks before his female delivered twin offspring. The operation disconnected both the main olfactory system and inputs from the vomeronasal organ. Recovery from anaesthesia was normal, and healing of the scalp incision site was uneventful. The male remained active and continued to feed normally. Olfactory discrimination tests subsequently demonstrated the effectiveness of the operation. The patterning of this male's copulatory behaviour clearly differed from that of intact animals. His female partner had apparently ovulated but failed to conceive, as revealed by measurements of circulating LH and progesterone. However, instead of the expected peak in ejaculatory mounts, the male copulated frequently throughout the entire period spanned by video tape records

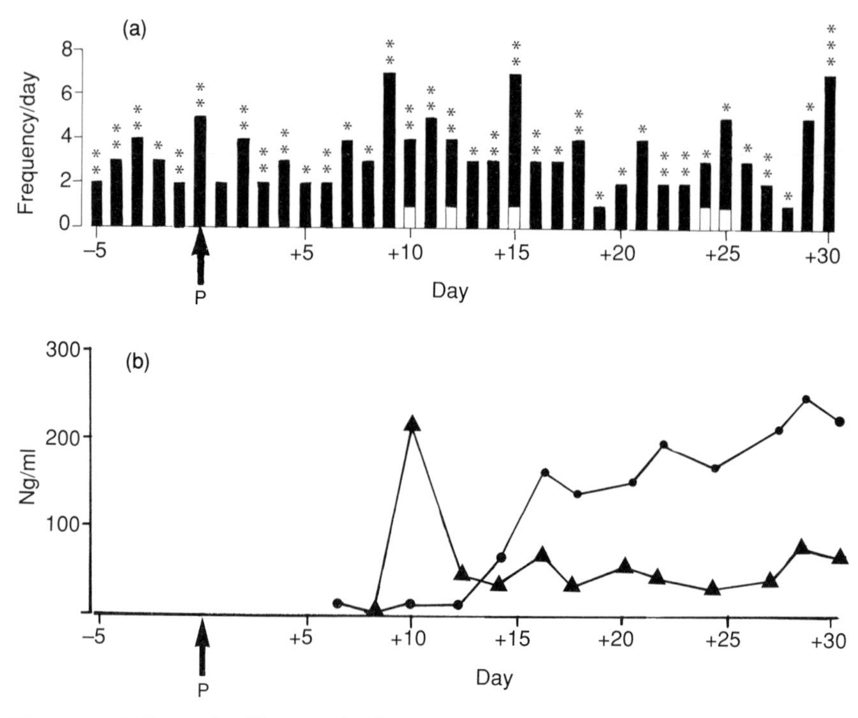


**Fig. 6.3** Bilateral olfactory bulb section in a male common marmoset: effects upon sexual interactions with the female partner before and after parturition. (a): Daily frequencies of mounts and attempted mounts are shown as histograms. Solid bars = male initiated mounts; open bars = female initiated mounts; * = an ejaculation. (b): Daily measurements of plasma luteinizing hormone (LH ▲) and progesterone (•) are graphed beginning on the sixth day after parturition. (P). Further details of methodology are given in the text.

(Fig. 6.3). This experiment and several others (Lloyd and Dixson, unpublished data) indicate that whilst olfactory inputs are not essential to activate the male's sexual and copulatory behaviour, they are important in allowing the male to co-ordinate these activities with respect to hormonal changes in the female during the post-partum period. Some females will accept copulation throughout the post-partum period if the anosmic male is sufficiently aroused and initiates mounts. The situation is analogous to that observed during short-duration paired encounters, where sexual arousal in male marmosets is sufficient to obscure possible effects of female olfactory cues (Dixson and Lloyd 1988).

The evidence reviewed above indicates that olfactory cues are important for female sexual attractiveness in captive common marmosets during the post-partum period. The same may also be true for other callitrichids which are said to exhibit 'post-partum estrus' (e.g. *Cebuella pygmaea*: Soini 1987*a*; *Saguinus œdipus*: Ziegler *et al.* 1987*a*). However, it is also clear from the videotaping and endocrine studies of captive *Callithrix jacchus* that changes in proceptivity and subtle aspects of receptivity also occur in some females. Rigorous behavioural measurements were required to record these changes in the females' behaviour. If continuous video monitoring had not been used, it might easily have been concluded that no changes in female behaviour or attractiveness occurred. This was the conclusion reached by Stribley *et al.* (1987) in their studies of *Leontopithecus rosalia*; but in that study behaviour was recorded for only 30 minutes each day. The same reservation should also apply to field studies, if, as is usually the case, only a small sample of the animals' sexual activity has been recorded.

## Relative testis size and mating systems

Building upon Parker's (1970) original hypotheses concerning 'sperm competition' in insects, it has been shown that relative testis size is greatest in primates and other mammals which have multi-male mating systems (Harcourt *et al.* 1981; Kenargy and Trombulak 1986). This is the case, for instance, in the chimpanzee, in which sexual selection has favoured the evolution of large testes capable of maintaining maximal numbers of spermatozoa per ejaculate. In the monogamous gibbon or polygynous gorilla, by contrast, relative testis size is small; this is consistent with lack of sperm competition between males. If female callitrichids mate with a number of males, rather than being strictly monogamous, the question arises as to whether selection for increased relative testis size has occurred in this group. In Harcourt *et al.*'s comparative studies they noted that relative testis weight in *Saguinus œdipus* was greater than expected for a monogamous primate. Unfortunately, data on testicular weights in callitrichids are not available in most species, particularly for free-ranging animals. Data relating to the lengths and widths of testes are more easily obtained and testicular volumes may be calculated using these measurements. Figure 6.4 shows a logarithmic plot of testicular volume versus body weight for 39 primate species, including seven callitrichids. This figure reproduced from an earlier report (Dixson 1987*a*) gives some indication of larger relative testicular volumes in *Cebuella pygmaea*, *Callithrix argentata*, and *Saguinus nigricollis*. However, this analysis has many limitations, including the small sample sizes available for certain species. Only one measurement of testicular volume was available for *Cebuella*, one for *C. argentata*, and two measurements for *S. nigricollis*. For none of these species were body weights and testicular volumes obtained from the same specimens, and so average

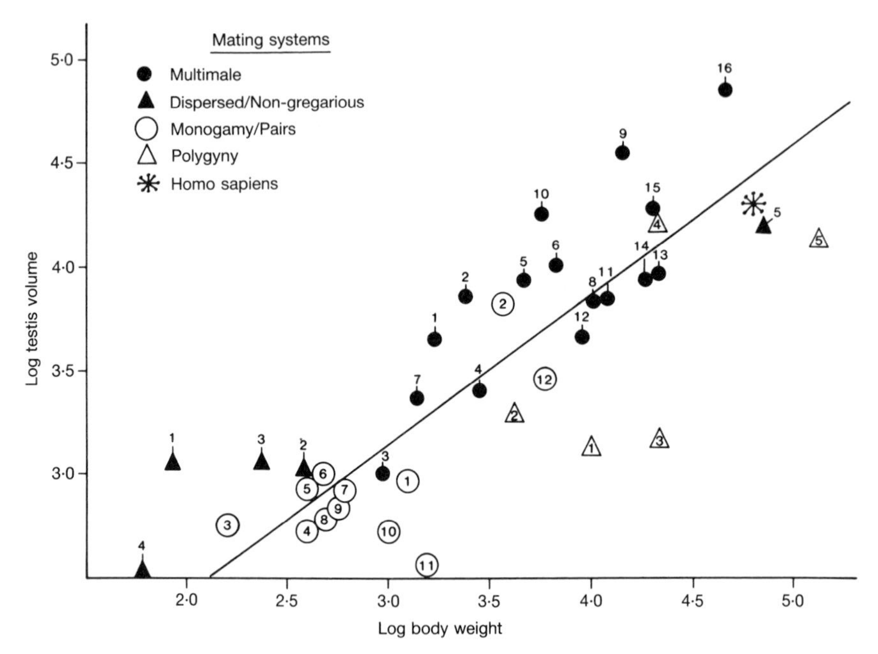


**Fig. 6.4** Logarithmic plot of mean testis volume versus body weight in 39 primate species. A principal axis line has been fitted to these data (slope 0.718). After Dixson (1987*a*).
*Multimate mating system* (●) 1. *Lemur fulvus*. 2. *L. macaco*. 3. *Saimiri sciureus*. 4. *Cebus apella*. 5. *Cercopithecus aethiops*. 6. *Allenopithecus nigroviridis*. 7. *Miopithecus talapoin*. 8. *Macaca nigra*. 9. *M. arctoides*. 10. *M. fascicularis*. 11. *Cercocebus atys*. 12. *C. aterrimus*. 13. *Papio anubis*. 14. *P. papio*. 15. *P. ursinus*. 16. *Pan troglodytes*.
*Dispersed or non-gregarious mating system* (▲) 1. *Microcebus murinus*. 2. *Cheirogaleus major*. 3. *Galago senegalensis*. 4. *G. demidovii*. 5. *Pongo pygmaeus*.
*Monogamous mating system* (polyandry in callitrichids?) (○) 1. *Avahi laniger*. 2. *Lemur variegatus*. 3. *Cebuella pygmaea*. 4. *Callithrix jacchus*. 5. *C. argentata*. 6. *Saguinus nigricollis*. 7. *S. midas*. 8. *S. oedipus*. 9. *Leontopithecus rosalia*. 10. *Aotus lemurinus*. 11. *Pithecia pithecia*. 12. *Hylobates agilis*.
*Polygynous mating system* (△) 1. *Colobus guereza*. 2. *Cercopithecus ascanius*. 3. *Theropithecus gelada*. 4. *Mandrillus leucophaeus*. 5. *Gorilla gorilla*.
*Man* (✻)

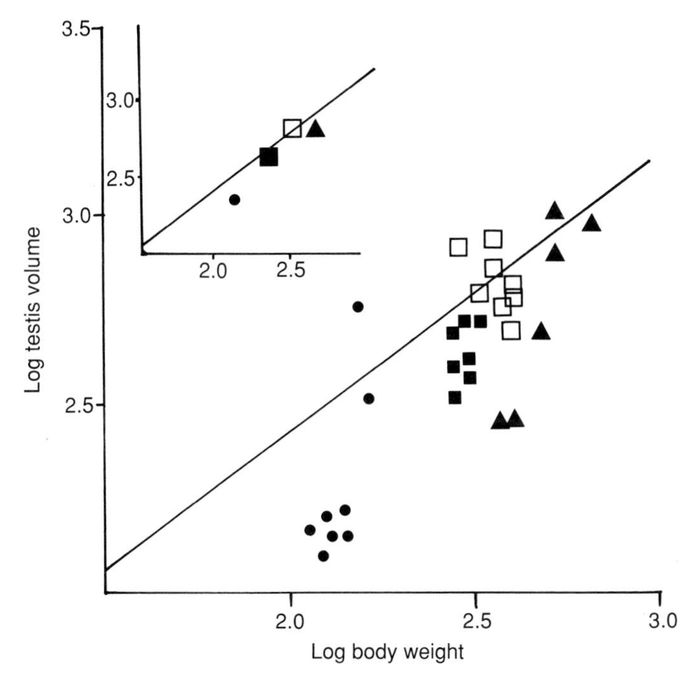


**Fig. 6.5** Logarithmic plot of mean testis volumes versus body weight in three callitrichid species. The lower graph shows values for individual specimens of *Cebuella pygmaea* (●); *Callithrix jacchus* (captive individuals (□):, free-ranging (■) and *Saguinus oedipus* (▲). The upper graph shows mean values for each species. On both graphs the same principle axis line as calculated for Fig. 6.4 (slope 0.718) has been inserted to facilite comparisons between the various plots. Data are previously unpublished measurements made by the author at the Anthropological Institute (Zurich) and MRC Reproductive Biology Unit (Edinburgh) and by Dixson, Anzenberger and Monteiro Cruz at Tapacurá (Brazil). A single specimen of *C. pygmaea* and five specimens of *S. oedipus* measured by Hershkovitz (1977) are also included.

body weights of adult males of each species were plotted on the graph. Adult body weights can vary enormously within callitrichid species and there are differences between captive and wild individuals, or even between separate populations in the wild (e.g. in *Saguinus mystax* and *S. fuscicollis*: Garber and Teaford 1986). Intraspecific variations in testicular volume are also pronounced, as can be appreciated by considering Fig. 6.5. Individual data on *Cebuella pygmaea*, *Callithrix jacchus*, and *Saguinus œdipus* have been plotted as well as mean values for each species. In all cases except one (the highest value for *Cebuella*), testicular volumes and body weights were obtained from the same individuals. Whilst a few male *C. jacchus* and

*S. oedipus* do have relatively large testes, the mean values for these species are low, as are those for *Cebuella*, and do not indicate effects of sexual selection due to sperm competition. It is also apparent that testicular volumes in captive *C. jacchus* exceed those for wild-trapped specimens, which raises questions concerning environmental and seasonal effects upon testicular function. A more detailed analysis using complete data from a range of callitrichid species would certainly be worthwhile. However, the available information does not indicate that selection for testicular enlargement has occurred in Callitrichidae in the same manner as for multi-male species such as the chimpanzee or various macaques and baboons.

## DNA 'fingerprinting' and mating systems

Traditionally, behavioural observations have been used to define mating systems in primates, but this approach has many limitations. It is virtually impossible to establish paternity under field conditions by using behavioural criteria alone. Observations on the composition of social groups can lead to wrong assumptions about the mating system. As an example, the 'one-male units' of certain *Cercopithecus* monkeys or *Erythrocebus* may not represent strictly polygynous mating systems, since influxes of additional, sexually active, males can occur from time to time (Cords 1987; Rowell 1988). Field studies of callitrichids have yet to determine whether polyandry is prevalent in breeding units and which male, or males, sire the offspring.

The discovery of hypervariable minisatellite sequences in the human genome has led to the development of DNA 'fingerprinting' techniques (Jeffreys *et al.* 1985*a,b*) which also have novel applications in fieldstudies (e.g. Wetton *et al.* 1987). Figure 6.6 shows DNA 'fingerprints' of captive family groups of common marmosets, compared to those of several unrelated human subjects. Some useful pointers for future fieldwork emerged from these studies. Firstly, human probe 33.6 is informative for determining relatedness of group members, but there is much band-sharing between animals. Maternal-specific and paternal-specific bands occur, but they are limited in number and this renders determination of parent–offspring relationships a difficult task. A second point of interest concerns the shared placental circulation between dizygotic twin fœtuses in callitrichids (Wislocki 1939) and resulting haemopoietic chimaerism of the offspring (Benirschke *et al.* 1962). DNA derived from the white blood cells of such twins yields virtually identical DNA fingerprints (Dixson *et al.* 1988). Whilst this may further complicate paternity determination using blood samples collected from free-ranging callitrichids, it should facilitate identification of twins

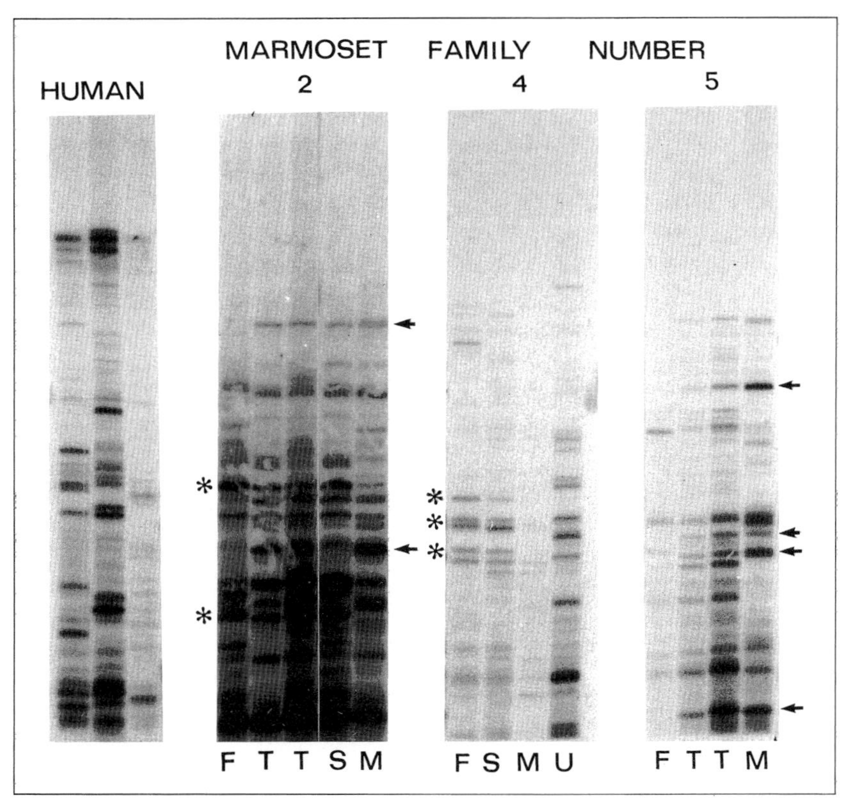


**Fig. 6.6** Autoradiographs showing hybridization patterns in 3 family groups of marmosets (No. 2, 4 and 5) and in 3 (unrelated) human subjects. F = Father; M = mother; S = single offspring; T = twins, U = unrelated male 361BK fostered into family 4. Examples of maternal-specific (←) and paternal-specific bands (*) are marked for the various groups. (After Dixson *et al.* 1988).

and provide information on whether co-twins remain in a group or migrate together between groups. Since trapping and retrapping of callitrichids can be accomplished in the wild (e.g. Scanlon *et al.* 1988 for *C. jacchus*) it should be possible to collect blood samples and attempt DNA 'fingerprinting' analyses in the near future.

DNA fingerprinting of free-ranging groups of *Callithrix jacchus* has now been carried out (Dixson, Anzenberger, Monteiro da Cruz, Patel, and Jeffreys) and results are reported in the Proceedings of the Second Schultz-Biegert Symposium *Paternity in Primates: genetic tests and theories* (ed. R.D. Martin, E.J. Wickings, and A.F. Dixson). Karger, Basel (1992).

# 7

# The social organization of marmosets: a critical evaluation of recent concepts

*Hartmut Rothe and Kurt Darms*

## Introduction

Only a few years ago it would not have occurred to most scholars of marmoset and tamarin behaviour to question the universality of sexual and reproductive monogamy in this family. Similarly, we knew of no objections to the transferability of data obtained from the more or less intensive study of, at the most, five to seven species (*Callithrix jacchus*, *C. humeralifer intermedius*, *C. flaviceps*, *Cebuella pygmaea*, *Saguinus fuscicollis*, *S. mystax*, *S. labiatus*, *S. oedipus*, and *Leontopithecus rosalia*: systematics according to Mittermeier *et al.* 1988*b*) to the whole taxon Callitrichidae, whose genera live in such widely differing habitats.

## Definitions

There is considerable confusion about the definition and correct use of such pivotal terms as monogamy and polyandry. Most authors do not differentiate between the sociographic and sociosexual levels when describing and interpreting social systems and structures (Kleiman 1977; Rutberg 1983; for discussion and references see Wickler and Seibt 1983). We consider this difference to be essential, and propose the following definitions.

The central feature of a (primate) society defined on the sociographic level is whether it has just one or more than one member of either sex (monogynandrous (pair), polyandrous, polygnous, polygynandrous), or whether it is isosexual (isogynous, isoandrous). The definition of a society on the sociosexual level: (1) includes only those members, irrespective of its size, which are sexually and reproductively active; (2) differentiates between males and females. For example, a society defined on the sociographic level as monogynandrous *must* be sexually monogamous on the male *and* female side and *can* be reproductively monogamous on the male and female side. In a sociographically polygynandrous society, the males and females can be sexually or both sexually and reproductively monogamous and/or polygamous when sexual interactions exclusively *within* the respective sociographically defined society are considered (for example, only one male and female are sexually active; only one male and female are reproductively active; more than one male is sexually active with only one female, etc.). Finally, in a sociographically polyandrous society, the males can only be sexually, or sexually and reproductively, monogamous, whereas the female may be monogamous or polygamous in both respects.

A third way to define societies concentrates on the genetic relations of their members. It is important to differentiate between: (1) family, a non-related male–female pair and offspring (nuclear family *sensu* McGrew 1986); and (2) group, a family plus at least one unrelated member (incorporative family *sensu* McGrew 1986), or society, containing, in the extreme case, only unrelated conspecifics. Transitions from families to groups, and vice versa, can occur through migrations or death. Lacking specific information on the members' genetic relationships and/or the history of the society, we should use the terms 'society' or 'social unit'.

## Sociographic aspects

Marmoset (and all other callitrichid) societies are polygynandrous groups or families (see Sussman and Garber 1987; Stevenson and Rylands 1988; Ferrari and Lopes Ferrari 1989). In sociographic terms, complete societies of marmosets (at least an unrelated male and female) are not monogamous. In this respect, no differences exist between captive and wild groups (Epple 1970; Rothe 1975*b*). Many field workers have stressed that groups/families contain more than one adult member of either sex, but the data from wild as well as captive groups/families refer to extremely skewed samples. Of approximately 160 natural societies (about 50 per cent of which were *C. pygmaea*) only about 50 have been observed more or less intensively, as opposed to several hundreds or even thousands in captivity.

## Demographic variables

Immigration and emigration may operate more rapidly and have greater effects on the demographic and genetic structure of a society than births and deaths (see for example, Scanlon *et al.* 1988). This is especially true if reproduction is confined to only one pair (see below), and if mating/reproduction is seasonal (for example, Rylands 1982; Hubrecht 1984; Soini 1987*b*; Stevenson, in Stevenson and Rylands 1988; Ferrari, in press), and infant mortality is high (Soini 1982*b*). Under laboratory conditions we have only limited access to the analysis of emigration and immigration (Koenig and Rothe 1990, 1991*a*). The initial trigger for these processes is hard to detect, and the same may apply to the field situation (Ferrari 1988*a*,*b*, personal communication). It is difficult to comment correctly and unequivocally on the causes of intragroup conflicts which might be linked to the limited or non-existing options of leaving or staying in the native family/group (Rothe *et al.* 1986; Darms 1987*a*). In the captive situation, the term 'eviction' is synonymous with being excluded from the family's/group's social network, the final act being performed by the keeper. Here we consider 'evictions' as emigration, assuming that conflicts in the wild, including severe fighting, would be replaced by a gradual peripheralization and/or early migration of one of the combatants (see also Baker 1987; Darms 1987*a*; Stevenson and Rylands 1988; Koenig and Rothe 1990).

## Immigration, emigration, turnover

Are marmoset (callitrichid) societies closed or open (permeable to strange conspecifics)? The question of migration rates, and the age and sex of dispersing individuals, is fundamental for an understanding of the mating system and social organization of callitrichids, but lamentably still poorly understood.

In the laboratory, it is usually impossible to integrate genetically related or unrelated (sub)adult strangers into established groups/families (Epple 1967, 1970; Rothe 1975*b*, 1979; Evans 1983; Sutcliffe and Poole 1984; Harrison and Tardif 1988), although it is possible with infants and juveniles (Kuester 1978; Pook 1978*b*; Rothe, unpublished observations; Wolters, personal communication), as well as with adults in incomplete societies (Rothe *et al.* 1987). Likewise, attempts to reintegrate evicted members into their natal family have failed with both sexes (Rothe, unpublished observations, but see Kleiman 1979 for *Leontopithecus*). In this last case, we conclude that eviction is tantamount to permanent separation, although a gradual peripheralization might not be irreversible (Soini 1982*b*; Darms 1987*a*).

In a long-term investigation, Koenig and Rothe (1990, 1991*a*) demonstrated a lack of emigration/immigration in infant/juvenile *C. jacchus* (2–10 months old). Those of different (adjacent) families did not form play groups (considered by Goldizen (1987*a*,*b*) as a strategy preceding migration), and all members showed agonistic or even hostile behaviour towards potential immigrants, even if already familiar. Koenig and Rothe (1991*a*) concluded that immigrants are welcomed only when of benefit to all group/family members. Brusek (1985), Morosetti (1986), and Radespiel (1990) observed a strong extragroup orientation in juveniles following weaning and/or after the birth of a new set of siblings, and Koenig (in Koenig and Rothe 1990) observed this to be more pronounced in twins than in singletons. There were, however, differences between twins, as well as an active avoidance of sociopositive interactions of juvenile immigrants with strangers. To summarize, therefore, in captivity intact marmoset groups/families are unwilling to integrate subadult/adult strangers, and we conclude that under these conditions births and deaths are the pivotal demographic variables determining group size and composition.

In the wild, friendly contacts between individuals of different social units do occur, but evidently rather infrequently (Maier *et al.* 1982; Rylands 1982; Stevenson, in Stevenson and Rylands 1988; Alonso and Langguth 1989). Hubrecht (1985) recorded 23 interactions between groups of *C. jacchus* at Tapacurá, Brazil, including a case of sexual behaviour between a female (10–15 months old) and one or more males of a neighbouring group (see, however, Soini 1982*b*). Relations between neighbouring groups will evidently depend on their history as well as the characteristics of the home ranges where they overlap. Rylands (1982), for example, observed that his *C. kuhli* study group was indifferent to one neighbouring group, actively interacted (chasing, long-calling, and displaying) for long periods with a second group, and would mix and interact socially with a third. Rylands (personal communication) believed that this might reflect a lack of a reason for disputes (no evident food sources in the area of overlap) in the first case, a dispute for an important food source in the second, and relatedness in the third (the neighbouring group was possibly newly formed, evidenced by an abnormal composition).

Regarding migration rates, Scanlon *et al.* (1988) and Monteiro da Cruz and Silva (1989), working at Tapacurá, reported a high turnover rate of 50 per cent and 30 per cent, respectively, for *C. jacchus*; a considerable instability in the demographic and genetic structure of the groups involved. However, a *C. jacchus* group studied by Alonso and Langguth (1989) in Paraíba was stable over 13 months, and Ferrari and Lopes Ferrari (1989) argued that the Tapacurá population is characterized by a very high density (15 times higher than elsewhere, see Stevenson and Rylands 1988) and might not be representative. Immigration/emigration events for other *Callithrix* species are evidently infrequent (see, for example, Ferrari (1988*a*,*b*) who studied *C. flaviceps* during 13 months, and Rylands (1982) who studied *C. humeralifer intermedius* for 12 months, reviewed by Ferrari and Lopes Ferrari 1989). For *Cebuella*, Soini (1982*b*) calculated a modal fluctuation rate of two independently locomoting individuals per year.

However, he estimated that 60 per cent of the group size changes were caused by the disappearance/reappearance of subadult/adult members, and 40 per cent by births and deaths. Ferrari and Lopes Ferrari (1989) argued that the comparatively small groups of *Cebuella* (smaller than is typical of *Callithrix*) and relatively frequent migrations may be indicative of species-specific mechanisms of dispersal related to a unique form of ranging behaviour (distinct territories separated by relatively large distances). Sussman and Garber (1987), referring mainly to the studies of Terborgh and Goldizen (1985) and Garber *et al.* (1984) on *Saguinus*, argued that migrations are regular events in callitrichid societies. Price (in press c) points out, however, that the *S. mystax* population studied by Garber *et al.* (1984) had been recently introduced, and that another study of the same species by Ramirez (1984) failed to find high levels of migrations. Ferrari and Lopes Ferrari (1989) recognized a fundamental difference in group stability between *Callithrix* and *Saguinus*. They emphasized the frequent changes in group composition reported for *Saguinus* (Dawson 1976, 1978; Izawa 1978; Neyman 1978; Garber *et al.* 1984; Terborgh and Goldizen 1985; Soini 1987*b*), and the relative stability of *Callithrix* groups (see, however, Price in press c, who argues that there is no difference). The 27 cases of migrations (18 males, nine females) documented by Terborgh and Goldizen (1985; Goldizen and Terborgh 1989) in 30 *S. fuscicollis* groups over seven years, provides a migration rate of less than one per group per seven years. It is evident that the data are as yet insufficient, but we believe that, while accepting a certain fluctuation in membership, results of the majority of field studies indicate that births and deaths, rather than dispersal events, are the decisive variables concerning the structures of families/groups.

The actual processes of emigration/immigration are still poorly known, and a crucial aspect regarding dispersal is whether immigrants/emigrants are related or not to the members of the society into which they immigrate or from which they emigrate. According to Goldizen and Terborgh (1989), 12 *S. fuscicollis* emigrants (of 18 whose origin and destination were known) were seen in an area adjacent to the home range of their native societies, and the remaining six migrated two or three home ranges away. Ferrari and Lopes Ferrari (1989) observed emigration of an adult male *C. flaviceps* into a neighbouring group and the subsequent formation of a new group, one to two weeks later, composed of three adult females from the study group and the two original adult males from the same neighbouring group. No intragroup aggression was documented during this process. In captivity, marmoset groups/families regulate their size and composition by evictions of subadult/adult members (Rothe *et al.* 1986; Stevenson and Rylands 1988; Darms 1989). The probability of being expelled, especially by same-sexed siblings, grows with increasing family size and thus with increasing age of the social unit, although there are considerable differences in the date and frequency of evictions (Rothe 1975*a*; Spichiger-Carlsson 1982; Anzenberger 1983; Darms 1987*a*,*b*, 1989). With an understanding of reproductive and demographic variables, we suggest that it is possible to obtain an estimate of the age of wild groups/families through their age/sex structure (Heltne 1978; Darms 1989). Thus, for example, a group/family with more than two adult members (including a breeding pair) was probably established at least 20–24 months previously. In this way, the size of the non-breeding subgroup is a rather precise reference to the duration of the group's/family's existence.

Waser (1988) provides several theoretical models on the probabilities relevant to philopatry and dispersal into adjacent home ranges. Assuming a moderate mortality rate, (intermediate home range turnover) and opportunistic philopatry (Model A. p.111), evidently not unrealistic for marmosets, then 'most juveniles will settle on natal or adjacent home ranges'. The ability to recognize kin is fundamental to avoid inbreeding in this model of relatively restricted dispersal (Harrison and Tardif 1988; Achilles 1991; Siebels 1991). In summary, we expect a close neighbourhood of related animals living in adjacent areas, and that animals involved in immigration/emigration are frequently related. This is especially so if turnover rates (including the replacement of the breeding female) are high. If the population is long established and population density is as high, as in Tapacurá, we may even expect an acceleration of these demographic processes and an enormous increase in the probability that animals living in adjacent areas are related.

To obtain a better understanding of the demographic processes and their influence on social organization, not only between genera but also between species and populations in different habitats, it is necessary to consider the definition of the group, the mechanisms causing dispersal (sibling competition, increase in group size), the inbreeding and outbreeding risks (relatedness of neighbours and group members), and the benefits and costs (predation, habitat quality, breeding opportunities, etc.). The consequences of high or low fluctuations of group membership remain obscure. However, we conclude that immigration of an unrelated animal is a rare event, thus exerting only a minimal effect on the demographic structure, and that genetically unrelated immigrants may play an important role as mating/breeding partners during the reproductive reorganization of a family/group following the death of the breeding male or female (Koenig 1987; Rothe and Koenig 1987; Koenig *et al.* 1988; Koenig and Rothe 1991*b*).

## Sociosexual aspects

A sociosexual description of the marmoset society has as a pivotal element the 'individual mating strategy'. Mating restricted to a single pair (reproductive monogamy) means that other animals of all age classes are not integrated into the sociosexual structure of the family/group.

### *Mating and breeding*

We first examine studies of captive groups. In naturally complete and unmanipulated families, both the founding male and female are, as a rule, monogamous (see, however, the definition of sexual monogamy). Alonso (1986) and Anzenberger and Simmen (1987) reported on a polygamous male, which bred with the alpha female and her highest-ranking daughter. The two females continued to be reproductively monogamous within a polygynmonandrous family. Abbott (1984) reported on two of 17 *C. jacchus* families in which the alpha male mounted a daughter, and on four of 17 families in which juvenile males mounted the alpha female. Epple (1970) mentions two brother–sister matings, one of which resulted in pregnancy. Crook (1988) describes a son–mother mating after the death of the father (the mother gave birth to non-viable twins). Rothe (1979) describes the eviction of the alpha male by the eldest son and subsequent breeding (six times) with his mother.

Even manipulated families of *C. jacchus*, *C. geoffroyi*, and *Cebuella* in which the father is replaced by a strange male (families turned into groups), resume sexual and reproductive monogamy after a transitional period of reproductive polygyny within a polygynandrous group during which the new male interacted sexually and impregnated more than one female, the alpha female and her eldest daughter (see also Abbott 1984; Rothe and Koenig 1991). In most cases, this 'transitional stage' ends more or less abruptly after the birth of the first or second set of infants of the subdominant female/daughter due to increasing aggressiveness between the two females. It is not always the mother who attacks the daughter, but may be the reverse (Box 1977; Vogt *et al.* 1978; Rothe and Koenig 1987, 1989, 1991; Rothe and Radespiel 1988; Koenig and Rothe 1991*b*). Mager (personal communication) observed both mother and daughter of a *Cebuella* group breed regularly for many years without major signs of aggressive competition between them. Alonso and Porfiro (1989) and Jaemmrich (1985) reported on the simultaneous breeding of two sisters in *C. kuhli* and *C. jacchus*, respectively.

Considering the enormous number of callitrichid societies with different life histories and demographic compositions which have been kept successfully under diverse laboratory conditions worldwide, and that these have shown an extremely low rate of departures from monogamy (Price and McGrew 1991; Rothe and Koenig 1991), it would be incorrect to consider this unambiguous result as an artefact of captive conditions. In our colony, for example, only one of about 70 naturally developed families/

groups of *C. jacchus*, *C. geoffroyi*, and *C. penicillata* have deviated from reproductive monogamy. Including only those families containing at least three adult members (41), the relative frequency of departures from monogamy is 2.4 per cent (Rothe and Koenig 1991).

At first sight, the situation is different in experimenter-formed groups of more than one adult and unrelated animal of either sex ('deliberate groups'—we prefer this term to 'artificial'). All members of such groups are sexually polygamous, but considerable differences exist concerning the frequency of sexual interactions (Epple 1970, 1975*a*, 1978*a*,*b*; Rothe 1975*b*, unpublished observation; Abbott and Hearn 1978; Abbott 1984, 1986). One pair clearly ranks on top, and each member of this pair has sexual intercourse with all other females and males, although most frequently with each other. Other members show less sexual behaviour both with each other and with the alpha pair. Abbott (1986) recorded 0.0–6.7 per cent of male mounts with subordinate females in deliberate groups (three males, three females). Subordinate males likewise showed little interest in subordinate females. In five of eight groups, only the alpha male achieved intromission and ejaculation and only with the alpha female. In three groups, the beta males also copulated, but achieved ejaculation in only 2.9 per cent of their mounts, whereas alpha males achieved 24 per cent. There is a clear tendency for the dominant pair to monopolize each other and prevent sexual behaviour both between him/her and subdominant members as well as between the latter (aristogamy *sensu* Bischof 1985). This is more marked for the dominant female, especially towards females, than for the dominant male (Rothe, unpublished observations; see also Abbott 1984, 1986). Thus, subordinate animals do not have full sexual access to the breeding male/female, especially during oestrus, and the mating system should be seen as functionally monogamous (see also Ferrari and Lopes Ferrari 1989).

Reproduction in deliberate groups is definitely limited to the highest-ranking female, whereas we cannot be sure if infants are fathered by more than one male; a possibility because spermatogenesis is not suppressed in subdominants (Rothe 1975*b*, 1979; Abbott *et al.* 1981). Abbott *et al.* (1981) recorded that three subordinate females out of 15 groups showed plasma progesterone levels greater than 10 ng/ml, but none of them preserved that level over the luteal phase (20–24 days). In a different investigation, also on deliberate groups, Abbott (1984) found all subdominant females to be sexually inhibited and infertile, whereas in eight of 17 families, the mother and one daughter ovulated. Suppression of subdominant females is more rigorous in deliberate groups than in families (Abbott 1984; Rothe, unpublished observations).

Reproductive monogamy does not exclude *a priori* the cohabitation of more than one same-sexed adult and unrelated conspecific, but in some cases in captivity serious intrasexual aggression occurs even within a few minutes after setting up a group with more than two individuals. Rothe (unpublished observations) tried to establish a group of five wild-born adult males and eleven females of *C. jacchus* (probably unrelated). Two males and five females were evicted on the first day, and a further three females had to be removed by the fifth day. The remaining six animals (three males and three females), lived as a group for a total of 111 days (in a cage of $75m^3$), when the highest-ranking pair was removed due to a serious increase in agonistic activity towards subordinates. The remaining two males and females lived together for nearly a year, until the death of the alpha female, after which the trio was dissolved. Another such 168 experiments have been performed involving four to six unrelated animals. The outcome was always the same: the quite rapid formation of a strict male and female hierarchy and the severe suppression of subordinates by the alpha pair.

Three aspects should be stressed here. Firstly, when a deliberate group was formed, dyadic sociopositive and coordinate interactions, as well as sexual behaviour, did not occur until the formation of male and female hierarchies. This process takes several minutes to several hours, depending on group size. In 152 of the 168 cases, the pair which won the first serious fight became the alpha pair, unchallenged for the duration of the experiment (one to four days). In 17 cases the alpha position changed (11 times with males, six times with females), and in a further four the alpha position changed twice (once with a male, three times with females), with the original alpha pair replacing the subsequent one (see also Abbott 1986).

Second, all subdominants, but especially the females, showed slight stress (permanent diarrhoea, hair loss, restlessness, weight loss, and exaggerated submissive behaviour) due to more or less regularly occurring attacks by the alpha pair, especially during the alpha female's postpartum oestrus (Rothe 1975*b*) and preceding birth. However, it should be mentioned that on some occasions sociopositive interactions (allogrooming, huddling) were more frequent between the alpha male and subordinates than between the alpha female and subordinates.

Thirdly, subdominant females in deliberate groups resumed ovulation much later than did female offspring of a natural family following their removal from the group/family (see also Abbott 1984, 1986). Furthermore, they subsequently had irregular pregnancies (miscarriage, stillbirth), and showed lactation failure and infant abuse (for example, cronism), and neglect. Despite the small data base on these aspects, the observations are consistent.

To summarize, deliberately formed groups do not show departure from reproductive monogamy. Reproduction is the privilege of the alpha pair. The pair-bond (whatever this may be, see Evans 1986; Bodemeyer 1990) is not formed until after the establishment of a hierarchy. The alpha animals generally restrict sociopositive behaviour to their partner and suppress and socially peripheralize the subdominants and unrelated conspecifics, even though they may eventually act as non-reproductive helpers (see below). The alpha-partner orientation of sociopositive interactions, and the socially peripheral position of the subordinates, raise doubts as to whether the term 'group' can appropriately be used to characterize these 'societies'.

## Departures from sexual/reproductive monogamy

Siess (1988) carried out a long-term experiment on the potential of adult and non-related *C. jacchus* to live in polyandrous (three males, one female) groups. The aim was to test the readiness of males to interact with each other in a sociopositive and non-agonistic manner; this must be considered a *sine qua non* for a co-operative polyandrous mating and breeding system. The males formed a strict hierarchy, even in the absence of a female, which resulted in the eviction of one, and the harsh suppression of another, by the dominant (see also Epple 1970, 1975*a*,*b*). The female chose the dominant male as a mate, but was not agonistic to the subdominant. Harrison and Tardif (1988) also argued that intense aggression between same-sexed individuals is difficult to reconcile with a breeding system based on co-operation. The dominant male tried to monopolize the female, and vice versa, whereas the subordinate was just tolerated, and not, therefore, incorporated into the mating/breeding system, evidently a prerequisite for reproductive polyandry (Siess 1988).

Compared to the laboratory, only limited information is available from the field. At present, valid conclusions on social organization which extend beyond the sociographic aspect cannot be drawn. Models, explanations, and hypotheses concerning the callitrichid mating system can be neither verified nor falsified.

As mentioned previously, a total of approximately 160 *Callithrix* and *Cebuella* societies have been observed for reasonable periods. A departure from sexual monogamy is reported in only five of these families/groups. These include (1) copulations by more than one male or female, although intromission and ejaculation were not certain (Rylands 1982, 1986*a*; Ferrari 1988*a*,*b*; Alonso and Langguth 1989; Ferrari and Lopes Ferrari 1989), (2) more than one breeding female per group (Soini 1982*b*; Scanlon *et al.* 1988; see also Ramirez 1984; Dietz and Kleiman 1986; Dietz and Baker 1991). Soini (1987*b*) reported mounting attempts by the subdominant male being impeded in *Cebuella*. However, only the dominant male copulated during the alpha female's oestrus, all attempts by the subdominant male being prevented by the higher-ranking (see also Siess 1988). Female promiscuity is possibly a means of enlisting male aid (Rylands 1982, 1986*a*; see below). The existence of more than two infants in each society is taken as definite evidence of two breeding females, but without observations of sexual behaviour it is impossible to say if more than one male is involved. Several other aspects require clarification, such as the relationship between the animals, the degree of affiliation to the social unit, and the duration of membership. Ferrari and Lopes Ferrari (1989) and Rylands (1982, 1986*a*) observed the replacement of the breeding female by another (in Ferrari's study group, the daughter), but both previous breeding females remained in the group/family, which continued to be reproductively monogamous.

According to field studies, departure from sexual and, especially, reproductive monogamy is a very rare event in wild marmoset societies (in only 3.1 per cent (sexual) and 1.9 per cent (reproductive) of all observed social units), thus showing no difference from the data on captive societies. Regarding *Saguinus*, Terborgh and Goldizen (1985) report on consort pairs in two *S. fuscicollis* social units. Ferrari (in press) argued that 'while such consortships rarely involve open aggression between males, the behavioural dominance of the breeding female could be the key factor determining which male (or males) is (are) able to copulate at the time of ovulation. Matings or "copulations" at other times, whether they involve intromission and ejaculation or not, would seem in these cases to represent only poor evidence of a polyandrous mating system' (p.10).

## . . . or exclusive sexual/reproductive relationships?

The majority of studies on marmoset behaviour and biology, both from the field and captivity, have revealed long-lasting privileged and non-restricted relationships between only two members of a family or group. This relationship is characterized mainly by a distinct spatial cohesion (short interindividual distance), which may be regarded as mate-guarding, especially during the (visual, olfactory, and acoustic) presence of strange and/or unrelated conspecifics (Anzenberger 1983; Darms 1987*a*; Soini 1987*b*; Juenemann 1990; Radespiel 1990; see, however, Terborgh and Goldizen 1985). Several authors use the term 'pair-bond' to characterize the particular relationship between the pair, or dominant male and female, of a family/group (Erickson 1978; Anzenberger 1983; Evans 1986; Darms 1987*a*; for references and discussion see Bodemeyer 1990). In our opinion, the concept of the 'pair-bond' is vague, poorly defined operationally (see also Kleiman 1985), and implies that it is based mainly on personal traits of the individuals concerned and not on a common or complementary social interest or strategy, the realization of which is not confined to specific individuals. Furthermore, it inhibits consideration of any other mating/breeding system, such as polyandry (see, however, Abbott and Hearn 1978). Radespiel (1990) has shown that a close spatial proximity between two group members, especially males, may not be due to a specific and reciprocal social affinity (to a pair-bond) but due to the quality/attractiveness of a given location in the home range (cage). Short interindividual distances may not always, therefore, reflect social affinity but (merely) social tolerance. Juenemann (1990) revealed a (possible) relationship between interindividual distances and 'personal interest' within triadic constellations, as well as efforts to regulate the proximity between specific family members (see also Darms 1987*a*,*b*). For example, the eldest son of a family exerts a great influence on the interindividual distance between his parents and eldest sisters. Radespiel (1990) observed comparable behaviour in the dominant male, regulating contact behaviour and frequency of his adult offspring with the dominant female.

Exclusive tasks and rights of the dominant male and female are apparently restricted to mating and reproduction, whereas other roles, for example, group defence, sentinel behaviour, and infant care, are also assumed by adult subdominant group members or offspring, predominantly adult males/sons (Darms 1987*a*; Radespiel 1990; Koenig and Rothe 1991*a*). Playing a specific role, however, seems to be related to age and rank and dependent on family size and composition (see also Rothe 1978*b*; Sommer 1980; Darms 1987*a*; Rothe *et al.* in review *a*). Excepting infant care, we have no indications that dominants prevent subordinates from assuming the above-mentioned roles, and it must, therefore, be in their interest to be relieved of these duties (Koenig and Rothe 1990, 1991*a*).

Reproductive exclusiveness requires mechanisms for reproductive/hormonal suppression of subordinates (see Abbott 1984, 1986) and/or incest avoidance (see also Sommer 1980; Amsler 1982; Erb 1983). Since it is rather unlikely for a marmoset society to contain exclusively unrelated members, individuals not only have to differentiate between 'known' and 'unknown', but also between 'related' and 'unrelated' conspecifics. Whereas parents generally avoid their offspring, and the latter each other, as mating partners, vacancies result in strange conspecifics being accepted immediately (for example, Bodemeyer 1990; Koenig *et al.* 1990). Experimental replacement of the dominant male can result in reproductive polygyny (see Abbott 1984; Rothe and Koenig 1987, 1988; Koenig and Rothe 1991*b*). Koenig *et al.* (1988) reported on three groups which had lost one parent, and in which breeding occurred between the remaining parent (twice the male, once the female) and an unrelated group member which had been integrated as an infant several months before the replacement of the breeding male. Both examples demonstrate discrimination of related and unrelated conspecifics, but the mechanism is not known (see also Achilles

1991). Darms (1987*a*,*b*) revealed that dyads of heterosexual and sexually mature siblings avoid proximity. Darms interpreted this as an economic strategy of incest avoidance on the one hand, and of signalling their non-mated status to potential partners on the other. Mated pairs show close proximity (Darms 1987*a*,*b*; Bodemeyer 1990; Juenemann 1990; Koenig and Rothe 1990; Radespiel 1990), thereby demonstrating their reproductive and dominant status. In deliberate groups, subdominant heterosexual dyads show close proximity to each other (Rothe, unpublished observations), since a special avoidance or advertising strategy is unnecessary. It remains unclear at the moment whether the so-called incest taboo is a mere side-effect of the latter strategy or not, or vice versa, or alternatively is a mere consequence of reproductive suppression (Abbott 1984, 1986; Evans and Hodges 1984; Hubrecht 1984; Heistermann *et al.* 1989). Abbott (1984) argued that social contraception effectively maintains the monogamous *status quo* of one breeding pair. The cycling of daughters might occur either as a result of a change in their relationship with their mother, perhaps prior to dispersal (see also Hampton *et al.* 1966; Epple 1967; Poole and Evans 1982), or precede a takeover as the breeding female. Hubrecht (1989) also demonstrated daughters ovulating and even becoming pregnant in captive families. No behavioural change was observed between mothers and daughters, only between the daughters and their younger sisters (aggression). Hubrecht also regarded ovulation by daughters as a preliminary to dispersal (see also Hubrecht 1985). We conclude that the term 'incest avoidance' incorrectly describes this specific aspect of the offspring's behaviour (for details see Bischof 1972, 1985; Darms 1987*a*).

The restriction of reproduction to the dominant pair involves the risk of temporal interruption or even termination of reproduction following the loss of the partner (Koenig *et al.* 1988). We have analysed 15 incomplete families in this respect (loss of the male three times, loss of the female 10 times, loss of both twice). In four cases inbreeding resulted (twice alpha male and daughter). In only one case had the infants definitely been sired before the death of the alpha female (about 50 days). In the remaining 11 families, reproduction was abandoned and they were dissolved due to severe fighting within a period of 2.5–24 months (for further details see Koenig *et al.* 1988). Families which had lost the alpha female showed a slight tendency to remain stable for longer than those which had lost the alpha male (Rothe 1978*b*).

Rothe (unpublished observations) split up a *C. jacchus* family into two subgroups of eight. One contained the parents and six (2:4) offspring, the other only siblings (2:6). Whereas the parental subgroup remained stable, all members of the 'decapitated family' fought with each other. In contrast to complete families/groups and deliberate groups, no separate male and female hierarchies were established. The oldest son dominated his siblings, expelled two sisters (one his twin) after five and 89 days, and his younger brother (then 13.5 months old) after 107 days, at which time agonistic behaviour ceased completely and quite abruptly. However, the remaining five (1:4) siblings showed a remarkable indifference to each other, even to their twins, and interactions (huddling, grooming) were very infrequent. Solitary behaviour (autogrooming, feeding, manipulating the cage equipment, etc.), on the contrary, increased enormously. The sibling group was dissolved 299 days after its inception.

No data are available on the reproductive reorganization of incomplete social units in the wild. It is difficult to know whether a wild callitrichid society is incomplete or not, although the absence of pregnant and/or lactating females, for example, might be an indication (see for example, Soini 1982*b*; Scanlon *et al.* 1988). We can, therefore, only speculate on the captive data. One of the most interesting results is the more or less immediate onset of destabilizing behavioural processes after the loss of a parent. Four important aspects should also be mentioned: (1) as a rule the remaining parent did not participate, at least initially, in the aggressive interactions, and did not interfere in the conflicts of other family members; (2) offspring of both sexes were involved in agonistic interactions, indicating that the reinstatement of a vacant alpha position is linked to an overall reorganization of the social structure, and not confined to the gender for which the alpha position needs to be re-established; (3) stability is guaranteed only by the presence of *both* dominant members; and (4) dependent infants delay the onset of reorganization processes until weaning (Koenig *et al.* 1988; Rothe, unpublished observations).

Considering the fact that the immigration of unrelated conspecifics is infrequent, the establishment of a new, non-incestuous, breeding pair might be the exception and dissolution of the entire family/group the rule. Preservation of stability guarantees reproduction, and thus increases reproductive success. Emigration of the breeding male or female before losing dominant status would, therefore, be disadvantageous. Scanlon *et al.* (1988) reported a case of immigration of a reproductively active female, but the stability of the group was not ascertained.

## Parental care: cause or result of a specific breeding system?

One of the principal advantages of monogamy is considered to be male investment in offspring (for review see Whitten 1987). However, according to Kleiman (1977), monogamy lacks clear-cut correlates; there is great variability in the male's investment, and some authors doubt the general legitimacy of linking the two aspects at all (Ferrari in press). Two hypotheses are discussed: the dominant male is primarily interested in caring for his infants (monogamy a consequence of paternal investment); and the dominant male is primarily interested in the breeding female (paternal care is a byproduct), and the key aspect is the male's guaranteed access to the female (for details see Goldizen 1987*a*,*b*). If monogamy has evolved because of the need for paternal care, it is necessary to ask why callitrichids have evolved high litter-mother weight ratios, twinning, and postpartum oestrus in the first place (Rylands 1982, 1989*a*; Pook 1984; Whitten 1987; Barlow 1988). Several authors have shown that infant care by the dominant male is not necessarily confined to genetically related infants (Box 1977; Abbott 1978; Rylands 1982, 1986*a*, 1989*a*; Cebul and Epple 1984; Cleveland and Snowdon 1984; Ferrari 1988*a*,*b*. Rothe *et al.* in review *a*; Wolters, personal communication), and if Scanlon *et al.* (1988) are right, even pregnant females may disperse, in which case, the probability of infants being raised by unrelated conspecifics is very high. If so, benefits through kin selection are unlikely to explain co-operative breeding (see also Tardif *et al.* 1984; Harrison and Tardif 1988). Rothe and Koenig (1988) observed a newly-established male provide a distinctly higher contribution to the care of his predecessor's infants than to his own. Furthermore, this male was later more involved in caring for infants he fathered with a subdominant daughter of the female than for the infants he sired with her mother (Rothe and Koenig 1991). These observations support the second hypothesis, that the primary object of the male's interest is the sexual partner and not the infants.

More complicated is the question of whether a specific mating/breeding system is required to guarantee optimal infant care. Goldizen (1987*a*,*b*) and Goldizen and Terborgh (1989) argued that *S. fuscicollis* pairs are unable to raise twins successfully without helpers, thus a polyandrous mating system would be the only and logical consequence. Goldizen's argument hinges on two observations: lone pairs are seldom seen in the wild, and were never seen breeding; and infant-carrying group members forage less, especially problematic for pregnant and/or lactating females. Goldizen and Terborgh also adopt the traditional view of callitrichids as obligate and prolific breeders, and not simply potentially prolific given optimal conditions. They unfortunately confuse two criteria: the occurrence of lone pairs at all on the one hand, and the apparent absence of attempts to breed on the other. The reasons for these two aspects may be distinct. According to Goldizen and Terborgh's (1989) Table 1, groups with no more than two individuals were seen in only one of the seven years of observation (1984): this is evidently, therefore, a rare phenomenon. The authors mention in this context the risk of dispersal of single individuals as one of the main reasons. Not attempting to breed, however, is a quite different aspect. Goldizen and Terborgh's (1989) argument that a pair foregoes breeding until joined by a helper implies a remarkable cognitive ability, as yet unproven. Lacking clear evidence of such a sophisticated 'birth-control' strategy, this argument is unconvincing and even implausible. There is no reason to assume that total abstinence from breeding would be more advantageous in increasing reproductive success than the use of all reproductive possibilities, for example raising one infant once a year (see Soini 1987*b*). Dawson (1976; Dawson and Dukelow 1976), Rylands (1982), and Goldizen and Terborgh (1989) have shown that the full reproductive potential is not realized under all circumstances (see also Hubrecht 1984), but may be related to the ecological conditions in the respective breeding season. Ferrari and Lopes Ferrari (1989) indicated a generic difference in this respect, with *Saguinus* and *Leontopithecus* generally breeding only once a year, and the marmosets twice, and link this to food availability during times of scarcity (the marmoset's capacity to explore gum sources).

Ferrari (in press) did not observe reduced activity in infant-carrying *C. flaviceps*, but rather intensified foraging (see, however, Price in press *a*,*b*). Ferrari does not accept Goldizen and Terborgh's (1989) view that lone pairs are unable to raise twins, nor the need to postulate an interrelationship of infant care and the mating/breeding system. Rylands (1982, 1986*a*) also argued that reproductive success is enhanced by helpers, but this has not been demonstrated. Ferrari (in press) attributes the scarcity of male–female pairs to a higher mortality in females, and thus to an imbalanced sex-ratio, but does not substantiate why females should have a higher mortality *per se*. Field data show no differences in sex ratio in the observed groups/families (Soini 1982*b*; Hubrecht 1984, 1985; Stevenson and Rylands 1988; Scanlon *et al.* 1988; see also Rothe *et al.* in press for an analysis of captive *C. jacchus*). Sex differences in mortality may be due to differential dispersal or age differences in dispersal of each sex, but both captive and field data have failed to show this. McGrew and McLuckie (1986) observed a tendency for philopatry in male *S. oedipus*. Koenig (in Koenig and Rothe 1990), Koenig and Rothe (1991*a*), and Rothe *et al.* (in review *a*, submitted *a* and *b*) observed that males/sons contributed more to infant carrying and group defence than females/daughters. This may indicate a greater collaboration and tendency to philopatry, but more data from the wild are required.

Captive male–female pairs have no problems in raising infants without additional helpers. Koenig and Siess (1986) and Lucas *et al.* (1927) even describe three cases (*C. jacchus*, *C. geoffroyi*, and *C. jacchus* × *C. penicillata*) of infant rearing by the mother only. In our colony (71 groups/families of *C. jacchus*, *C. geoffroyi*, and *C. penicillata*—325 sets of infants), there is no significant indication of decreasing mortality with increasing number of helpers (see also Rothe *et al.* in press, in review *b*). On the contrary, the presence of too many 'helpers' may be disadvantageous or even dangerous, due to frequent handling by older and sometimes inexperienced siblings, as well as an increase in intervals between suckling caused by conflicts between the mother and elder offspring, mainly adult daughters. Rothe and Winter (unpublished observations) found that newborn infants in large (more than 15 helpers) *C. jacchus* families were not suckled for 5–6 hours during the first day because the mother was unable to retrieve the infants from lower-ranking individuals (see also Tardif *et al.* 1984, 1986; Harrison and Tardif 1988).

Garber *et al.* (1984) argued a positive correlation between infant survival and the number of helpers in wild *S. mystax* (see also Moya *et al.* 1980). Soini (1982*b*) assumed a high perinatal mortality in wild *Cebuella* which could result in misinterpretation concerning this aspect. Likewise, other factors such as infant rearing experience, perinatal behaviour, and the physical condition of the mother and infants should also be considered (see Rothe 1973; 1974; 1975*a*, 1978*a*; Stevenson 1976*b*).

Unrelated helpers which have been recruited may be replaced by the juvenile-subadult offspring after about one year. Hence the question arises as to whether or not a polyandrous mating/breeding system is really so beneficial. Due to dizygotic twinning, twins may have different fathers. This attribute is regarded as a crucial stimulus for a second male to join a pair and form a reproductive polyandrous system (Rylands 1982, 1989*a*). Ferrari (in press) argued 'whereas polyandrous mating may be tolerated by unrelated tamarin males as long as their group contains insufficient helpers, the situation in the marmosets may be far more similar to that seen in lions (Emlen 1984) in which case the co-operative breeding of siblings would not be linked to group size but to other factors such as the availability of males' (also argued by Rylands 1982, 1989*a*). Terborgh and Goldizen (1985) report on a social unit of *S. fuscicollis* containing two brothers, but only one was sexually active.

Another critical point is the availability and 'recruitment' of helpers, as well as the synchronous emigration of at least two unrelated conspecifics to establish a breeding pair/group. During 21 months, Terborgh and Goldizen (1985) witnessed the formation of (only) three bisexual social units (one polygynandrous, two polyandrous). Hence group formation by group/family segments from different social units is evidently rather rare. Ferrari (1988*b*) witnessed a similar group formation by segments in *C. flaviceps*. But the probability of group formation by single emigrants would seem to be higher than that by more than one emigrant. If we follow the arguments of Goldizen and Terborgh (1989) strictly, at least three animals have to co-ordinate their actions without knowing of each other's plans or strategies (group interactions are too seldom to enhance the chance for members of different societies to get acquainted). According to Goldizen and Terborgh (1989), adult offspring in *S. fuscicollis* delay their own reproduction and prefer to stay in their native groups/families, for example, because of a shortage of vacant breeding positions and suitable habitat and in order to minimize the as yet unsubstantiated risks of dispersal. The authors indicate, in this case, a minimum probability of group or pair formation. The field data do not substantiate the assumption of Goldizen and Terborgh (1989) that lone pairs were unable to raise infants, and that paternal care determines the mating/breeding system (or vice versa). These authors show only that pair/trio formation in *S. fuscicollis* is a rare event, for whatever reason. Indirectly, they do admit, however, that *S. fuscicollis* pairs are able to rear infants on their own ('. . . even if the parents were capable of doing all of the infant-carrying, the helper's aid might allow the parents to conserve energy, survive longer and/or breed again sooner than would otherwise be possible'; p.297). French (1983) provides a similar argument. According to Soini (1982*b*), pair formation by emigrating individuals in *Cebuella* is a regular, but gradual, process, and there is no reason to assume less predation risk for the pygmy marmoset than for *S. fuscicollis*. Even if we concede major species differences in the energetic investment in infant-rearing (Tardif and Harrison 1986), there is no sound reason to postulate the complete inability of single pairs of any species to raise infants.

Ferrari and Lopes Ferrari (1989) presume group formation in *C. flaviceps* by two sets of twins from different groups to be the most probable, but not exclusive, way for offspring to obtain breeding positions (see also Baker 1987). 'Fraternal' polyandry of marmoset males might be seen as an extension of a monogamous pattern, especially in the case of twin siblings. For marmosets, polyandry as a breeding strategy would be more closely related to factors such as population density and availability of breeding females than to group size. Where polyandry occurs it will be fraternal, thus the degree of direct intrasexual competition for reproductive status will be relatively low, with kin selection playing an important role in the evolution and maintenance of the societies (Ferrari and Lopes Ferrari 1989; see also Rylands 1982, 1989*a*). This hypothesis, however, still requires confirmation. We have no further information on the development of the newly formed group in Ferrari's study site, nor of the genetic relations of the native societies. More data are also needed to judge whether communal emigration and group formation is the rule or just one option. Rothe (unpublished observation) recorded the four oldest male offspring of a large *C. jacchus* family (18 members in two spacious rooms) withdrawing from the family centre over a period of many months. For several hours each day, they interacted exclusively with each other. This might be interpreted as the initial stage of peripheralization/emigration of the male subgroup. Darms (1987*a*), Groeger (1988), Rothe and Radespiel (1988), and Radespiel (1990) observed sexual/soliciting behaviour between the breeding male/female and females/males, as well as between male and female offspring, of two neighbouring *C. jacchus* families at the 'home range border', away from the family centre and other members, and in situations when they believed themselves to be alone (see also Hubrecht 1984, 1985). They seldom or never showed this behaviour in the presence of family members, but rather demonstrated indifference or even reacted/interacted aggressively toward/with the unrelated neighbours. Later on, however, agonistic/aggressive behaviour was also directed towards siblings of either sex, or the encounter was abandoned by withdrawal, especially when one or both approached. This strategy was also followed up by the male and female offspring which had been evicted from their native families but could keep contact with both their native and neighbouring families (see Rothe and Radespiel 1988). Within a relatively short period after eviction, a hierarchy had been established between the twin brothers and twin sisters, respectively, and no sexual behaviour was shown in the presence of the twin. The change in behaviour of the consort partners in the presence of another family member compared to the situation when alone clearly contrasts with the assumption of Goldizen and Terborgh (1989) that pairs will not breed until a helper has been recruited. It also contrasts with the hypothesis of fraternal polyandry (Rylands 1982, 1989*a*; Ferrari and Lopes Ferrari 1989). Our observations on *C. jacchus* indicate that conspecifics (same or opposite-sexed siblings) should be considered as competitors rather than helpers.

## Co-operation: *cui bono*?

Co-operation, predominantly in the form of infant care, is unquestionably a prominent aspect of the callitrichid social organization (review in Riedman 1982; Stevenson and Rylands 1988; Price and Evans in press; Rothe *et al.* in review *a*, submitted *a* and *b*). The questions arise, however, as to whether co-operation/helping is a *sine qua non* for the functioning of this social organization, whether it is necessary to guarantee benefits for the individual or all members of the society, or even whether it is an essential element of callitrichid social organization? Helping can be explained by: (1) gain in rearing experience (Epple 1975*a*,*b*, 1978*a*,*b*; Kaspereit 1977; Lange 1977; Hoage 1978; Cleveland and Snowdon 1984; but see Tardif *et al.* 1984); (2) the contribution to inclusive fitness via kin selection (Hamilton 1964; West Eberhard 1975); (3) delayed benefits through reciprocal altruism (Trivers 1971; Axelrod and Hamilton 1981). The fact that a female can rear up to three litters within a year with the aid of two males, as observed in *C. humeralifer*, would certainly fulfil the condition that altruistic acts could be reciprocated in a short space of time (Rylands 1982). Ferrari (in press) considers the provision of infant care as a form of submissive behaviour towards the breeding female to ensure permanence in the group, also suggested by Rylands (1982). The alpha female is considered to require assistance because of such factors as the high fetal/infant weight in relation to the mother, obligate twinning, postpartum oestrus, and high energetic costs of lactation (Ingram 1977, 1978*a*; Kleiman 1977, 1985; Eisenberg 1977, 1981; Leutenegger 1980; Rylands 1982, 1986*a*, 1989*a*; Pook 1984; Sussman and Garber 1987; Tardif *et al.* 1990; Price in press *a* and *b*; Rothe *et al.* in review *a*; submitted *a* and *b*).

According to field (Soini 1982*b*; Ferrari, in press) and captive data (review in Rothe *et al.* in review *a* and *b*, submitted *a* and *b*), the presence of non-reproductive helpers is not a prerequisite for breeding success (see above). The answer to the first of the questions above is, therefore, no. The second question is more difficult to answer, since it incorporates several aspects. One to which great importance is ascribed concentrates on the number of helpers and the degree of help. In a long-term study on infant-carrying in *C. jacchus*, Rothe *et al.* (in review *a*, submitted *a* and *b*) have shown that enormous differences exist in the contributions of family members, and only one or two individuals may be considered to help substantially ('qualified helpers') (see also Rylands 1986*a*; but Kleiman 1985). The number of (adult) family/group members is by no means, therefore, equal to the number of helpers, which are in the minority, nearly always the same individuals, and predominantly male (sons) (see also Price and McGrew 1990; but Garber *et al.* 1984; Sussman and Garber 1987). These helpers often contribute more to infant care than the dominant male (see also Box 1977; Gottschling 1984; Stevenson and Rylands 1988; Alonso and Langguth 1989), which is, therefore, no measure of breeding status. We conclude that assessment of the real amount of relief to the breeding pair or female is very difficult, both qualitatively and quantitatively. What is a real contribution to infant care? Are we allowed, for example, to regard a 5 per cent reduction of carrying by the mother as substantial relief, especially considering that additional conspecifics consume resources?

Another critical point in this context is timing of effective help. Our long-term studies of infant-carrying (Rothe *et al.* in review *a*, submitted *a*) have forced us to reevaluate the assumed efficiency of the marmosets' (and other callitrichids') helper system, chiefly for the following reasons:

(1) During the most critical period of the infant's life (the first few days) other family/group members do not substantially contribute to carrying: the mother prevents others, including the dominant male, although this is variable (see also Engel 1985*a*,*b*).

(2) Helpers contribute most when the infant is still dependent but already moving around alone and taking solid food, and when the absolute carrying frequency is already substantially reduced.

(3) During weaning, when the infant has already gained considerable weight, the dominant male and female are very often the predominant or even exclusive carriers (there is, however, large intra- and intergroup variation and dependency on the presence of adult male offspring).

(4) The breeding animals' contribution to infant carrying is generally higher than that of each of the other members, irrespective of group/family size and the number of potential helpers.

(5) The breeding female's contribution to infant-carrying shows a clear dependency on the presence of at least one adult male non-reproductive helper (adult son) (see also below), but is otherwise the least variable and least influenced by group/family size, age/sex structure etc. (for details see Rothe *et al.* in review *a*).

(6) The breeding female contributes more to infant care and carrying, especially in the first three weeks, than would be necessary for its survival.

(7) A slight reduction in the alpha male's infant carrying is compensated by an increase in the female's contribution, and likewise a considerable reduction of the dominant males contribution is balanced in groups/families without additional helpers by the alpha female (see also Yamamoto *et al.* 1987).

(8) In larger groups/families, compensation is provided by the mother and, as a rule, one male non-reproductive helper, hence again only two members take the main burden, and an increase in the contribution of a helper does not automatically relieve the mother, but nearly always another helper.

(9) Substantial relief of the parents/alpha female from infant carrying does not occur until the family/group contains a fully adult son, at least two years, therefore, after pair formation.

(10) In captivity, the more space available for a family/group and/or the greater the opportunity to interact with strange conspecifics, the lower is the helper's carrying performance, including that of the alpha male.

There is, all in all, an advantage for the breeding members/alpha female when one or more qualified helpers are present, but it is by no means so substantial as to link successful infant care and survival to the mating/breeding system, since in nearly all situations when helping would be most crucial to infant survival, it is ultimately the alpha female who has to master the situation (see also Tardif *et al.* 1990).

Regarding the value of infant carrying for breeding experience, Rothe *et al.* (submitted *a* and *b*) have shown that substantial helping is restricted to just one or two individuals, and not from the first day after birth. Engel

(1985*a*,*b*) demonstrated that the alpha female prevents inexperienced family members from carrying and handling the newborn (see also Kaspereit 1977; Lange 1977; Morosetti 1986). Increasing family size or the presence of qualified helpers decreases infant carrying by juveniles and subadults (Rothe *et al.* submitted *a* and *b*). Winter and Rothe (unpublished observation) observed excessive infant care by experienced helpers which eventually required intervention. These few examples indicate that: (1) assistance in rearing siblings is not the crucial incentive for most non-reproductive helpers to stay in their native family; (2) infant care need not be learned directly (see also Tardif *et al.* 1984), but rather depends on social competence in general, obtained in a complete family up to subadulthood.

Why, therefore, do non-reproductive helpers stay in the family/group? Do they prefer to stay, or are they forced to do so due to intra- and extra-group constraints? The second question also applies to helpers, who although admittedly increasing their reproductive success by inclusive fitness, would do better if they dispersed. One of the most frequently cited arguments as to why individuals stay is the option to 'inherit' breeding status in their native group, or part of the home range by annexing (Emlen and Vehrencamp 1985; McGrew and McLuckie 1986; Sussman and Garber 1987; Rylands 1989*a*). Predation risk is also cited.

The rare occurrence of inheritance, the rather long tenure of the breeding pair (Rylands 1982, 1986*a*; Soini 1982*b*, 1987*b*, 1988; Hubrecht 1984, 1985; Ferrari 1988*a*; Stevenson and Rylands 1988), and the fact that only one position in either sex class may be replaced, argue against inheritance as an adaptive strategy, quite apart from the possibly deleterious effects of inbreeding. The presumed higher predation risk for a single animal has not yet been substantiated. Furthermore, this hypothesis contrasts with the fact that family/group members may forage alone at a considerable distance from each other (Fonseca *et al.* 1980; Coimbra-Filho *et al.* 1981*a*; Hubrecht 1984; Muskin 1984*a*,*b*), and that most emigrations, in the wild and captivity, involve just one animal (Soini 1982*b*, 1987*b*, 1988; Rothe *et al.* 1986; Darms 1987*a*, 1989; Goldizen and Terborgh 1989; Stevenson and Rylands 1988; see, however, Ferrari and Lopes Ferrari 1989). We consider social contact to be a crucial resource for each individual, since it guarantees social competence (for example, Darms 1987*a*). Deprivation of social contact might be the motivation for dispersal as soon as a social alternative arises (see also Soini 1982*b*), and the family may offset the consequences of their peripheralization/emigration by promoting contact between peripheral/emigrating conspecifics of adjacent societies (Hubrecht 1984).

To summarize, inheritance, annexing an enlarged home range, learning how to handle infants, and the presumed predation risks of dispersal cannot be considered to be essential for philopatry, and most important is the option for social contact. Qualified help in raising infants is restricted to just one or two non-reproductive helpers, and their help is indeed of advantage; proximately mainly for the breeding pair, ultimately for all family/group members, assuming the non-qualified helpers (less than 10 per cent of carrying) leave their native unit as soon as possible. Their contribution may be helpful, but not necessarily advantageous to the remainder. We do not know, however, whether helping to rear infants is a byproduct of privileged relationships with the breeding animals, or whether privileged access causes increased carrying. The acquisition of privileged relationships is probably a matter of seniority and/or hierarchical status and authority over the (twin) sibling(s) (for example, Darms 1987*a*). Effective non-reproductive twin helpers have won their twin fight (Rothe, unpublished observation).

## Conclusion

It is important to differentiate between the sociographic unit and functional unit when considering the definition of a family, group, or social unit. The same individuals are not always involved in different contexts such as breeding, infant rearing, and group and territorial defence, and we can assume that the larger the society the greater will be the difference between sociographic and functional units. It is inaccurate to simply calculate, for example, the number of helpers merely from group/family size when considering prerequisites for reproductive success.

To date, little is known about the interrelationships between functional units of different sociographic units. Turnover, for example, may have no consequences for the functional breeding or infant-rearing unit as long as the respective animals do not compete or co-operate in this context. The analysis of functional aspects requires consideration of who *really* co-operates or interacts with whom in a given context, and not simply which individuals could eventually interact or co-operate. Proximate causes influencing individual co-operation are still not well understood (for example, Pryce 1988; Koenig and Rothe 1991*a*). The decision to co-operate and/or how much to invest in co-operation, according to common or conflicting interests (Chase 1980), is probably influenced by a number of variables such as age, seniority, sex, and the presence of infants, amongst others (for example Koenig and Rothe 1991*a*).

There is a considerable literature on the theoretical aspects of such as mating/breeding systems, dispersal patterns, group size, parental/paternal investment, and co-operative rearing, but neither theory nor model sufficiently covers or explains all aspects of the social system and organization of primates. Theoretical dogmatism will not solve the problem in question, only the evaluation of empirical data, descriptive and experimental, from both the field and captivity. It is evident that the gaps in the data are still enormous, rendering final conclusions and interpretations as yet impossible. Long-term data on more social units in natural habitats are urgently required, remembering that strategies such as mating, breeding, and dispersal in a given context and habitat may be deleterious for another under similar conditions. The subsummation of social units to a 'statistical unit' may be misleading or even false. Finally, intensive field studies (for example, Rylands 1982; Ferrari 1988*a*,*b*; Ferrari and Lopes Ferrari 1989; see Rylands and Faria, this volume) have shown that, even within the genus *Callithrix*, considerable differences exist regarding such aspects as habitat use, population density, feeding strategies, seasonality of resources, and mating/breeding, and care should be taken in generalizing across not only genera, but also species and populations.

## Acknowledgements

The authors are indebted to L. Achilles and A. Koenig for their valuable comments and technical assistance, and to A.B. Rylands for critical comments and a revision of an earlier text.

# 8

# Flexibility and co-operation as unifying themes in *Saguinus* social organization and behaviour: the role of predation pressures

*Nancy G. Caine*

## Introduction and historical survey

For most of the history of behavioural primatology, New World monkeys were treated as the poor cousins of the Old World monkeys and apes. Despite some important and intriguing early research on captive (e.g. Hampton *et al.* 1966) and free-ranging (e.g. Carpenter 1934) platyrrhines, the data from these investigations figured only rarely in the development and testing of theory in primate behaviour. Considered to be primitive and simple, and difficult to observe in the wild, New World monkeys were not regarded as good candidates for most behavioural research. It seemed to be the general consensus (with important exceptions) that there was little to be gained from studying social organization and behaviour in species that were far removed from the human lineage in terms of both time and space (Kinzey 1986).

Around 1980, investigations of New World monkeys began to appear in the published literature with much greater regularity than before. A number of field studies in Central and South America were initiated, and the data began to accumulate. Far from being dull and predictable, each genus of platyrrhine seemed to offer some surprise: female group transfers, mixed species associations, tool use, co-operative breeding, and more (Strier 1990). Recognizing the role that these new data were coming to play in the development of theory in primatology, the International Primatological Society dedicated its 1988 Congress to 'New World Primates, New Frontiers'.

From the beginning, the Callitrichidae was the more maligned of the two New World families. Being small and 'squirrel-like', the five callitrichid genera received even less attention than the cebids, among which the squirrel monkeys (*Saimiri*), at least, were relatively well known. Some intriguing field work prior to 1980 (e.g. Moynihan 1970) attracted the attention of a small group of primatologists who had interests in tamarins and marmosets, but it was with the publication of *Five New World Primates* (Terborgh 1983), in which the behaviour of two species of free-ranging tamarins (*Saguinus imperator* and *S. fuscicollis*) was described in detail, that a wider audience was attracted. When Sussman and Kinzey (1984) and Terborgh and Goldizen (1985) suggested that tamarins practice a variable, co-operative mating system that can even include polyandry, the old image of the primitive, simple callitrichid was dealt its final blow (see also Garber *et al.* 1984). In its place was an incomplete but exciting new framework for understanding callitrichids, one that focused on behavioural ecology and communal relationships within groups.

It may or may not be coincidental that, at about the same time that primatology began to take notice of New World monkeys, it also began to reopen the issue of predation as an important selection pressure on primate social behaviour. Considered by many to play a crucial role in the evolution of animal sociality, predation was nonetheless largely ignored by primatologists in the 1960s and 1970s, when discussions of intragroup phenomena such as social rank and communication dominated the literature. In the 1980s important new dialogues opened on the relative roles of inter- and intragroup competition and safety from predators in the determination of group size and structure (e.g. Wrangham 1980; Terborgh and Janson 1986). These discussions followed an enormous amount of research in optimal foraging theory applied to the behaviour of birds, rodents, and other animals (cf. Stephens and Krebs 1986). Terborgh's (e.g. 1983) and Goldizen's (e.g. 1987*b*) observations of the tamarins in Manu Park, Peru, led them to conclude that the natural history of tamarins has been and continues to be shaped in a significant way by predation pressures. Others are less certain of the extent of predation's role in callitrichid social evolution (e.g. Garber 1988*a*), but no one dismisses it.

In this chapter I begin by providing support for two assertions about tamarin (genus *Saguinus*) social behaviour. These are (1) that the fundamental elements of tamarin social life are co-operation, tolerance, and adaptability, and (2) that predation is among the most important, if not the most important, selection pressure influencing the social behaviour and group structure of tamarin species. In concluding my paper I present the argument that the co-operative, adaptable behaviour displayed by tamarins is best understood as a response to high predation risks and the consequent dependence upon group mates.

## Variability in the social life of *Saguinus*

The genus *Saguinus* includes at least 11 species and some 33 forms (Hershkovitz 1977). Tamarins range from Panama to central Brazil (the tamarin genus *Leontopithecus* is endemic to southeastern Brazil), and may be found in areas of primary or secondary growth. Many show a preference for edge habitat. Some of the differences in microhabitat and macrohabitat are at the level of species; others are at the level of subspecies or even group (e.g. Dawson 1978). Sussman and Kinzey (1984) conclude that the ability to exploit a variety of habitats 'is the hallmark of callitrichid ecology'. (p.425)

Comprehensive field studies of the social behaviour and ecology of tamarins are rare, and even with the recent proliferation of investigations there is still a woeful lack of data from the wild. This, more than anything, continues to hamper our understanding of tamarins. Table 8.1 shows those species for which there are field data regarding social behaviour. The best data come from four species: *S. fuscicollis*, *S. mystax*, *S. oedipus*, and *S. geoffroyi*. There are some data, but mostly regarding mixed-species associations, for *S. labiatus* and *S. imperator*. For species such as *S. inustus*, *S. bicolor*, and *S. leucopus* there is virtually nothing published of which I am aware. The limitations of even some of the better field studies include lack of individual identification, possible multiple sampling of the same groups, and lack of data on genetic relatedness. This final weakness is particularly serious in light of the continuing interest in polyandrous matings and co-operation among males in tamarin groups. A critical question, of course, is whether or not this tolerance can be explained by kin selection, an issue to which I will return later.

Controlled and quantified studies of tamarin social behaviour in captivity have also produced data on only a few species. *S. fuscicollis* and *S. oedipus* have been more extensively studied than their congeners; there are a few studies of *S. mystax* and *S. labiatus*. The most common topics of study in the laboratory have been vocal and olfactory communication, and issues relating to reproduction (e.g. reproductive suppression of females). Snowdon and his colleagues and students (e.g. Snowdon 1986) have provided us with a rich collection of work on callitrichid vocalizations, and Epple (1987*c*) has been producing data on tamarin (mostly *S. fuscicollis*) scent-marking behaviour for over two decades. Common in the literature are 'intruder studies', in which a strange conspecific is presented to one or more tamarins under controlled conditions (e.g. Epple 1978*b*; French and Snowdon 1981). The goal of these studies has been to elucidate the variables that regulate territoriality, adult pair bonds, and sexual attraction. Table 8.2 lists some other topics that have received attention in laboratory studies.

We have learned a great deal from captive studies, including the fact that there is surprising behavioural variability within and across species. One must be aware that differences in housing, husbandry routines, and experimental protocols undoubtedly contribute to variability in results across captive studies (Kleiman 1985). However, the observed variability

**Table 8.1** Field studies of the social behaviour of *Saguinus* species. This list does not include some ongoing or unpublished investigations, nor does it include studies that were primarily census-taking or non-social (e.g. feeding habits) in orientation. Some of the time frames and group sizes are estimates. A number of the authors cited have multiple publications derived from their respective field studies; I have made reference only to representative work.

| Species | Details of study duration | Reference |
|---|---|---|
| *S. geoffroyi* | Nineteen (?) months. Five groups of individually marked animals | Dawson (1978) |
| | Intermittent observations of 35 captive and 125 wild tamarins over the course of nine years | Moynihan (1970) |
| | Four months. Three groups | Lindsay (1980) |
| *S. fuscicollis* | Over five years on several marked and unmarked groups | Terborgh/Goldizen (e.g. 1983; 1987*a*) |
| | One group for 15 months | Soini (1987) |
| | 156 days. One group intensively; 13 others less intensively | Yoneda (1984) |
| | One mixed species group (*S. labiatus* and *S. fuscicollis*) for 10 months | Heymann (1990*a*) |
| | One mixed-species (*S. mystax* and *S. fuscicollis*) group for three months | Garber (1988*a*) |
| | Two mixed-species (*S. mystax* and *S. fuscicollis*) groups for 13 months | Norconk (1990) |
| *S. mystax* | Three months, ? groups | Castro and Soini (1978) |
| | One group for 27 days; 18 others less intensively | Garber *et al.* (1984) |
| | Not specified | Ruth (1987) |
| *S. nigricollis* | Eighty-eight days. Ten groups | Izawa (1978) |
| *S. imperator* | Fifteen months. Primarily one group (social behavior not emphasized) | Terborgh (1983) |
| *S. labiatus* | Five month study of *S. labiatus-S. fuscicollis* associations | Pook and Pook (1982) |
| | Six months. Ten groups | Yoneda (1981) |
| | Six-month study of *S. labiatus-S. fuscicollis* associations | Buchanan-Smith (1990) |
| *S. oedipus* | Two years. Three groups | Neyman (1978) |
| | Two trapped and released groups | Savage *et al.* (1990) |
| *S. midas* | One week | Thorington (1968) |
| *S. bicolor*, *S. leucopus*, *S. inustus* | None published (?) | |

**Table 8.2** Examples of behavioural topics studied in captive groups of *Saguinus*

| Species | Behaviour | Reference |
|---|---|---|
| *S. fuscicollis* | Scent marking | Epple and Smith (1985) |
| | Intruder studies | Epple (1978*b*) |
| | Reproductive suppression | Epple and Katz (1984) |
| | Vocalizations | Moody and Menzel (1976) |
| | Cognition | Menzel and Menzel (1979) |
| *S. oedipus* | Scent marking | French and Cleveland (1984) |
| | Intruder studies | French and Snowdon (1981) |
| | Reproductive suppression | French *et al.* (1984) |
| | Vocalizations | Cleveland and Snowdon (1982) |
| | Parenting | Tardif (1984) |
| | Mating strategies | Price (1990*b*) |
| | Social organization | McGrew and McLuckie (1986) |
| *S. labiatus* | Vigilance | Caine (1984, 1986, 1987) |
| | Vocalizations | Masataka (1987) |
| | Infant care | Pryce (1988) |
| | Aggression | Schaffner (1991) |
| *S. mystax* | General social behaviour | Box and Morris (1980) |
| | Mating systems | Malaga (1985) |
| *S. geoffroyi* | Communication | Moynihan (1970) |
| | Male–female relations | Skinner (1986) |
| *S. midas* | General social behaviour | Omedes and Carroll (1980) |
| *S. imperator* | Dominance | Knox (1990) |
| *S. inustus*, *S. leucopus*, *S. bicolor*, *S. nigricollis* | No published data (?) on social behaviour in captivity | |

may arise more fundamentally from individual adaptability. Before I return to this issue, I shall describe some of the behavioural variability that has been recorded in studies of captive and free-ranging *Saguinus*.

Until very recently, monogamy was assumed to be the keystone of tamarin (indeed, all callitrichid) social organization. Most of the social behaviour observed in tamarins was presumed to be a consequence of their monogamous lifestyle. There was evidence as early as the mid 1960s that monogamy was perhaps not the best descriptor of tamarin social and sexual life, yet most investigators focused on the aspects of the data that fit the pre-existing paradigm. In part the fault lay in the data themselves; many of them could be interpreted within the framework of more than one conceptual scheme. For example, in captivity, trios of *S. fuscicollis* (e.g. Epple 1972), *S. oedipus* (e.g. Hampton *et al.* 1966), and *S. mystax* (Malaga 1985), in which there are two males and one female, usually do very nicely. The female commonly copulates with both males and, while male–male fights do sometimes occur, the trios are often very stable and non-aggressive. This would appear to be polyandry. But there is frequently a stronger relationship (i.e. more interactions of various sorts, including copulations) between the female and one of the two males (e.g. Epple 1972; see also Soini 1987*b*, on wild *S. fuscicollis*), a situation that more closely resembles monogamy. In the wild, group compositions that fail to conform to a monogamous pair model were detected 10 years ago by Neyman (1978) and Dawson (1978) for *S. oedipus* and *S. geoffroyi*, respectively. Both investigators found a great deal of change in group membership over relatively short periods of time, with both males and females migrating. This would seem to indicate something other than an extended family lifestyle, but it is possible that the migrant adults were offspring of a monogamous pair who were in transitional periods of group membership. Recent data from *S. fuscicollis* (Terborgh and Goldizen 1985) and *S. mystax* (Garber *et al.* 1991) shed doubt on the proposition that the immigrants are always related, but we need much more data on genetic relationships to be certain. Also notable is the fact that, for most species in which multiple adults have been observed in a group, the ratio of adult males to adult females is greater than 1 (e.g. Terborgh and Goldizen 1985 for *S. fuscicollis*; Neyman 1978 for *S. oedipus*; Garber *et al.* 1991 for *S. mystax*). This ratio does not always exceed values that would be predicted by chance, but, significantly, it is usually not less than 1 (but see Ramirez 1984 for *S. mystax*).

In captivity, *S. fuscicollis* (e.g. Epple 1978*b*) and *S. oedipus* (French and Snowdon 1981) pairs may behave aggressively to novel conspecifics during intruder studies. This would be the predicted response if monogamy is assumed. But among *S. oedipus* it is only the males who react with aggression, and only towards other males. *S. oedipus* females anogenital mark with great frequency at these times, but it is unclear exactly what such marking means (French and Snowdon 1981; Harrison and Tardif 1989). In *S. fuscicollis* the response to an intruder is generally agonistic and sexually monomorphic, but there is also a tremendous amount of variability in the response across individuals (Epple 1978*b*). In *S. labiatus*, pair–pair introduction studies by Coates and Poole (1983) were uneventful. Wild tamarins of at least four species (*S. fuscicollis*, *S. oedipus*, *S. geoffroyi*, *S. mystax*) have been observed to come and go from troop to troop with frequency and apparent ease (see below). Hence the notion that aggression toward potentially competitive adults maintains monogamous pair-bonds is, at the very least, an oversimplification, although the intruder study data have important implications for species differences and for our understanding of intrasexual competition (French 1986).

In *S. fuscicollis* (Epple and Katz 1984) and many, but not all, groups of

*S. oedipus* (French *et al.* 1984; Tardif 1984), daughters do not cycle when housed with their mothers. In *Leontopithecus rosalia*, daughters cycle but do not copulate (French and Stribley 1987). Once removed and paired with a male, however, they rapidly become fertile. This would seem to reduce incest and mother–daughter competition, thereby promoting the stability of the family. But it is also consistent with a polyandry interpretation, especially when considered in light of the fact that there is no known comparable endocrine suppression for tamarin males (Sussman and Garber 1987). Furthermore, in the wild only one female per group is reproductive. This is true of all species for which there are relevant data, although there are also occasional (and important) exceptions to this general rule (Terborgh and Goldizen 1985 for *S. fuscicollis*; Ramirez 1984, Garber *et al.* 1991 for *S. mystax*).

Finally, there is extensive paternal care in all species of tamarins studied to date, with the possible exception of *S. nigricollis* (Izawa 1978). Paternal care is a typical feature of monogamy (Kleiman 1977). But the preponderance of more than one male per group and the extensive infant care by each and all of these males mimics the co-operative breeding observed in some bird species (cf. Goldizen and Terborgh 1989). Unless this is simply a matter of brothers sharing a limited resource, the willingness of tamarin males to coexist peacefully in the company of a single reproductively active female tests the limits of current theory in primate sociobiology.

In actuality, making a choice among the various interpretations regarding tamarin mating systems is probably unnecessary, because the list of *Saguinus* species in which some combination of monogamy, polygyny, and polyandry are all observed is likely to grow. The evidence regarding tamarin mating patterns is, therefore, becoming weighted in favour of a variable breeding system (Goldizen 1987*a*). It appears that tamarins can be monogamous, polygynous, or polyandrous, depending on some combination of social and ecological factors that is as yet not well understood.

The variability that we are now recognizing in the mating systems of tamarins is also characteristic of individuals within groups. On a range of traits from ovarian cyclicity (e.g. Brand and Martin 1983 for *S. oedipus*) to reactions to intruders (Epple 1987*b* for *S. fuscicollis*) to infant care (e.g. Pook 1984) there are marked differences among individual monkeys. Callitrichid researchers commonly appear to accept this variability as a fact of tamarin life, but remarkably little attention has been paid to its significance. Table 8.3 shows some of the studies for which individual and or group variability was explicitly mentioned by the author(s) as being striking. Clearly, then, we must conclude that one characteristic of the social behaviour of *Saguinus* species is adaptability. Apparently, individuals and groups can adapt to dynamic social and ecological factors in such a way that very different social patterns emerge. Species differences are

**Table 8.3** Studies that make explicit reference to individual and/or group variability in the social organization or behaviour of *Saguinus* species

| Species | Subject | Captive/wild | Reference |
|---|---|---|---|
| *S. fuscicollis* | Responses to novel conspecifics | captive | Epple and Alveario (1985) |
| | Group size and composition | wild | Terborgh and Goldizen (1985) |
| *S. mystax* | General social behaviour | captive | Box and Morris (1980) |
| | Group size and composition | wild | Garber *et al.* (1984) |
| *S. labiatus* | Vigilance | captive | Caine (1984) |
| | Group size | wild | Freese *et al.* (1982) |
| *S. oedipus* | Exploration of new 'territories' | captive | McGrew and McLuckie (1986) |
| | Anogenital marking | captive | French and Snowdon (1984) |
| | Group size and composition | wild | Neyman (1978) |
| | Mating behaviour | captive | Brand and Martin (1983) |
| | Infant care | | Tardif *et al.* (1990) |
| *S. geoffroyi* | Group size | wild | Moynihan (1970) |
| *S. inustus* | Group size | wild | Freese *et al.* (1982) |
| *S. nigricollis* | Group size and composition | wild | Izawa (1978) |
| *S. imperator*, *S. leucopus*, *S. bicolor*, *S. midas* | No relevant data sources (?) or not enough samples to generalize | | |

blurred by intraspecific group differences, and consistent group differences, if any, can be overwhelmed by individual differences (French and Snowdon 1984).

## Co-operation and tolerance: recurring themes

What, then, constrains a tamarin's social behaviour? That is, what limitations are there on its otherwise flexible sociality? Can we identify the proximate and ultimate mechanisms that operate on its development and adult expression?

A trait that characterizes all *Saguinus* species studied to date is 'co-operative' (Neyman 1978; Soini 1987*b*; Goldizen 1988; Snowdon and Soini 1988). Perhaps foremost on the list of the many co-operative behaviours exhibited by tamarins is infant care. In the species studied to date, all adult members of *Saguinus* groups carry infants, usually from within a few hours or a few days of birth. Mothers typically carry infants relatively infrequently, sometimes taking them back only to nurse. In *S. labiatus*, group members actually compete to carry infants (Pryce 1988), and individuals will follow one another around in a persistent attempt to attract the infant(s) to them (personal observation).

There appear to be substantial costs associated with carrying an infant or, even more so, twin infant tamarins. There is significant metabolic expense when carrying so much weight (Kirkwood and Underwood 1984; Goldizen 1987*a*), and there is reduced mobility, especially when carrying twins, that may interfere with travel (Price in press *a*). In *S. oedipus* (Price in press *a*), *S. labiatus* (Caine personal observation) and *S. fuscicollis* (Goldizen and Terborgh 1986), tamarins eat less when they are carrying infants.

Not only do tamarins cooperate to carry infants, but they also allow infants to take food from them (e.g. Caine personal observation for *S. labiatus*; Izawa 1978 for *S. nigricollis*; Terborgh and Goldizen 1985 for *S. fuscicollis*). In some species, adults actively offer food to infants (e.g. Feistner 1984 for *S. oedipus*). Food sharing by *S. oedipus* adults is not motivated by satiation or lack of interest in the food; Feistner and Chamove (1986) found that the food items most likely to be shared with infants are items that are also most preferred by adults. In addition, the data of Mayer *et al.* (1992) on foraging styles indicate that captive tamarin (*S. labiatus*) adults are far more likely than squirrel monkeys (*Saimiri*) to tolerate each others' presence at a newly discovered food source.

Infant care is not the only realm in which tamarins display co-operative behaviour. For example, when travelling and foraging, group members move in a cohesive manner (e.g. Izawa 1976 for *S. nigricollis*; Goldizen

1987*b*, Yoneda 1984*b* for *S. fuscicollis*; Lindsay 1980 for *S. geoffroyi*; Garber 1988*a* for *S. mystax*) and, at least in *S. labiatus*, individuals take turns acting as sentinels for one another (Zullo and Caine 1988). In *S. labiatus*, food calls are given when particularly palatable food is found, even if only in small amounts (Addington *et al.* 1991). Presumably, these calls attract group mates to the food source. At least some species of tamarins also produce 'monitoring' calls (sometimes called cohesion calls or contact calls) (Dawson 1978, Lindsay 1980 for *S. geoffroyi*; Neyman 1978 for *S. oedipus*; Moody and Menzel 1976 for *S. fuscicollis*). According to Caine and Stevens (1990), these are calls that are emitted in no one particular context and elicit no one particular response. Rather than serving to directly regulate space, monitoring calls allow individuals to keep track of the general whereabouts of their group mates, thereby maintaining intragroup cohesiveness and permitting co-operative ventures (such as vigilance or transferring an infant, in the case of tamarins). Caine and Stevens (1990) found that, in captive *S. labiatus*, monitoring calls are given almost continuously throughout the day, ceasing only when all members of a group have entered their nest box for the evening. Rates of monitoring calls increased when one member of the group was temporarily out of sight and when the possibility of losing visual contact with a group member was experimentally increased.

High rates of aggression and strictly enforced status relationships need not preclude co-operation within groups, and may even be a prerequisite for some sorts of co-operation in some species (e.g. the progression orders of baboons; see Rhine and Westlund 1981). However, aggression and status enforcement hamper the sharing of social roles and pre-empt other, more affiliative behaviours. Aggression within free-ranging groups of tamarins is rare, a fact that has been noted by numerous field researchers (see Table 8.4). In captivity, aggression does occur, but it tends to 'erupt' in punctuated bouts, e.g. between juvenile twins (personal observation of *S. labiatus*; see also Rothe *et al.* 1988 for *Callithrix jacchus*) or between mother and daughter (Epple 1975*a*, *S. fuscicollis*), that are preceded by or interspersed with much longer periods of peaceful interaction. A four-month study of aggression in four groups of captive *S. labiatus* showed no cases of injurious aggression in almost 176 group-hours of observation. Most of the aggression that was observed lasted for less than 20 seconds and consisted only of head shakes and teeth chatters (Schaffner 1991).

Dominance hierarchies within tamarin groups are usually indiscernible by conventional means, apart from the priority of access that one female (and her mate, in monogamous family groups) may have to copulation opportunities. Knox (1990), in a careful analysis of agonistic behaviour in *S. imperator*, concluded that hierarchical rank structures do exist, but do not adequately characterize *S. imperator* social organization. Caine

**Table 8.4** Explicit references to a general lack of aggression within *Saguinus* groups. Many field and laboratory investigations fail to mention within-group aggression at all, perhaps because it is so infrequently observed. In captivity, punctuated bouts of aggression are exhibited by some individuals during 'intruder' studies; 'expelled' individuals may also suffer aggression in the laboratory

| Species | Captive/wild | Reference |
|---|---|---|
| *S. fuscicollis* | wild | Goldizen (1989) |
| | referring to interspecific interactions within mixed groups of wild *S. labiatus* and *S. fuscicollis* | Buchanan-Smith (1990) |
| *S. geoffroyi* | wild | Moynihan (1970) |
| *S. mystax* | captive | Box and Morris (1980) |
| | wild | Heymann (personal communication) |
| *S. labiatus* | captive | Coates and Poole (1983) |
| | captive | Schaffner (1991) |
| *S. oedipus* | captive | Price and Hannah (1983) |
| *S. nigricollis* | wild | Izawa (1978) |
| *S. midas* | captive | Omedes and Carroll (1980) |
| *S. imperator*, *S. inustus*, *S. leucopus*, *S. bicolor* | No data available (?) on within-group interactions | |

(unpublished data) found no correlation between the orders with which individuals in *S. labiatus* groups took a prized piece of food on consecutive days. Furthermore, the fact that so many social tasks are shared by all members of a group (e.g. infant care and vigilance) suggests a certain egalitarianism that is unusual in primates.

De Waal (1986) argues that, after a bout of aggression, primates must engage in reconciliatory behaviour in order to reaffirm status relationships. However, a study of post-conflict behaviour in *S. labiatus* revealed no evidence of reconciliation following aggression (Schaffner 1991). If de Waal is correct in assuming that reconciliation is primarily related to the reaffirmation of relative social status, the absence of such behaviour in red-bellied tamarins suggests a social organization based on non-hierarchical relationships with groupmates.

Despite any quarrels that have taken place during the day, all members of a tamarin group pack themselves into a tight ball, limbs entwined, to sleep (Caine *et al.* 1992). This is true for all species and all groups (in the field or in captivity) in which tamarin sleeping habits have been observed and discussed. In my experience with *S. labiatus*, only if an individual is undergoing a true group expulsion is it not included in the huddle. Allowing such close physical contact is probably indicative of the generally tolerant nature of tamarin social interactions.

Tamarins are also very co-operative with respect to anti-predator behaviour. The use of sentinels has been documented in captivity (Zullo and Caine 1988 for *S. labiatus*) and noted in the wild (Moynihan 1970, Dawson 1978, Lindsay 1980 for *S. geoffroyi*; Goldizen 1987*b* for *S. fuscicollis*). Bartecki and Heymann (1987) witnessed a *S. fuscicollis* group co-operatively mobbing a snake. High levels of visual vigilance are frequently reported by field researchers (e.g. Goldizen 1988 for *S. fuscicollis*), and there is some evidence for a co-operative division of roles with regard to surveillance of the environment (e.g. Caine 1986). More will be said about anti-predator adaptations below.

Finally, we know that tamarins cooperate not only intraspecifically, but interspecifically. *S. fuscicollis* (Terborgh 1983), *S. imperator* (Terborgh 1983), *S. mystax* (Castro and Soini, 1978), and *S. labiatus* (Yoneda 1981) all form mixed species associations in which the two groups co-ordinate activity and movement, sometimes with remarkable precision and persistence. Certain long-distance vocalizations are used to locate the corresponding groups in the morning and territorial interactions with neighbouring groups include both species (each interacting with its intraspecific rival; Buchanan-Smith 1990, *S. labiatus* and *S. fuscicollis*). Buchanan-Smith (1990) observed joint mobbing of a tayra (*Eira barbara*) by *S. fuscicollis* and *S. labiatus*. There is a good deal of unresolved speculation regarding the function of these interspecific associations, but there is no dispute over the efficiency of the proximate mechanisms that co-ordinate the associations (Norconk 1990).

What is perhaps most remarkable about the co-operative behaviour of tamarins is that we now have reason to believe, as mentioned above, that co-operation can take place outside the context of kin selection or perhaps even without the benefit of long-term associations. Tamarins of at least four species have been observed to enter and exit groups routinely, and, in at least five species, temporary intraspecific mergers of tamarin groups have been observed (Izawa 1978 for *S. nigricollis*; Thorington 1968 for *S. midas*; Castro and Soini 1978 for *S. mystax*; Moynihan 1970 for *S. geoffroyi*; Soini and Soini 1982, cited in Snowdon and Soini 1988 for *S. fuscicollis*). These observations suggest that the degree of relatedness within tamarin groups might be relativly low, although we need more data on genetic relatedness to be sure (Garber *et al.* 1991). Harrison and Tardif (1988) found attenuated aggression between related males, relative to unrelated males, in an intruder study of *S. oedipus*. The authors conclude that kinship may underlie co-operative behaviour among males in free-ranging populations. Accordingly, polyandrous matings might be mostly fraternal, but Terborgh and Goldizen (1985), at least, have observed otherwise. More recently, Savage *et al.* (1990) found unrelated (determined through genetic analyses) individuals in both of the groups of wild *S. oedipus* that they examined. Likewise, engagement in infant care appears not to be limited to parents and siblings, since all adult males and females in a group will carry babies. Furthermore, and perhaps most importantly, even in primate species where within-group degree of relatedness is known to be high (e.g. Hylobatidae), the range and extent of co-operation, tolerance, and social adaptability does not equal that of tamarins (cf. Leighton 1987).

## Predation pressures and the evolution of *Saguinus* sociality

What selection pressures have led to the co-operative nature of tamarins? To answer this question one must first distinguish the terms 'social' and 'co-operative'. There are significant benefits associated with being social (i.e. living in a group), including predator detection and avoidance, access to mates, and defence of resources. Each of these advantages of group life probably applies to tamarins. However, a species can be social without being especially co-operative. Furthermore, many species exhibit co-operation in particular contexts (e.g. territorial defence) without behaving co-operatively in others (e.g. infant care). The best functional explanations for tamarin sociality will be those that account for the fact that cooperativeness takes a generalized form in *Saguinus*.

I propose that the ability and willingness of tamarins to behave co-operatively is largely a consequence of predation pressure. Predators are ubiquitous in the habitats of tamarin species. While anecdotal reports of predation on tamarins are all we have in most cases, such anecdotes are remarkable for their consistency across and within species. Nowhere have I found an exception to the observation that tamarins are virtually terrified of raptors flying overhead (e.g. Moynihan 1970, Lindsay 1980 for *S. geoffroyi*; Soini 1987*b* for *S. fuscicollis*; Hampton *et al.* 1966 for *S. oedipus*). Goldizen (1987*b*) reports one raptor attack per week per tamarin group at Manu Park. Heymann (1990*b*) studied the responses of three groups of *S. mystax* and *S. fuscicollis* to avian predators. He found that there were alarm events every 2–3 hours, over half of which were in response to raptors. Most of the responses to avian threats were 'strong': rapidly retreating to a lower level of the tree, running from the periphery of a tree to the tree trunk, hiding under leaves, or, most dramatically, falling to the ground. Following an observed attack by an ornate hawk-eagle (*Spizaetus ornatus*), the study group remained still for almost four hours. For several days following the attack the monkeys stayed lower than usual in the trees and appeared to be nervous.

In addition to predation by birds, there are sightings of tamarins being eaten by snakes (Heymann 1987 for *S. mystax*), tayras, and ocelots (cf. Goldizen 1987*b*). Caine and Weldon (1989) found that captive *S. labiatus* are able to discriminate between the fecal scents of predatory and non-predatory Neotropical mammals. Scents from felids elicited avoidance and, in some cases, alarm calls. In response to predators tamarins may mob (Bartecki and Heymann 1987 for *S. fuscicollis*; Moynihan 1970 for *S. geoffroyi*), freeze (e.g. Izawa 1978 for *S. nigricollis*) or alarm call. At least two species of tamarins have different alarm calls for terrestrial and aerial predators (Epple 1975*a* for *S. fuscicollis*; Neyman 1978 for *S. oedipus*).

When retiring to the sleeping site each evening the tamarins employ a variety of adaptations to reduce vulnerability to nocturnal predators. These include torpor (Moynihan 1970 for *S. geoffroyi*), retirement before dusk (Yoneda 1981 for *S. fuscicollis* and *S. labiatus*), increased vigilance (Caine 1984, 1987 for *S. labiatus*) and careful selection of sleeping sites (Dawson 1979 for *S. geoffroyi*; Neyman 1978 for *S. oedipus*; Caine 1990 for *S. labiatus*). Caine *et al.* (1992) found that captive *S. labiatus* chose sleeping boxes that were most concealed, highest from the floor, and covered from above. When forced to sleep in an unpreferred type of box, the tamarins significantly increased their rates of vigilant scanning. As Sussman and Kinzey (1984) put it, 'the roosting habits of *Saguinus* seem to be predator-proof'. (p.33).

While behaviours such as mobbing certainly require co-ordinated effort, I believe that it is vigilance for predators, rather than response to them, that demands and has promoted co-operative social behaviour in tamarins. Behavioural biologists have reasoned that the simplest way that group life promotes anti-predator benefits is by providing more eyes and ears to detect danger. A number of studies of bird flocks have shown that individual rates of vigilance are reduced when there are group mates who are also being vigilant (e.g. Powell 1974). Simply put, individuals can spend more time eating and less time looking around when responsibility for vigilance is shared. Of some interest here is a test of Pulliam's (1973) mathematical model of optimal foraging by Elgar and Catterall (1981). These authors suggest that, at least for house sparrows (*Passer domesticus*), the benefits increase up to a group size of about seven animals; after that there is little additional gain, and there may be costs associated with increased feeding competition. Tamarins are rarely observed as loners, and average group sizes rarely exceed more than seven or eight; a number that has also been proposed, however, in relation to efficient use of food resources (Pook 1984).

Scanning the environment for predators need not be highly co-ordinated within groups, but it certainly can be. In my studies of captive *S. labiatus* (Caine 1984, 1987) I have found remarkable levels of vigilance in the form of visual scanning of the environment. In the presence of a mildly threatening object that was partially hidden the tamarins performed 'vigilance checks' (Caine 1986). These checks were performed by all group members but only one tamarin checked at a time, thereby minimizing redundant effort. Red-bellied tamarins also use sentinels under a variety of circumstances, including foraging (Zullo and Caine 1988). While comparable controlled studies have not been performed using other *Saguinus* species, similar claims for scanning and sentinel use have been made in a number of field- and laboratory-based publications (Moynihan 1970, Dawson 1978 for *S. geoffroyi*; Goldizen 1987*b* for *S. fuscicollis*).

It is probably safe to say that co-operation, whether for the sake of infant care, vigilance, or foraging, requires some degree of tolerance; tolerance, in turn, demands some behavioural flexibility. Given its likely susceptibility to predators, the worst scenario for a tamarin might well be to find itself alone, or unintegrated into a group. In my experience with captive *S. labiatus*, lone adults accept an unfamiliar mate almost instantly. In one case a female I acquired who had lived alone and in sensory isolation from conspecifics her entire adult life (eight years) rested in full body contact with her new mate only hours after he was introduced to her. When I introduced two subadult females to an unrelated and unfamiliar lone adult female there was not only no aggression or obvious discomfort among them, but the three assumed the classic tamarin sleeping huddle on the first night following their introduction. Goldizen (1987*b*) writes that, 'during several months, a *S. imperator* group consisting of only one male and one female shared its territory with an *S. fuscicollis* group. The female *S. imperator* was quite habituated to the observer's presence, while the male was not. When the *S. fuscicollis* group was being observed, the female *S. imperator* normally traveled in the midst of the *S. fuscicollis* group, while her mate remained 50 to 100 meters away, apparently too scared to approach more closely.' (p.37). This female, it seems, was reluctant to leave the safety of the larger group, even if it meant choosing another species over her own mate. A group of tamarins might admit new members, even at the expense of some mating competition, in order to reap the benefits of additional vigilance. This, in turn, requires social adaptability, and it may account for the differences we have observed between and within individuals, and the corresponding variability in group structure within species.

The adaptability of tamarins has primarily been discussed in the context of variable matings systems (see above). Goldizen and Terborgh (1986) have provided a hypothesis that predicts shifts in breeding demography over time. They argue that tamarins can be monogamous or polyandrous, depending on the number of adult offspring available to help with the infants. That is, given that a female alone cannot eat enough to carry and nurse two babies by herself, she needs at least a mate and probably other helpers, too. If there are older siblings, they can be the 'sitters' and the adults can be monogamous. If not, the female and her mate will accept additional adult males, all of whom copulate with the female (see also

French and Inglett 1989 on *Leontopithecus*). Given the chance that any infants could be his, each adult male will invest in the offspring. Additionally, because breeding vacancies for males (and females, for that matter) are probably rare, sharing a female with one or more males is an acceptable compromise (Goldizen and Terborgh 1989).

Ferrari and Lopes Ferrari (1989) compared the social organization and behaviour of marmosets (*Callithrix*) and tamarins. Marmosets, by virtue of specialized dentition, are consummate gum feeders, which provides them with a relatively stable, predictable food source year-round. Tamarins consume gums, but cannot easily initiate gum glow (Garber 1984*b*), and thus their feeding ecology is more opportunistic. Compared to tamarins, marmosets have smaller home ranges, larger group sizes, and less variable social structures. Tamarins, according to Ferrari and Lopes Ferrari, demonstrate variable group size, group composition, mating behaviour, etc., primarily because their food sources are unstable.

In Ferrari and Lopes Ferrari's account, feeding ecology drives tamarin sociality toward flexibility, but this explanation cannot by itself explain the levels of co-operation that tamarins exhibit across social settings. In Goldizen and Terborgh's (1986) scenario, infant care requirements explain tamarin mating systems and account for co-operation and flexibility, too. But I would like to press the case a bit further. Goldizen and Terborgh (1986) found that tamarins do not usually eat while they are carrying an infant. Price (in press *a*) documented this phenomenon in captive *S. oedipus* and I have observed it in captive *S. labiatus* also. Given that it may be clumsy to eat with one or two heavy babies on one's back, the hesitancy to eat is still surprising, considering that carriers need to eat all the more if they are handling so much extra weight. What we can safely postulate is that a dorsally clinging baby is a prime target for a raptor; furthermore, the carrier, too, is particularly vulnerable given the increased weight and consequent lack of agility. Heymann (1990*b*) found that a free-ranging group of *S. mystax* responded with alarm calls to significantly more events after the birth of an infant. Boinski (1987*b*) found that squirrel monkey (*Saimiri oerstedi*) neonates are quite vulnerable to predation when riding on their mothers' backs (which they do almost constantly during the first two months of life). In eleven months she saw 41 predation attempts by toucans, eagles, and hawks; in two of the three successful attempts the victims were neonates. Thus a carrier must be particularly vigilant while charged with babies, and vigilance is incompatible with looking under leaves for insects or prying apart a fruit (Goldizen 1987*a*). Price (in press *a*) found that *S. oedipus* do not scan more when carrying than when not carrying, but she attributes this to the observation that carriers remain more quiet and hidden from view while carrying babies. In any case, there must be a number of carriers if everyone is to eat and the infants are to be properly cared for.

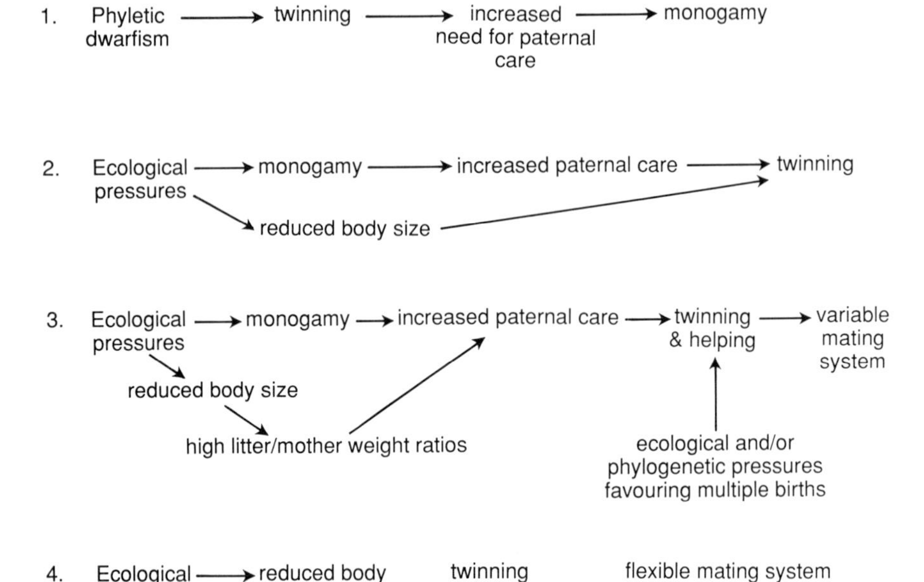


**Fig. 8.1** Models of the evolution of *Saguinus* social organization. Models 1, 2, and 3 are after Leutenegger (1980), Pook (1978*a*, 1984), and Goldizen (1990), respectively. In model 4, agreeing with Eisenberg (1978) and Rylands (1989*a*), I assume that the particular ecological pressures that selected for reduced body size in callitrichids are related to the occupation of an insectivorous niche. The evolution of twinning may be a strategy to maximize reproductive output in the face of heavy predation (Rylands personal communication), in contrast to the allometric scaling/obstetrical hypothesis forwarded by Leutenegger (1980). Rylands (1989*a*) further argues that opportunistic exploitation of secondary growth (Rylands 1986*b*, 1987) has driven callitrichids toward an 'r' strategy that includes polyandry. Garber *et al.* (1991), in contrast to many others (e.g. Goldizen 1990), propose that monogamy may not be an ancestral breeding condition, at least for *S. mystax*.

My argument is that, whereas the need for multiple infant carriers may be the basis for a variable social organization, the foundation of that need may be directly related to predation pressures. This view is presented in Fig. 8.1, where there are four possible models of the evolution of social behaviour in tamarins, the last of which represents my own position.

Along with their hypotheses regarding the evolution of variable mating patterns in tamarins, Goldizen and Terborgh (1989) also propose an explanation for the evolution of delayed breeding and care of young by older siblings within tamarin groups. They present evidence that breeding vacancies in tamarin populations are limited, and dispersal is risky. A sibling can increase its inclusive fitness under these conditions if it remains in the natal group and assists its relatives. This analysis is compatible with my own, insofar as the dangers of dispersal are largely a function of predation risks when travelling alone (cf. Cheney and Seyfarth 1983).

I believe that the need for co-operative vigilance also helps to explain the remarkable tolerance of adult males for polyandrous life. If polyandry is tolerated only because the parents need help in carrying babies, should we not find that, in a group with more than one adult male, one of the two or more males routinely ejects (or attempts to eject) the others from the group when an older sibling(s) is mature enough to help with the new babies? I have seen such behaviour reported only once (Vogt *et al.* 1978 for *S. fuscicollis*), although it may happen more often. Goldizen (1987*a*) does report a negative correlation between the presence of older offspring and number of mature males in *S. fuscicollis*. In captivity, *S. fuscicollis* males show less interest in strange females than they do in their female partners (Epple 1990), yet one would think that males might welcome opportunities to interact with strange females if mating competition is potentially high. One would also expect a great deal of competition among males for access to copulations, yet the data show little or no such competition in either captive or wild groups with more than one male (e.g. Epple 1972 for *S. fuscicollis*; Hampton *et al.* 1966 for *S. oedipus*; Goldizen 1987*a* for *S. fuscicollis*; Heymann personal communication for *S. mystax*; Caine personal observation for *S. labiatus*). Even females, who have a good deal to lose if they allow other potentially reproductively active females to enter or remain in their groups, sometimes do so (e.g. Goldizen and Terborgh 1986 for *S. fuscicollis*; Garber *et al.* 1984, 1991 for *S. mystax*). Thus it seems that there must be strong proclivities towards tolerance, even if it sacrifices reproductive success in the short run. My position is that co-operative tendencies that evolved in response to predation pressures may have aided in or even pre-adapted tamarins for the development of very flexible social systems, including systems permitting facultative polyandry.

There are, nonetheless, circumstances under which tamarins are notoriously intolerant. In the wild, territorial encounters between groups are usually quite aggressive. Similarly, as mentioned earlier, captive pairs of *S. fuscicollis* and *S. oedipus* are unfriendly to novel conspecifics in the context of experimental introductions. Captive *S. fuscicollis* females are also likely to aggress against their adult daughters, sometimes to the point of fatal injury (Epple 1975*a*). Hampton *et al.* (1966) reported mother–

daughter aggression in his group of *S. oedipus*, although Tardif (1983) and Savage *et al.* (1988) found no such overt fighting in their study groups, and I have observed no mother–daughter aggression in *S. labiatus*. Taken together these data indicate intrasexual competition that is at least occasionally strong, especially between females, and an integrity of group and space (cf. Terborgh and Stern 1987). These aspects of tamarin behaviour are hard to reconcile with reports that, in the wild, group transfers are easy and frequent, and aggression is rare within groups. Clearly we must learn more about the circumstances that influence and direct the establishment of relationships among individual tamarins. Undoubtedly the ages, parity, and relatedness of group members, as well as food availability, season, and other ecological variables, affect how, when, and if tamarins respond aggressively to extra- or intragroup individuals in the wild (Sussman and Garber 1987; French 1986). Species differences will undoubtedly emerge, as well. Already we know that *S. oedipus* and *S. fuscicollis* differ in response to intruders in captivity: *S. oedipus* males, for example, are more aggressive than *S. oedipus* females to intruders of the same sex (French and Snowdon 1981). This is not the case for *S. fuscicollis* (Epple 1978*b*). Perhaps the male intrasexual competition that is predicted by polyandry is more intense and hence more observable in *S. oedipus*. However, neither Neyman (1978) nor Hampton *et al.* (1966) reported male–male aggression in wild or captive groups of this species.

## Conclusions

My argument can be summarized as follows. There is a clear unifying theme of co-operation, tolerance, and flexibility in and across tamarin species. The observed behavioural variability across individuals, groups, and species is by no means random or unguided, however. Many of the differences are more quantitative than qualitative, and, more than anything else, they seem to serve the goal of cohesion and co-ordination within groups. I propose that the primary purpose of this adaptability is to permit coping with ecological and social vagaries in whatever way best provides protection from predators. Resource defence/procurement and infant care requirements must pose powerful selection pressures on tamarins, but predation pressure may be even more fundamental. In fact, predation pressure may be partially responsible for the degree of infant care that is required by tamarins. Anti-predator considerations may also impose constraints on foraging behaviour. For example, it is possible that tamarins cut short their foraging time in order to travel to safe sleeping sites before dusk (Moynihan 1970; Caine 1987).

I also contend that low rates of overt aggression, despite theoretically intense intrasexual competition, are related to predation pressures. In tamarins, reproductive competition seems to take on forms that minimize fighting. Reproductive suppression via endocrinological means, for example, effectively maintains the breeding rights of a single female without the duress or danger of open competition. Likewise, the overt conflicts that sometimes occur in captivity tend to be swiftly (if violently) resolved, often through rather sudden expulsion of a group member. For tamarins, a premium is placed on co-operation, which is in turn undermined by hierarchies and fights.

In conclusion, I propose that predation pressures have played a critical role not simply in determining that tamarins live in groups, as has been suggested by Terborgh and Janson (1986), but also in guiding the types of social interactions and relationships that have been observed within tamarin groups. It would be wrong (and indefensible) to argue that predation pressures have operated to the near exclusion of all others; selection surely operates multidirectionally. I do, however, believe that the role of predation has been underestimated in analyses of primate sociality at the level of within-group relationships and social processes. Tamarins provide a case in point.

## Acknowledgements

An earlier version of this paper was presented at the 1988 Congress of the International Primatological Society in Brasília, Brazil. I am grateful to Jeff French for filling in for me there. Anthony Rylands made useful suggestions regarding Fig. 8.1. Thanks to A.D.K. for allowing me essential albeit rare moments of uninterrupted work on this project.

# 9

# Communal infant care in marmosets and tamarins: relation to energetics, ecology, and social organization

*Suzette D. Tardif, Mary L. Harrison, and Mary A. Simek*

## Introduction

The need for helpers in infant care has been hypothesized as shaping the mating system and social structure in callitrichid primates. Recent field studies suggest that substantial variation may exist in mating patterns and social structure both within and between callitrichid species. Disagreements remain as to whether the modal mating system is monogamy or polyandry. However, the inability of a single adult to raise multiple offspring of a relatively high total litter weight by herself is frequently cited as causally related to both monogamy (Hershkovitz 1977; Leutenegger 1980) and polyandry (Garber *et al.* 1984; Terborgh and Goldizen 1985; Sussman and Garber 1987).

The energetic demands resulting from the rearing of two infants of a relatively large weight may include the increased foraging demands resulting from lactation as well as the energetic and mechanical burden that infant carrying represents. Goldizen (1987*a*) has proposed that, in the wild, the energetic demands of feeding and lactation in saddle-back tamarins will limit the mother's and father's carrying to approximately 20 per cent and 40 per cent respectively. Price (in press *a*) reports that, even in a captive setting, the behaviour of cotton-top tamarins which are carrying infants reflects probable costs associated with this activity; animals spent significantly less time feeding, foraging, moving, and engaging in social activities when they were carrying infants than when they were not.

Given the presumed importance of infant care in shaping social structure, we attempted to determine whether parameters of infant care vary across callitrichid species and, if so, how those differences may relate to differences in ecology and social organization.

We compared the following parameters of infant care across callitrichid species:

*Infant carrying*:
- duration of infant carrying;
- latency to first carry by a non-mother;
- the effect of helpers upon parental infant carrying.

*Infant provisioning*:
- identity of provisioners;
- type of food shared;
- the development of feeding independence.

Most of the data to be discussed were collected from the study of captive groups at the Marmoset Research Center in Oak Ridge, Tennessee, USA, and from published studies of infant care in other callitrichid colonies. For most parameters, data from four species were available: the common marmoset (*Callithrix jacchus*), the saddle-back tamarin (*Saguinus fuscicollis*), the cotton-top tamarin (*Saguinus oedipus*), and the golden lion tamarin (*Leontopithecus rosalia*). Data on additional species are presented when available.

## Infant carrying

Three features of infant carrying will be compared across species; these are duration of infant carrying, latency of carrying by non-mothers, and the effect of helpers on parental carrying of infants. Arguments regarding the relation of these features to energetics and feeding will be presented.

Figure 9.1 illustrates the percentage of time that infants were carried for weeks 4–10 for three species of tamarin and two species of marmoset. Data were obtained from the literature and by direct observation of three species at Oak Ridge. Data from multiple studies were combined, for *C. jacchus* (Ingram 1977; Locke-Haydon and Chalmers 1983; Tardif *et al.* 1986), *S. oedipus* (Cleveland and Snowdon 1984; Tardif *et al.* 1986), and *S. fuscicollis* (Vogt *et al.* 1978; Tardif, unpublished). One source was available for *L. rosalia* (Hoage 1978) and *C. argentata* (Buchanan-Smith 1984). Prior to week 4, infants are generally carried over 90 per cent of the time by all species. However, after that time, there are obvious species differences, with tamarins carrying infants for a larger proportion of time than do marmosets. Comparable published data on *Cebuella* were not available, but a study by Wamboldt *et al.* (1988) suggests that *Cebuella* infants are carried approximately 30 per cent of the time at week 6 and 20 per cent at week 8, a pattern similar to *Callithrix*.

Because the amount of time that infants are carried could be affected by a number of variables such as cage size and group composition, we examined infant carrying in two pairs of matched groups of each of two species

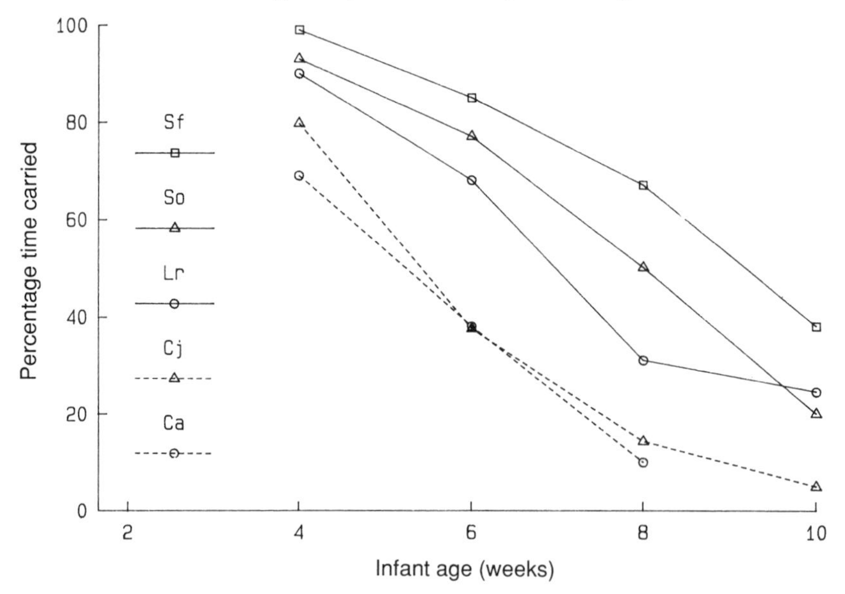


**Fig. 9.1** Percentage of time that infants were carried during weeks 4–10 for the following species: Sf, *Saguinus fuscicollis*; So, *Saguinus oedipus*; Lr, *Leontopithecus rosalia*; Cj, *Callithrix jacchus*; Ca, *Callithrix argentata*. All data were derived from studies of captive animals. See text for specific sources.

(*S. oedipus* and *C. jacchus*) and two matched trios of each of three species (*S. oedipus*, *S. fuscicollis* and *C. jacchus*) housed under identical conditions. Each pair or trio of groups was matched for group composition and history as closely as possible. Each group was observed for three 20-minute sessions per week for eight weeks to determine infant-carrying patterns. Further information on this study can be found in Tardif *et al.* (1986).

Table 9.1 indicates the percentage of time infants were carried in each group during weeks 3–4, 5–6, and 7–8. There was a great deal of inter-group variation; the group with the oldest helpers (Trio #4) carried the highest percentage of time. However, even with this variation, the tamarin species always carried infants a higher percentage of the time than did the marmosets during weeks 7–8; the carrying time of the two tamarin species was similar.

Data from field studies of *Leontopithecus rosalia* (A. Baker, personal communication) and *S. fuscicollis* (P. Garber, personal communication) indicate that in both these species, infants are carried for shorter periods of time in the wild than in captivity. Field studies of *S. oedipus* (A. Savage, personal communication) indicate similar patterns to captive *S. oedipus*. While we know that the actual length of time that infants are carried in the

**Table 9.1** Percentage of the time that infants were carried in matched comparisons of three species. Sf, *Saguinus fuscicollis*; So, *S. oedipus*; Cj, *Callithrix jacchus*

| Pair/ trio No. | No. of infants | Infant age (wks) | Percentage of time Infants Carried | | |
|---|---|---|---|---|---|
| | | | Sf | So | Cj |
| 1 | 2 | 3–4 | — | 100.0 | 95.5 |
| | | 5–6 | — | 80.3 | 57.5 |
| | | 7–8 | — | 52.4 | 5.4 |
| 2 | 1 | 3–4 | — | 100.0 | 94.4 |
| | | 5–6 | — | 68.8 | 12.4 |
| | | 7–8 | — | 32.4 | 0.6 |
| 3 | 2 | 3–4 | 100.0 | 100.0 | 100.0 |
| | | 5–6 | 90.2 | 94.8 | 22.8 |
| | | 7–8 | 66.1 | 81.6 | 3.8 |
| 4 | 1 | 3–4 | 100.0 | 100.0 | 100.0 |
| | | 5–6 | 91.1 | 100.0 | 97.6 |
| | | 7–8 | 61.0 | 72.5 | 34.8 |
| $\bar{X}$ | | 3–4 | 100.0 | 100.0 | 97.4 |
| | | 5–6 | 90.6 | 85.9 | 47.6 |
| | | 7–8 | 63.5 | 59.7 | 11.1 |

laboratory and in the field may not be equivalent, consistency of the pattern of carrying behaviour within species across captive conditions suggests that species differences will be confirmed in the field.

The significant differences in carrying behaviour observed in the laboratory might easily be explained by a corresponding difference in the growth-rate of infants. If, for example, *Callithrix jacchus* infants attain a significantly greater proportion of their adult weight earlier than do *Saguinus oedipus* infants, marmoset parents or helpers may cease to carry infants earlier because they are physically unable to continue. Difference in carrying times between callitrichid species have been noted by others, and it has been suggested that they reflect the slower maturation of the larger tamarin species and the faster maturation of the smaller marmoset species (Kleiman 1977; Leutenegger 1980).

Figure 9.2 provides a detailed comparison of growth in two species displaying differing periods of infant care. Figure 9.2a shows the actual growth curves based on body weight for the first 24 weeks of life in *Callithrix jacchus* and *Saguinus oedipus*. Raw growth data were taken from MRC animals, from literature reports on individual animals or groups of animals (Hershkovitz 1977; Hearn 1983), and from additional weights on hand-raised infants (G. Epple and L. Watson, personal communication). Mean weights for each species were estimated for target ages at four-week intervals by interpolation. To avoid the underestimation of actual mean

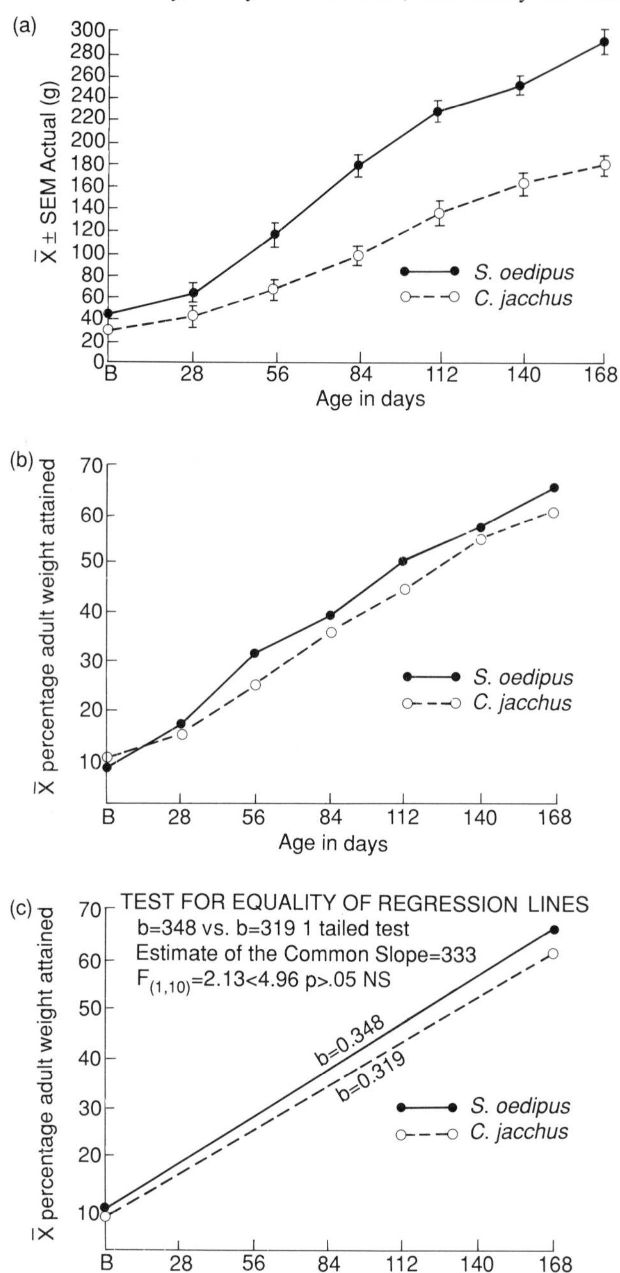


**Fig. 9.2** Growth curves for *Callithrix jacchus* and *Saguinus oedipus* from birth to 170 days. (a) actual weight gain; (b) growth as percentage of adult weight attained; (c) regression line describing growth attained curves.

weights that might be introduced by the inclusion of sickly or dying infants, only individuals for which a four-month weight was reported were included in these calculations.

Since large differences exist between species in terms of adult body size, actual growth curves are not comparable. Therefore, the mean values for target ages in each species were divided by the mean adult weight reported for individuals or calculated for groups of each species. Figure 9.2b shows the growth curves of the same species expressed as the percentage of adult weight attained at four-week target age intervals for the first 24 weeks of life. It is evident that when the effects of body size are eliminated, the two species show growth curves that are close approximations of one another.

Growth from birth to maturity in mammals is generally best represented by a complex curve. Kirkwood (1985) has described the pattern of growth of body weight in callitrichid primates as a simple decelerating exponential curve. However, as can be seen in Figure 9.2b, during the portion of the curve under consideration here growth is essentially linear.

To compare the rates of growth between species statistically, the growth attained curves for each species were further smoothed by least-squares linear regression. Correlation coefficients between the actual value of $Y$ and the predicted value of $Y$ were in excess of 0.974 for both species. Figure 9.2c shows the regression lines describing the growth attained curves for *Callithrix jacchus* and *Saguinus oedipus*. The slope of each regression line is the growth velocity of the sample. Slopes for the two species are represented by the values 0.319 and 0.348 respectively.

A test for equality of these two regression lines by the method of Sokal and Rohlf (1981) yielded an estimate of the common slope of 0.333 and an F value with one and ten degrees of freedom of 2.13. At the 0.05 level of probability, a null hypothesis of no difference between the slopes of the two regression lines cannot be rejected. When the effects of differences in adult body size are eliminated, the growth rates of the two species are not significantly different.

Figure 9.3 provides regression lines describing growth attained for *S. fuscicollis* and *Leontopithecus rosalia*, in addition to *C. jacchus* and *S. oedipus*. When the infant carrying patterns illustrated in Fig. 9.1 are compared with the growth attained curves for these four species in Fig. 9.3, it is clear that the significant differences in carrying behaviour in marmosets and tamarins observed in the laboratory cannot be explained by corresponding differences in the growth rates of infants in the species.

An environmental factor which is related to observed species differences in infant care is the ranging patterns of free-ranging populations. Table 9.2 indicates the mean daily path length and home range size for four species. As a measure of daily distance travelled, home range sizes are suggestive of differences between species. However, home range size

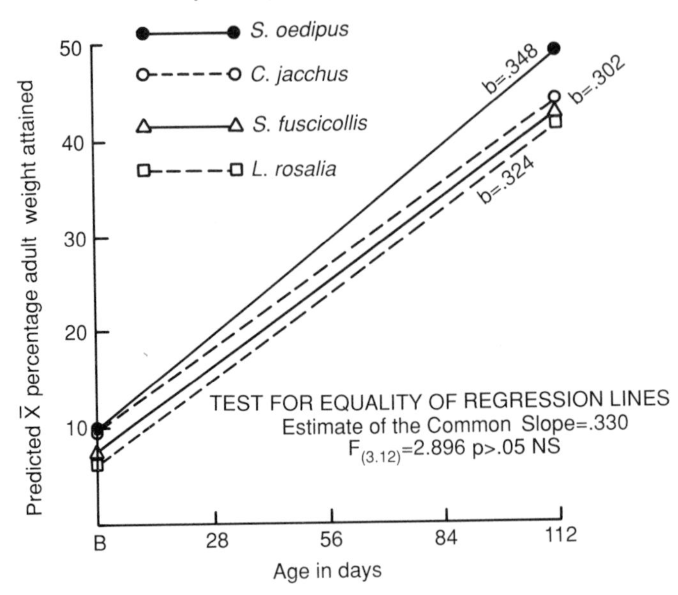


**Fig. 9.3** Regression lines describing growth attained for *C. jacchus* (○), *S. oedipus* (●), *S. fuscicollis* (△), and *L. rosalia* (□).

**Table 9.2** Comparison of home range, path length, and travel velocity for species discussed in text

| Species | Mean home range (ha) | Path length (m/day) | Travel velocity (mean m/hour) | Reference |
|---|---|---|---|---|
| *Callithrix jacchus* | 4.0 | 704 | 64 | Hubrecht (1985) |
| *Callithrix kuhli* | 10.0 | 974 | — | Rylands (1989*b*) |
| *Saguinus fuscicollis* | 40.0 | — | — | Garber (1988*a*) |
| | — | 2325 | 232 | Pook and Pook (1982) |
| *Saguinus oedipus* | 9.0 | 1750 | 136 | Neyman (1978) |
| *Leontopithecus rosalia* | — | 1496 | — | Peres (1986*b*) |
| *Leontopithecus chrysomelas* | 36.0 | 1792 | — | Rylands (1989*b*) |

frequently varies within species with seasonal availability of resources and habitat equality. Also, within-species variation in home range use renders this value meaningless in cross-species comparisons.

Daily path length is the best absolute measure of distance travelled. All values are derived from data presented in the literature and represent the results of tracking tagged individuals by radiotelemetery. However, examination of published reports revealed substantial variation in the number of hours per day that individuals were monitored. Accurate comparison of the relative energy expenditure on travel between species must include both a measure of distance and a measure of time. Therefore, travel velocity, or the distance travelled per day divided by the number of hours of activity per day, is used here. This value, expressed as mean metres travelled per hour of activity, is listed in Table 9.2.

This table indicates there are clear differences between a marmoset pattern of travel and a tamarin pattern. The mean metres per hour travelled by tamarins is about twice that of marmosets. A two-sample comparison of means performed on data presented by Hubrecht (1985) for *Callithrix jacchus* and by Neyman (1978) for *Saguinus oedipus* yielded a T value of 3.56. Using a one-tailed test with three degrees of freedom, a null hypothesis of no difference between the mean metres per hour travelled by the two species can be rejected at the 0.05 level of probability. Therefore, for the significant difference in the patterns of carrying behavior seen in the laboratory we do see a significant difference in the pattern of probable energy expended in travel between the two species under free-ranging conditions.

In Fig. 9.4, the mean percentage of time infants are carried at eight weeks in the laboratory is plotted against the mean daily path length travelled in the field for four species. While the sample size is too small for correlation statistics, the trend of increased infant-carrying time with increased daily path length is clearly apparent. The extended infant-carrying behaviour displayed by tamarins may represent an adaptation to a foraging pattern that is based on long-distance travel.

Another feature of infant care which may vary relative to energetic burden is the latency of non-mothers to carry infants. Of the four callitrichid genera, *Leontopithecus* displays a longer period of maternal retention of the infants. Kleiman (1977) noted that non-mothers first carried golden lion tamarin infants at 9–17 days; in the other genera, fathers may carry infants on day 1. Kleiman hypothesizes that this difference is explained by the larger body size (and smaller infant/adult weight ratio) in *Leontopithecus*, relative to other callitrichid species; however, as previously demonstrated, the infant/adult weight ratio appears to be similar across callitrichid species. It is still possible that the mechanical burden of infant carrying is less for the larger-bodied species. In addition, there is not

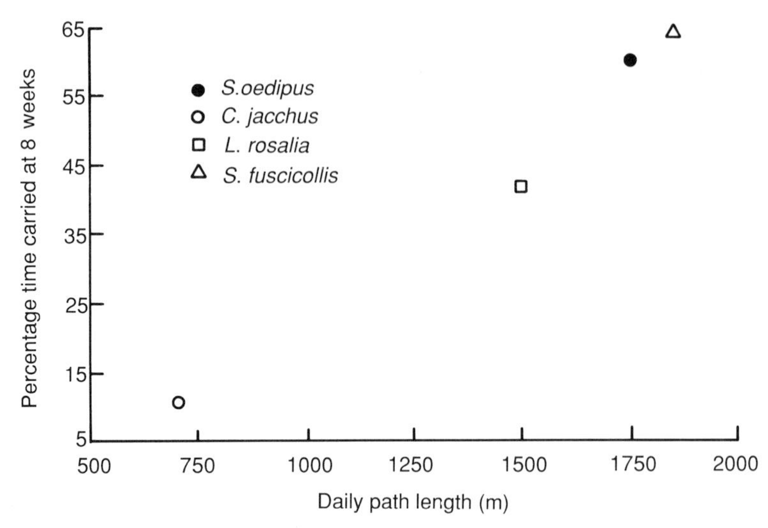


**Fig. 9.4** Relation between percent carrying time for 8-week-old captive infants and mean free-ranging daily path length for four species. See Fig. 9.3 for key to species.

a clear relation of latency to carry by non-mothers to body weight across genera; *Saguinus, Cebuella*, and *Callithrix* all display infant carrying by non-mothers as early as day 1 and generally within week 1 (Epple 1975*b*; Ingram 1977; Cleveland and Snowdon 1984; Tardif *et al.* 1990; Yamamoto, this volume). Numerous factors such as litter size, parity, and group size might also affect the age at which infants are first carried by non-mothers; this again makes comparisons across studies difficult.

Infant weight (or lactation) burden does affect the latency to carry by non-mothers within species for the golden lion tamarin and the cotton-top tamarin. In one study, golden lion tamarin mothers relinquished twin infants on day 9–11, while singleton infants were not relinquished to other carriers until day 14–17. A similar result has been found for cotton-top tamarins, with non-mothers first carrying twins on day 3 (median), while singletons were first carried by non-mothers on day 6.5 (Tardif *et al.* unpublished). These results suggest that, within a species, some aspect of the energetic burden of infant care probably does determine the role of non-mothers in infant care.

Goldizen (1987*a*) proposes that helpers beyond the father are necessary for the survival of offspring. Garber *et al.* (1984) have demonstrated a statistically significant relation between the number of adult males and the number of surviving infants in free-ranging groups of *Saguinus mystax*. In the light of these arguments, we examined how the parents' behaviour is

**Table 9.3** Effect of helpers on parental infant carrying in four callitrichid species. –, reduction: 0, no effect

| Species | Mother | Father | Reference |
|---|---|---|---|
| *C. jacchus* | — | 0 | Ingram (1977) |
| | 0 | — | Tardif *et al.* (1986) |
| *S. fuscicollis* | 0 | 0 | Epple (1975*b*) |
| *S. oedipus* | — | 0 | Cleveland and Snowdon (1984) |
| | — | 0 | Ziegler *et al.* (1990) |
| | 0 | — | Wolters (1978) |
| | 0 | — | Tardif *et al.* (1990) |
| | 0 | — | McGrew (1988) |
| *L. rosalia* | 0 | 0 | Kleiman *et al.* (1988) |

affected by the presence of helpers in captivity. Table 9.3 summarize data on the effect of helpers on parental carrying time across species. The data are ambiguous regarding the effects of helpers on parental carrying. In two species parental carrying is reported to be reduced by the presence of helpers, but the identity of the parent whose carrying is reduced varies from study to study.

A number of factors may account for these highly variable results. They include the following:

1. *Length of study.* Whether observations were conducted over four versus 12 weeks, for example, will drastically affect the results. In most species, mothers are most extensively involved in infant carrying in the early weeks and their participation declines as that of other group members increases (Hoage 1978; Ingram 1977).

2. *Individual differences.* In Wolters' study of cotton-top tamarins (1978), as well as in our own (Tardif *et al.* 1990), individual differences in female lactational abilities were identified. Differences such as this have clear effects on infant carrying patterns and could sway results, particularly in a study involving only 3–5 sets of parents, as many studies do.

3. *Definition of helpers.* Cleveland and Snowdon (1984), as well as others, have found that offspring less than 14 months of age generally carry infants infrequently. Therefore if most of the helpers in a study are juveniles, their effect on parental carrying may be minimal.

4. *Costs of carrying.* Finally, the costs of infant carrying may be less in captivity than in the wild. Costs may also vary from colony to colony if infant carrying has different effects on access to food, for example.

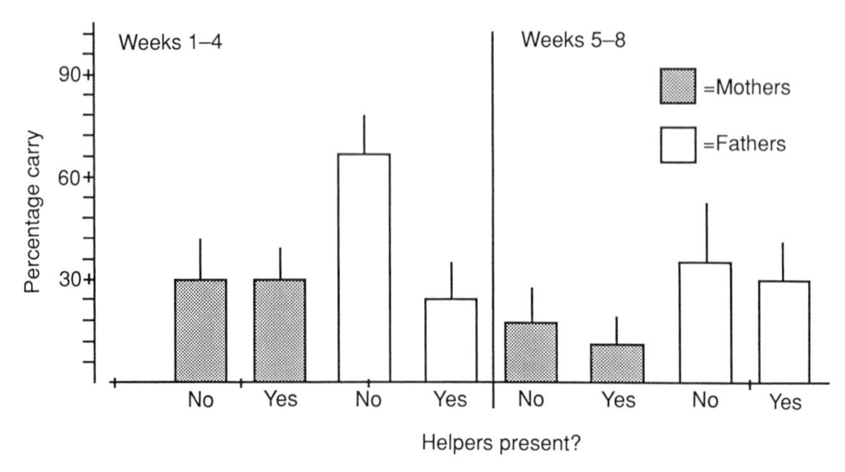


**Fig. 9.5** Mean percentage of time mothers (solid bars) and fathers (open bars) spent carrying infants relative to infant age (weeks 1–4 vs. 5–8) and presence of helpers.

We examined infant-carrying patterns in 36 *S. oedipus* litters from 23 captive groups to determine the effects of various demographic factors (e.g. infants number, infant age, group size) on different caregivers (Tardif *et al.* 1990). Examination of a relatively large population under standardized housing conditions provides the opportunity to explore the effects of helpers with minimal interference from confounding variables. Figure 9.5 presents the mean percent carrying time with 95 per cent confidence interval for mothers and fathers in groups with or without helpers. A group with helpers was defined as one in which individuals other than the parents did at least 15 per cent of the infant carrying. In weeks 1–4, fathers carried significantly less in groups with helpers than in groups without. The highest percentage of carrying overall was done by fathers in groups with no helpers. No other differences were significant. In weeks 5–8, helpers had no significant effect on maternal or paternal carrying. However, mothers overall carried significantly less than fathers. These data suggest that, generally, mothers limited their carrying more than fathers, and that fathers were more affected by the presence of helpers.

## Infant provisioning

While carrying infants is probably the most obvious aspect of infant care in callitrichids, provisioning infants with solid food is another important aspect of infant care. Callitrichids are unique amongst primates in the

**Table 9.4** Characteristics of food sharing across callitrichid species

| Species | Active sharing reported | Percentage active sharing | Preferred items shared | Reference |
|---|---|---|---|---|
| *Cebuella pygmaea* | yes | — | — | Feistner and Price (in press) |
| *Callithrix jacchus* | yes | 4.3 | less | Simek (1988) |
| | no | — | | Feistner and Price (in press) |
| *C. geoffroyi* | no | — | — | Feistner and Price (in press) |
| *C. argentata* | no | — | — | Feistner and Price (in press) |
| *S. labiatus* | no | — | — | Feistner and Price (in press) |
| *S. oedipus* | yes | 10.0 | more | Feistner and Chamove (1986) |
| | yes | 12.3 | same | Simek (1988) |
| | yes | 23.3 | — | Davis and Richter (1980) |
| | yes | 28.0 | — | Feistner (1985) |
| *L. chrysomelas* | yes | — | — | Feistner and Price (in press) |
| *L. rosalia* | yes | 94.0 | less | Brown and Mack (1978) |
| | — | — | less | French *et al.* (1985) |

degree to which group members provide solid food to youngsters. It has long been recognized that infants take food from other group members. However, Brown and Mack (1978) described a set of behaviours in golden lion tamarins, including eye contact, food presenting, and a characteristic food call, which were all associated with a lack of resistance to transfer, indicating that golden lion tamarins actively share food. This active food sharing has been observed in other callitrichids (Feistner and Price in press; see Table 9.4), though the percentage of transfers which appear to be active varies substantially across species, being lowest in *Callithrix* and highest in *Leontopithecus*.

In most callitrichid species, infants are provisioned almost entirely when they begin to eat solid food, then gradually develop into independent feeders. We hypothesized that species differences in food provisioning would parallel carrying with marmosets feeding independently earlier than tamarins. Unfortunately the data in the literature were not directly comparable across studies in the way that they frequently were for infant carrying. Simek (unpublished) conducted a comparative study of infant

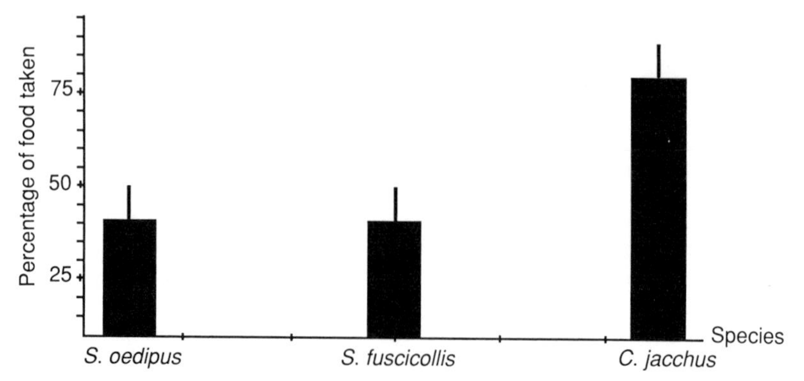


**Fig. 9.6** Mean percentage of food taken independently during weeks 5–8 for three species.

**Table 9.5** Comparison of identity of infant provisioners in four species

| Species | Who provisions infants | Reference |
|---|---|---|
| *C. jacchus* | Siblings > father > mother | Simek (1988) |
| *S. oedipus* | Father | Wolters (1978) |
| | Father > siblings > mother | Davis and Richter (1980) |
| | Siblings > father = mother | Simek (1988) |
| *S. fuscicollis* | Siblings > mother ≥ father | Cebul and Epple (1984) |
| *L. rosalia* | Father = mother (?) | Hoage (1982) |

provisioning from 5–16 weeks in the cotton-top tamarin, saddle-back tamarin, and common marmoset. By week 9, all three species were taking a similar percentage of their food independently. However, during weeks 5–8 *C. jacchus* did take a larger percentage of its food from the environment and a smaller percent from provisioners (see Fig. 9.6). *L. rosalia* is similar to *Saguinus* in taking no more than 50 per cent of food independently at 2 months of age (French, unpublished).

The literature suggests that, as with carrying, all individuals are involved in provisioning infants but no clear species differences are evident as to who, if anyone is the primary provisioner (see Table 9.5). The only consistency worth mentioning is that across species mothers were generally not the primary provisioners of infants. As the mother is the only group member with the added energetic demand of lactation, it may be that she is less willing to relinquish food to infants.

Most anecdotal accounts of sharing in the wild have been of infants taking insects from other group members. It may be that this type of sharing is simply more visible; however, Feistner and Price (in press) suggest that food transfers are increased when rare, highly preferred food items are present. Captive studies suggest that food items may be shared more or less relative to their desirability (see Table 9.4). In golden lion tamarins and in common marmosets preferred food items are less likely to be shared, while in cotton-top tamarins preferred items are shared equally or more often. This difference might be related to the prolonged dependence in the cotton-top tamarin relative to the other two species. One feature which differentiates food in the wild much more than in captivity is difficulty of acquisition. This is a feature which might make insect sharing more prevalent in the wild.

Provisioning of the infants may play a substantial role in how the social groups is formed and how it functions. Provisioning parallels infant carrying in being more intense and of longer duration in tamarins than in marmosets. Future results both from the field and from experimental manipulation of feeding situations in captivity can be useful in defining the mechanisms which determine who shares food and what food items are provisioned.

## Conclusion

Species differences in infant care are apparent within the Callitrichidae. Specifically, *Callithrix* carry and provision infants for a shorter period than do *Saguinus* species. *Leontopithecus* display an intensity of infant care that is intermediate between *Callithrix* and *Saguinus*. These differences are not linked with adult body size or growth rate but are correlated with ranging patterns in the wild.

On the basis of these preliminary data, we hypothesize that among callitrichid primates there is a direct relationship between the resource type and foraging pattern used by a species and the intensity and duration of direct infant care that are required by that species to raise viable offspring. While the relation indicated in Figure 9.4 does not imply a causal relation between these variables, we hypothesize that the two may be causally related in the following fashion. Species such as *Callithrix jacchus* depend in the wild upon a spatially clumped and temporally continuous dietary resource in the form of exudates (see references in Table 9.6; Ferrari and Lopes Ferrari 1989). Infants may forage independently earlier simply because the foraging path length is short. In *Saguinus* species, dietary resources (fruits and flying insects) are widely dispersed both spatially and temporally, and infants may have to be carried longer if they are to remain with the group over a long daily foraging path.

The intermediate position of *Leontopithecus rosalia* is less clearly explained since specific data on diet and foraging patterns in the wild are scarce for this species. However, Rylands (1989*b*) has provided extensive field data on the closely related *Leontopithecus chrysomelas*. In comparing

**Table 9.6** Comparison of extent of exudate feeding in species discussed in text

| Species | Exudate feeding | Reference |
|---|---|---|
| *C. jacchus* | 30 per cent daily activity | Maier *et al.* (1982) |
| *C. kuhli* | 31–34 per cent plant feeding | Rylands (1989*b*) |
| *S. fuscicollis* | 7.6 per cent feeding time | Garber (1988*a*) |
| *S. oedipus* | Occasional | Neyman (1978) |
| *L. rosalia* | Not available | |
| *L. chrysomelas* | 3–11 per cent plant feeding | Rylands (1989*b*) |

*L. chrysomelas* to the sympatric *Callithrix kuhli*, Rylands noted that *L. chrysomelas* exploited a larger home range (40 hectares versus 10 hectares for *C. kuhli*), fed substantially less frequently upon exudates (3–11 per cent of plant feeding records versus 31–34 per cent of the plant feeding records for *C. kuhli*), exploited larger insect prey at specific locations (bromeliads), and used a greater diversity of widely dispersed food resources than did *C. kuhli*. These variables can be presumed to affect daily path length, and the intensive infant provisioning exhibited by *Leontopithecus* species may function as a substitute for infant carrying.

The species differences in duration and intensity of infant care could result in differences in the importance of helpers across species. They suggest that additional helpers, beyond the mother and one male, may be less critical to infant survival in marmosets than in tamarins. These hypothesized relations between foraging/travel patterns, infant care and social structure could be assessed through acquisition of specific information on the energetic demands of infant carrying in a controlled captive setting. We hypothesize that the behavioural differences described here do indeed reflect species differences in the energetic demands of infant care; however, the energetic burden of infant care in different species and under different foraging regimes must be more directly measured to test this hypothesis.

## Acknowledgements

The authors thank Gisela Epple, Devra Kleiman, and Lyna Watson for generously providing additional data on infant growth; Paul Garber, Anthony Rylands, Andrew Baker, and Anne Savage for their comments on infant-carrying patterns in free-ranging animals; Fred Alexander, Richard Tardif, and Owen Lovejoy for helpful comments on the manuscript; Susan Russo for assistance in preparing the figures. Research was supported by NIH Grant RR-02022 to Suzette Tardif.

# 10

# From dependence to sexual maturity: the behavioural ontogeny of Callitrichidae

*Maria Emília Yamamoto*

## Summary

This chapter reviews data on behavioural ontogeny in the four callitrichid genera, and presents more detailed data on five *Callithrix jacchus* infants. Callitrichids are unique in the intensity of their relations to infants. The newborn, usually twins, may be carried from the first day by group members other than the mother. Weaning occurs from 9–13 weeks, when most of the food ingested is obtained through sharing or stealing. Agonistic behaviours may appear, as well as scent-marking. At the juvenile stage (beginning at 4–7 months) 'twin fights' may occur; particularly between same-sexed twins, determining status differences. At the sub-adult stage (beginning at 9–14 months) the young animal has the size and appearance of an adult, has mastered most of the adult behavioural repertoire, and achieves puberty. Although puberty takes place at this time the young does not conceive. At the adult stage (beginning at 12–21 months) sexual maturity is attained. Approximately every five months a new set of twins is born. Interactions with infants depend on the animal's age and previous experience. The behavioural ontogeny shows many similarities among the callitrichid genera, although some important differences call for further investigation.

## Introduction

Callitrichids present a combination of characteristics that make them extremely interesting subjects for the study of ontogeny and behavioural development. These are:

1. Twinning, unique among anthropoid primates. The twins present a high relative fetal weight, varying between 14 and 23 per cent (Leutenegger 1973).

**Table 10.1** Age, in months, of infant, juvenile, sub-adult, and young adult callitrichids of the four genera. *Cebuella*, data from Soini (1988); *Callithrix*, adapted from Ingram (1977) and Stevenson and Rylands (1988); *Saguinus oedipus*, data from Cleveland and Snowdon (1984); *Leontopithecus rosalia*, adapted from Hoage (1982)

| Stage | *Cebuella* | *Callithrix* | *Saguinus* | *Leontopithecus* |
|---|---|---|---|---|
| *Infant* | 1–5 | 1–5 | 1–7 | 1–4 |
| *Juvenile* | 5–12 | 5–10 | 7–14 | 4–9 |
| *Sub-adult* | 12–16 | 10–15 | 14–21 | 9–12 |
| *Young adult* | +16 | +15 | +21 | +12 |

2. Reproduction restricted to the dominant female, even when there are other potentially reproductive females in the group. The exclusivity of reproduction may be associated with suppression of ovulation in subordinate females (Snowdon 1990).

3. Sharing of infant care by the father and older siblings (Cleveland and Snowdon 1984; Hoage 1982; Ingram 1977).

4. Social group that, at least in captivity, may be formed only by kin, and that, in the field, may have a high degree of kinship (Ferrari and Lopes Ferrari 1989).

Indeed, many studies have been undertaken on this topic, mainly in captive settings, and although questions still remain, some important characteristics and species differences have emerged. The purpose of this paper is to examine these data, comparing the four genera, and to present a more detailed description of the behavioural ontogeny of *Callithrix jacchus*.

## Behavioural ontogeny: comparisons between genera

Before presenting the comparison, a word of caution. A compilation such as this presents the difficulty of putting together data that differ widely, having been collected by several researchers under various husbandry regimes, and sometimes scoring behaviours that have been differently defined. Besides this, some genera and species have been studied in more detail than others, especially in captivity. Although much of the variability in the data is due to real generic and species differences, some is the result of these other factors.

At least four developmental stages are recognized in the Callitrichidae: infant, juvenile (or adolescent), sub-adult and young adult. Table 10.1 presents the ages related to each stage for the four genera, adapted from the references cited, in order to have a more uniform classification.

Callitrichid infants are usually twins, and at birth weigh 13–15 g (*Cebuella*; Soini 1988), 22–38 g (*Callithrix*; Stevenson and Rylands 1988), 34–38 g (*Saguinus*, only two species; Leutenegger 1973), or 52–75 g (*Leontopithecus*, only *L. rosalia*; Hoage 1982). A few seconds after birth infants are able to grab the mother's fur, and *Callithrix jacchus* infants reach the breast 2–20 minutes after delivery (Rothe 1978*a*). Price (1990*a*), nevertheless, did not observe suckling up to one hour after delivery in 14 births in *Saguinus oedipus*. Unhealthy infants do not receive any special attention, and if they fail to cling they are allowed to drop. If the infant is unable to move or vocalize the mother shows very little interest in it, abandoning it to die (Rothe 1975*a*; Price 1990*a*). In *Callithrix jacchus* the father may carry the infants immediately after birth (Rothe 1974, 1975*a*) but in *Saguinus oedipus* this seem to be the case only for primiparous mothers (Adler 1988; Price 1990*a*).

Figure 10.1 presents the most noteworthy changes taking place during the infant stage (birth to weeks 22–30). This is the most extensively studied period, and it is when most of the behaviours of the animal's repertoire make their first appearance. The four blocks represent carrying, feeding, play, and agonistic behaviours. Hatched bars represent the first appearance of behaviours, the length indicating variability due to differences between species, individuals, or studies.

Until week 3 infants are carried all the time by the various family members (*Cebuella*, *Callithrix*, and *Saguinus*) or exclusively by the mother (*Leontopithecus*). There is, of course, a high degree of variation among genera and species. Some authors suggest that there is a preferential period of carrying by the mother in some species during the first few weeks (Cebul and Epple 1984; Cleveland and Snowdon 1984). Beginning around week 4, infants show increasing interest in their physical environment, leaving caregivers spontaneously and touching, smelling, licking, and gnawing objects. The first social interactions begin while the infants are still on the carrier's back, through grooming and twin contact (Izawa 1978; Kleiman *et al.* 1988; Snowdon and Soini 1988). Although frequently leaving the carrier's back, infants are carried until the weeks 8–12, when they are rarely seen 'on'. There is evidence that independence rates are different for tamarins (*Leontopithecus* and *Saguinus*) and marmosets (*Callithrix* and *Cebuella*), the first showing a longer carrying period than the former (Tardif *et al.* 1986, this volume). Weaning occurs from the weeks 8–15, although infants begin to taste solid food, either directly or through food sharing or stealing, much earlier. Sharing may be either passive (all species) or active, the latter observed in *Saguinus oedipus* (Cleveland and Snowdon 1984), *Leontopithecus rosalia* (Brown and Mack 1978; Hoage 1982) and *Leontopithecus chrysomelas* (Feistner and Price in press), *Cebuella pygmaea* (Feistner and Price in press), and *Callithrix flaviceps* (Ferrari 1987).

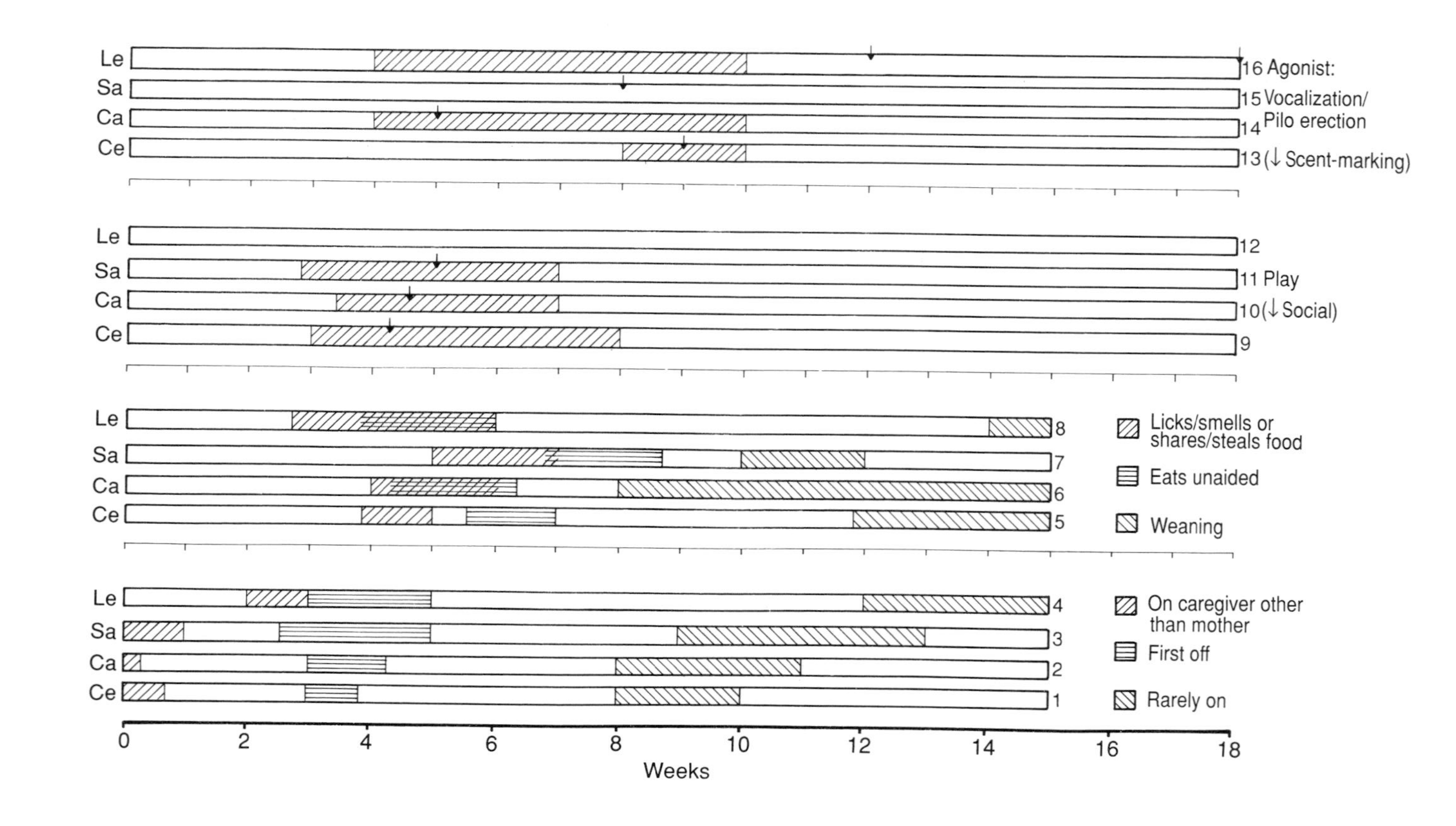
INFANT STAGE – FIRST APPEARANCES
Le
Sa
Ca
Ce
16 Agonist:
15 Vocalization/
14 Pilo erection
13 (↓ Scent-marking)
12
11 Play
10 (↓ Social)
9
8
7
6
5
Licks/smells or shares/steals food
Eats unaided
Weaning
4
3
2
1
On caregiver other than mother
First off
Rarely on
0
2
4
6
8
10
12
14
16
18
Weeks

Food sharing and stealing seem to be an initiation to self-feeding, since a little later (weeks 4–8) infants are eating unaided. By that age, social interactions with other group members occur frequently, through grooming, social play, and physical contact. Solitary play appears before social play, and is probably a way of getting to know and explore the physical environment, as well as developing the infants' physical abilities. Solitary play described by Izawa (1978) in *Saguinus nigricollis* is a good example of a behaviour that functions as a precursor and a preparation for later abilities, in this case specifically to catch insects. Some agonistic behaviours, such as partial pilo-erection, 'ehr-ehr' vocalizations, and tufts flattened, may appear in low frequencies, along with anogenital scent-marking.

The other stages of behavioural ontogeny (juvenile, sub-adult, and young adult) have not been studied so extensively as the infant stage, and changes reported are, with a few exceptions, of a quantitative rather than a qualitative nature. In other words, at these stages behaviours appear more frequently and consistently, and in a functional context.

At the juvenile stage (beginning at 4–7 months) the animal weighs about 75–77 per cent of adult weight (Hoage 1982; Snowdon and Soini 1988). The trend toward interacting with other group members besides parents is emphasized. The young marmoset/tamarin is now able to solicit grooming and groom others dexterously (Hoage 1982; Snowdon and Soini 1988; Stevenson and Rylands, 1988). Social play becomes rougher, and includes other siblings frequently, and the father occasionally. Also at this stage, 'twin-fights' may occur, which determine differences in later status (Stevenson and Poole 1976; Kleiman 1979; Sutcliffe and Poole 1984). The early occurrence of these fights is attributed to an adaptive mechanism: serious injuries are avoided while the animals still retain their milk dentition (Sutcliffe and Poole 1984). A major change in group composition usually takes place around this time: the birth of new infants. Juveniles show interest in young infants, sniffing, nuzzling, grooming, and sometimes

---

**Fig. 10.1** First appearances of behaviours related to carrying, feeding, play and agonism during infant stage in *Cebuella* (Ce), *Callithrix* (Ca), *Saguinus* (Sa) and *Leontopithecus* (Le).

1. Data for *Cebuella* from Larsson *et al.* 1982; Whitten 1987; Soini 1988.
2. Data for *Callithrix* from Box 1977; Buchanan-Smith 1984; Hearn 1983; Locke-Haydon and Chalmers 1983; Pook 1978*a*; Stevenson 1976*a*, 1978; Stevenson and Rylands 1988; Sutcliffe and Poole 1984; Voland 1977; Yamamoto 1990.
3. Data for *Saguinus* from Buchanan-Smith 1984; Cebul and Epple 1984; Cleveland and Snowdon 1984; Hampton *et al.* 1966; Izawa 1978; Pook 1978*a*; Stevenson 1976*a*; Wolters 1978.
4. Data for *Leontopithecus* from Buchanan-Smith 1984; Hoage 1978, 1982; Kleiman and Mack 1980; Rathbun 1979; Whitten 1987.

carrying and rejecting them. When infants get a little older, juveniles play and share food with them (Ingram 1977; Hoage 1982; Cebul and Epple 1984). Patterns related to sexual behaviour may also occur, such as mounting younger siblings and performing genital investigation in siblings and parents (Abbott 1978; Hoage 1982).

At the sub-adult stage (beginning at 9–14 months), the young animal is 82 per cent of adult weight and has mastered most of the adult behavioural repertoire. Anogenital scent-marking appears regularly, and sternal marking may be observed for the first time (Kleiman and Mack 1980; Hoage 1982; French and Cleveland 1984). Agonistic patterns of behaviour are exhibited, including pilo-erection, tuft-flicking, and arch-walking (Rathbun 1979; Hoage 1982). If a new set of twins is born, the sub-adult participates much more in infant carrying than he or she did as a juvenile (Ingram 1977; Cleveland and Snowdon 1984). Puberty is attained, with ovulation in females and an increase in testis size in males (*Callithrix jacchus*; Abbott 1978; Abbott and Hearn 1978), but the young animal does not conceive. There used to be some controversy over the occurrence of ovulation in daughters that remain in the family group. Recently, specific differences have been detected, which were probably responsible for some of the data discrepancies (see Abbott *et al.* this volume). In *Callithrix jacchus* 50 per cent of the daughters ovulate (Abbott 1984); in *Saguinus oedipus* no daughters ovulate (French *et al.* 1984; Ziegler *et al.* 1987*b*; Savage *et al.* 1988); in *Leontopithecus rosalia* there is apparently no suppression of ovulation (French and Stribley 1987); in *Cebuella pygmaea* Snowdon (1990) reports simultaneous pregnancies in mother and daughter living in the same group, suggesting that at least one of the daughters was not suppressed. Besides clear species differences, the females' status relative to twin or older sisters could also account for discrepancies in the measures of ovulation (Evans and Hodges 1984). A decrease in social contact may happen in some cases, leading to social isolation and peripheralization of the animal (Ingram 1978*c*; Pook 1978*a*; Price and Hannah 1983; Sutcliffe and Poole 1984; Yamamoto personal observation).

Callitrichids are considered as adults at an age varying from 12 to 21 months, when they attain adult size and appearance. At this time they reach sexual maturity, becoming capable of reproduction (Abbott 1978). However, while remaining in the family group, they show no behavioural changes, or any evidence of being sexually mature. Only when paired do they show changes in sexual, aggressive, and marking behaviour (Epple and Katz 1980; Epple 1981*b*; Tardif 1983; Epple and Alveario 1985). Sexual dimorphism in agonistic behaviour to conspecific intruders has been shown in some species, such as *Saguinus oedipus* (French and Cleveland 1984), *Callithrix jacchus* (Araújo and Yamamoto in press) and *Leontopithecus rosalia* (French and Inglett 1989). These may be related to the efficiency

**Table 10.2** Size and composition of families studied. M, male; F, female. The first infant in each family is the focal animal

| Family No. (size) | Father | Mother | Juvenile | Adult | Infant |
|---|---|---|---|---|---|
| 1 (4) | 05 | 06 | — | — | 1 F<br>1 F |
| 2 (4) | 29 | 02 | — | — | 1 F<br>1 M |
| 3 (4) | 27 | 40 | 1 M | — | 1 M |
| 4 (5) | 23 | 52 | — | 1 M | 1 F<br>1 F |
| 5 (6) | 55 | 58 | 1 M<br>1 F | — | 1 M<br>1 F |

of reproductive suppression by the dominant animals (Snowdon 1990; French and Inglett 1991).

## The first 22 weeks in captive *Callithrix jacchus*

The second part of this paper contains data from five *Callithrix jacchus* infants from five different captive families, living in the Primate Centre at the Federal University of Rio Grande do Norte, Natal, Brazil. The size and composition of the families are presented in Table 10.2.

The infants were observed through a one-way mirror during the first 22 weeks of life (infant stage), three hours weekly until the sixth week, and subsequently three hours every other week. Behaviours were recorded on a check sheet as they occurred (frequencies), or using stop-watches (durations). The facilities and feeding schedules are described elsewhere (Arruda *et al.* 1986).

During these 22 weeks, the following quantitative and qualitative data were collected:

1 time 'on';

2. frequencies and durations of rejections;

3. frequencies of the infants' attempts to climb on to caregivers, successful or otherwise, providing an index of success (successful attempts/total attempts × 100);

4. approaches and leaves of infants from caregivers and, conversely, caregivers from infants, which provides an index of infants' relative

responsibility for the maintenance of proximity (see Hinde and Atkinson 1970) in which a negative index indicates the responsibility of the caregivers, and a positive index the responsibility of the infants;

5. time spent in proximity (less than 15 cm) to caregivers;
6. time spent in physical contact with caregivers or twin;
7. time spent in grooming, either as groomee or groomer;
8. time spent in social play;
9. time spent in solitary play;
10. the first appearance of, for instance, scent-marking and agonistic behaviours.

The results are presented in four blocks: 2–4 weeks, 5–10 weeks, 12–16 weeks, and 18–22 weeks.

*Weeks 2–4*

This was a period of considerable dependence on caregivers. Figure 10.2 shows that infants spent most of their time being carried (96 per cent), and that caregivers were extremely tolerant, accepting their attempts to climb on, and rejecting the infants infrequently. Time 'off' carriers was rare, and watched closely by the father. During the brief periods 'off', the infants began exploring their physical environment through solitary play.

*Weeks 5–10*

This period marked the beginning of the process of independence. Although still carried for a considerable percentage of time (29.5 per cent, Fig. 10.2), infants began to leave the caregivers' back spontaneously, for brief periods. This incipient independence was not due only to the infants' initiative. Caregivers were less tolerant, accepting fewer of the infants' attempts to climb on (Fig. 10.2) and increasing the frequency of rejections (Table 10.3), forcing infants to get off and to stay off. The decrease in durations of rejections from week 2 to week 6 suggests a decrease in the intensity of the caregiver's rejection, but given the increase in their frequencies it seems more likely that infants responded to rejections more promptly in this second period than they did in the first, learning to cope, and, given their growing independence, becoming more resigned. There was a positive correlation between the frequency of rejections and the frequency of attempts to climb onto caregivers. This is clearly shown in Table 10.3: both frequencies increased from week 2 to week 6, and decreased from then on, suggesting that the infants sought hard for care until week 6, even being

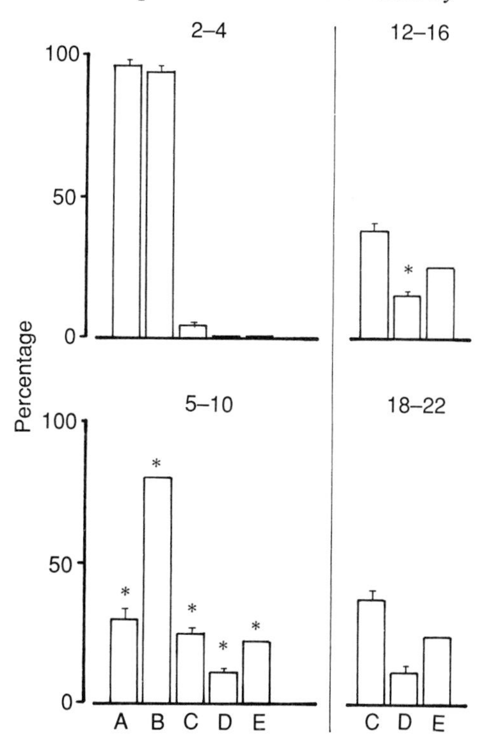


**Fig. 10.2** Care and interactions of infants to caregivers during weeks 2–4, 5–10, 12–16, and 18–22. The following behavioural categories are represented: A, time 'on'; B, proportion of successful attempts to climb; C, time spent in proximity; D, time spent in physical contact; E, percentage of approaches minus percentage of leavings by the infant. Percentages represent means of all caregivers.
* $p < 0.05$ for Student's $t$ test (A, C, and D) or Wilcoxon signed ranks test (B and E), between each period and the preceding one.

**Table 10.3** Sum of the frequencies and mean duration (in seconds) of rejections, and sum of attempts to climb and percentage of success, from 2 to 10 weeks

| Weeks | Rejections | | Attempts to climb | |
|---|---|---|---|---|
| | Frequency | Duration | No. of attempts | Percentage success |
| 2 | 11.0 | 15.4 | 2.0 | 100.0 |
| 4 | 49.0 | 10.9 | 77.0 | 78.0 |
| 6 | 124.0 | 6.8 | 147.0 | 42.2 |
| 8 | 51.0 | 1.7 | 27.0 | 50.0 |
| 10 | 30.0 | 2.0 | 8.0 | 0.0 |

**Table 10.4** Mean and range (in days) of first appearance of seven developmental markers in infant *Callithrix jacchus*

| | Mean (days) | Range (days) | Week |
|---|---|---|---|
| First day off | 15.4 | 11–20 | 2 |
| Solitary play | 19.6 | 11–25 | 2 |
| First solitary food | 28.6 | 25–34 | 4 |
| Social play | 31.6 | 25–49 | 4 |
| Twin fight | 53.5 | 33–73 | 7 |
| Scent mark | 62.8 | 33–73 | 8 |
| Last suckling bout | 61.2 | 54–67 | 8–10 |

frequently rejected, and subsequently became more independent, and probably more involved in other activities.

Part of time 'off' was spent near (26.2 per cent) and/or in physical contact (10.8 per cent) with caregivers (Fig. 10.2). Through weeks 5 and 6, however, the time spent in contact and proximity did not express the infants' preference, due to as yet underdeveloped motor abilities, which made it difficult for infants to approach caregivers. Nonetheless, despite their difficulties, infants approached caregivers more than they were approached by them (positive index of 22.4 per cent; Fig. 10.2). At the end of the period, around weeks 8–10, weaning occurred (Table 10.4), but infants tasted solid food much earlier (around the week 4), and were able to eat by themselves from weeks 5–6 (Table 10.4). Periods 'off' were spent in social and solitary activities, such as grooming and play (Fig. 10.3). The most frequent activity was solitary play (8.5 per cent), but being groomed and social play were also present.

*Weeks 12–16*

This third block is a period when the infants are physically quite independent. Carrying and nursing has ceased, and they eat solid foods unaided. Although quite active (Fig. 10.3), infants spent a considerable amount of time near (38.2 per cent) or in physical contact (14.8 per cent) with caregivers (Fig. 10.2). They approached the caregivers more during this period (26.0 per cent) than they did previously, suggesting that the increase in proximity and physical contact was due to the infants' initiative.

*Weeks 18–22*

This period did not present many changes when compared to weeks 12–16. Relationships with caregivers remained quite stable, but there were some differences in the distribution of the infants' activities. Solitary play, in the

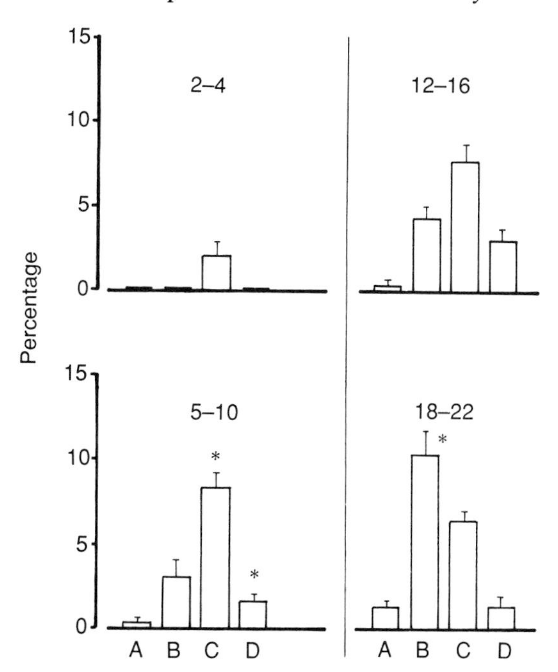


**Fig. 10.3** Grooming and play presented by infants during weeks 2–4, 5–10, 12–16, and 18–22. The following behavioural categories are represented: A, infant as groomer; B, infant as groomee; C, solitary play; D, social play.
* $p < 0.05$ for Student's $t$ test, between each period and the preceding one.

preceding periods the most frequent activity, was replaced by receiving grooming (10.2 per cent), due to an increase in the time spent in grooming, and not to a decrease in time spent in solitary play (see Fig. 10.3).

Table 10.4 shows the mean and range in days of some developmental markers for the five infants. The first independent behaviours appeared around the second and fourth weeks: the first day infants were seen 'off', in exploration (solitary play), as well as in social play, and the first solid food, which was either taken from the hands or mouths of older animals, or from the dish, where infants licked and gave little bites at food items. Agonistic behaviours first appeared directed at the twin (around week 7) and a little later, circumgenital scent-marking was also observed. Weaning occurred around weeks 8–10, when infants were quite able to eat by themselves.

From the weeks 2 to 22, infants showed a striking change in their dependence on caregivers, which was total at the beginning, but only partial at the end of the infant stage. This trend did not however, represent, the

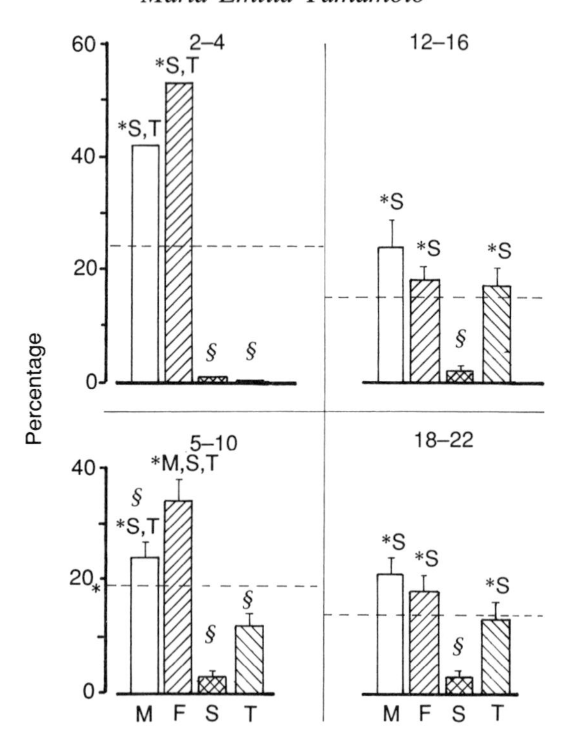


**Fig. 10.4** Social interaction of infants to mother (M), father (P), older siblings (S), twin (T), and mean interaction (hatched line), during weeks 2–4, 5–10, 12–16, and 18–22.
* $p < 0.05$ for Student's $t$ test; comparison between family members.
§ $p < 0.05$ for Student's $t$ test; comparisons between family members and mean interaction.

direction of the infants' relations with all family members. From the earliest stages infants were differentially cared for by family members, and also showed a preference for certain caregivers.

Figure 10.4 shows the mean interactions (sum of time spent carrying, grooming, and in physical contact) of infants with family members during the four periods. The broken line represents the mean of the interaction with all family members throughout each period. The mean provides an expected interaction, were all partners to interact equally with infants. The greatest mean interaction was observed during the first period, with significant decreases in the subsequent periods. The comparison of the mean interaction with that of each family member showed that there was a differential interaction with family members during each period, and also that the relative importance of family members changed through the 22 weeks that the infants were observed.

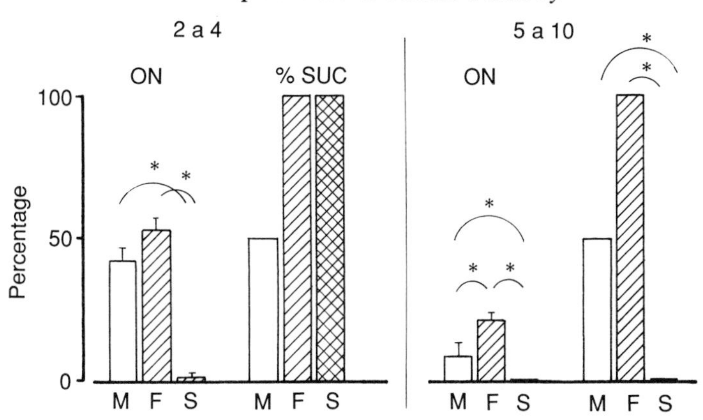


**Fig. 10.5** Percentage of time that infants were carried and proportion of successful attempts to climb on to the mother, father and older siblings. * $p < 0.05$ for Student's $t$ test (time 'on') or Wilcoxon signed ranks test (attempts to climb).

During weeks 2–4 the main partners were the father (53.0 per cent) and the mother (42.1 per cent) who interacted with infants at above the mean levels, and more than older siblings and twin. In the second period, mother and father remained the most frequent partners, although the father was the only one to interact with infants above the mean level. The twin, although interacting at below the mean level, showed a significant increase when compared to the first period. During the third period, infants interacted with mother, father, and twin at very similar levels, and only older siblings remained at below the mean level. This same pattern repeated itself in the fourth period.

It is clear that the relative importance of family members differed during the four periods. The only ones that did not show any alterations in the amount of interactions with infants were older siblings, maintaining a low level of interaction throughout the entire infant stage. The mother showed a significant decrease in the amount of interaction from the first to the subsequent periods; the father showed decreases from the first to the following three periods, and also from the second to the next two; and the twin increased its interaction from the first to the subsequent periods.

An examination of the behaviours involved in the interactions clarifies the reasons for these differences. Mother and father carried infants for equivalent amounts of time, and were tolerant of the infants attempts to be carried during weeks 2–4 (Fig. 10.5). Older siblings hardly carried infants at all, and the high tolerance index is probably due to infrequent attempts by the infants to climb on to siblings when compared to the father and mother (average number of attempts during the period: on

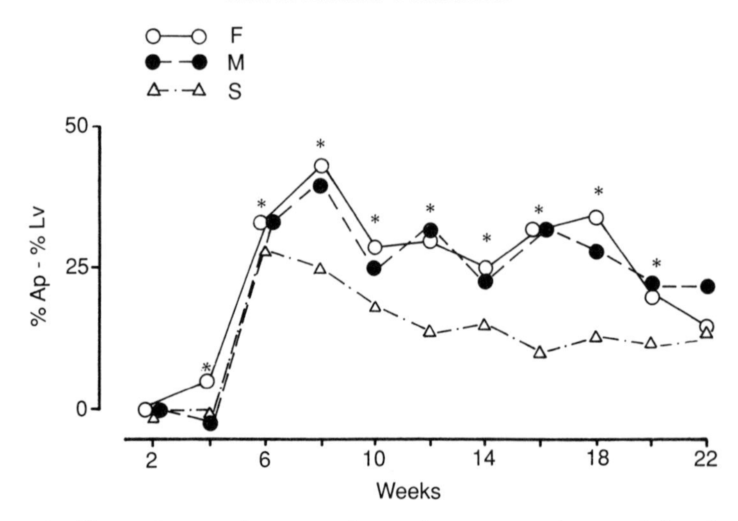


**Fig. 10.6** Percentages of approaches minus percentages of leavings by infants to father (F), mother (M), and older siblings (S) from weeks 2–22. * $p < 0.05$ for Student's $t$ test between mother and father, and older siblings and father (week 4), and parents and older siblings (remaining weeks).

mothers, 38; father, 34; on siblings, 7). During the second period (5–10 weeks), the father became the main carrier, although time 'on' was significantly less than in the first period. Infants sought the mother (82) as much as the father (99) during this period, suggesting that the greater time 'on' father was due to his higher tolerance (Fig. 10.5).

Data on the percentage of approaches by infants minus the percentage of leavings (Fig. 10.6) show that when infants were capable of independent locomotion (weeks 5–6) they differentially sought the parents rather than siblings. This preference was also clear in the time of proximity and physical contact (Fig. 10.7). From the fifth week, infants spent more time in proximity to parents than to older siblings, and more time in contact with parents and the twin than with siblings. Twin physical contact was high from the second period, and was the most frequent type of interaction between twins.

The most frequent grooming partner during the first period was the father (Fig. 10.8), but from the 18th week on he was replaced by the mother, mainly because she spent more time grooming the infants. Little grooming was observed between the twins and between twins and older siblings.

Data on social play partners is presented in the form of frequencies, together with the total duration of play episodes (Fig. 10.9). It appeared around the weeks 4–5, and the most frequent partner was the twin. Older

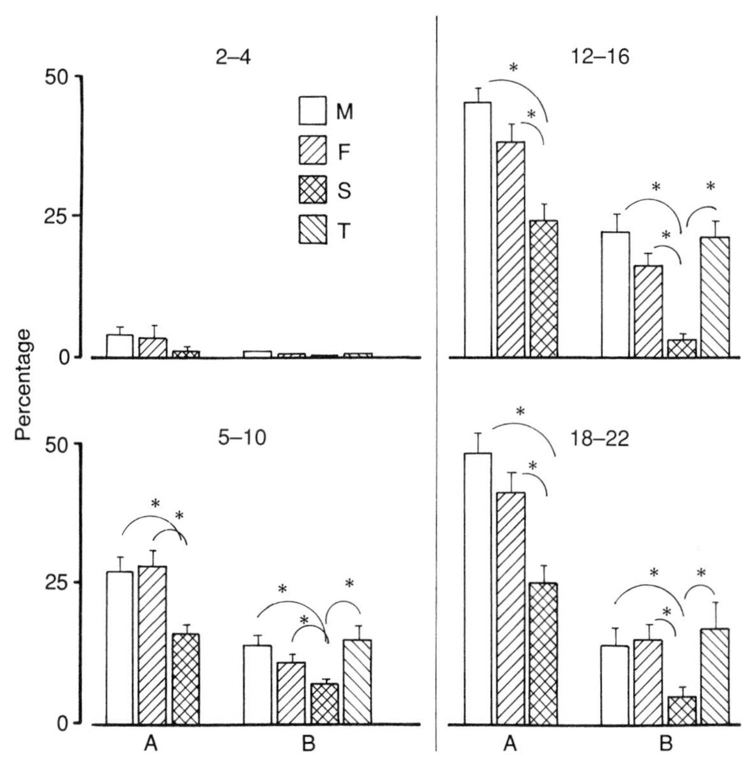


**Fig. 10.7** Percentage of time that infants spent in proximity (A) and physical contact (B) to mother (M), father (F), older siblings (S), and twin (T) during weeks to 2–4, 5–10, 12–16, and 18–22.
* $p < 0.05$ for Student's $t$ test.

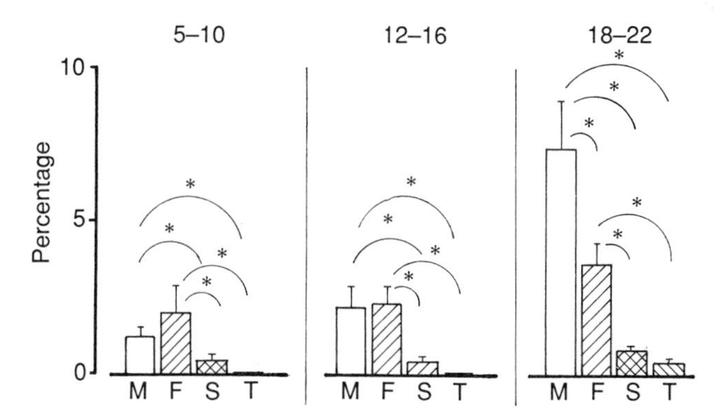


**Fig. 10.8** Percentage of time that infants spent in grooming interactions with mother (M), father (F), older siblings (S) and twin (T), during weeks 5–10, 12–16, and 18–22.
* $p < 0.05$ for Student's $t$ test.

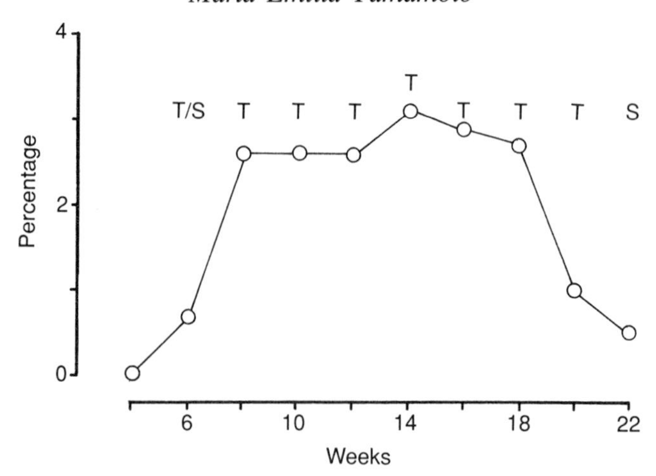


**Fig. 10.9** Percentage of time that infants spent in social play and main play partner (S, older siblings; T, twin) from weeks 4–22.

siblings were the main partners for play during the sixth and 22nd weeks. When only families with older siblings were considered, they were the main partners in weeks 6, 8, 14, 16 and 22. This result, as suggested by Cleveland and Snowdon (1984) for *Saguinus oedipus*, may indicate that older siblings have an important role in initiating the young infant to play. However, in families without older siblings, the most frequent partner was the twin in all weeks, suggesting that other members of the family do not take over that role in the absence of older siblings.

## Discussion and conclusions

The behavioural ontogeny of Callitrichidae presents very clear changes, which are most evident during the infant stage. Although changes from the end of infancy to sexual maturity are not, perhaps, as evident or numerous, they are nonetheless equally remarkable. All callitrichid infants present a decrease in time 'on' and a corresponding increase in time 'off' from birth until weaning. During time 'off' infants engage in social and solitary activities, such as grooming and play. Food sharing and stealing facilitate the transition from suckling to independent feeding, in a mechanism similar to that reported for Old World monkeys (Nicolson 1987), and also help the young animal to have a balanced nutritional intake (Ferrari in press). From the middle to the end of the infant period infants are already skilled groomers, quite independent, and able to move for themselves for long distances, as has been reported in field studies (Izawa 1978; Rylands

1981; Alonso 1984; Goldizen 1987*a*; Soini 1988). Differences in infant development in field and captive settings are difficult to detect, since field studies do not allow such detailed data as those obtained with captive animals. However, Stevenson and Rylands (1988) point out that the carrying period is extended in the field, and that juveniles do not participate in infant carrying. Also, some studies report differences in infant care for wild and laboratory-born parents kept in captive settings (Ingram 1978*b*; Silva *et al.* 1987), suggesting that captivity may have some influences on care patterns (see also Tardif *et al.*, this volume).

The other stages of development in callitrichids are more difficult to compare, since they have not been so well described. The birth of a new set of twins usually occurs at each stage, and the individual behaves toward them according to its age—more play and food sharing when younger, and carrying and providing contact as it gets older. Experience in the rearing of infants seems to increase future reproductive success of marmosets and tamarins (Ingram 1977; Epple 1978*a*; Cleveland and Snowdon 1984), and this could be the reason for extended family living, although the importance of experience is more clear for some species than others (Tardif *et al.* 1984). Puberty may bring social isolation, and sometimes even active eviction of the young animal from the group, especially if it is dominant. Kleiman (1979) suggests that the alpha pair suppresses, through aggression, high levels of assertiveness in their offspring. Observations in our colony, nevertheless, suggest that sometimes the subordinate animals, especially females, may suffer aggression from parents and siblings. These differences could be related to differences in fertility suppression (Snowdon 1990; French and Inglett 1991), in group composition or degree of kinship, or to opportunities of reproduction of subordinate animals (Ferrari and Lopes Ferrari 1989; Rylands 1989*a*). Field data report that groups receive individuals of various ages (Dawson 1978; Neyman 1978; Scanlon *et al.* 1988), and may have supernumerary animals not related to the alpha pair (Hubrecht 1984; Goldizen and Terborgh 1986; Rylands 1986*a*; Goldizen 1987*a*). Those transient or supernumerary animals could be sub-adults that chose to leave or were evicted from their natal groups. Reproduction for such animals depends on good opportunities, which could arise through the death or migration of the dominant animal or splitting of the group (Rylands 1989*a*).

Data available for *Callithrix jacchus* are mostly in agreement with the general pattern for callitrichid ontogeny. In this study, the father was the main caregiver in all five families, and parents carried infants more than older siblings. Some studies have indicated the father or adult male as the main carriers in the first weeks (*Callithrix*: Locke-Haydon and Chalmers 1983; Sommer 1984; *Saguinus*: Epple, 1975*a*; Goldizen 1987*a*), and others report a preferential period of carrying by the mother at the very beginning of life (*Cebuella*: Larsson *et al.* 1982; *Callithrix*: Buchanan-Smith 1984; *Leontopithecus*: Hoage 1978, 1982; Kleiman *et al.* 1988). Nonetheless, most studies report a high variability among families, which suggests that, except for the well-documented exclusive period of carrying by the mother in *Leontopithecus*, there seems to be no uniform pattern. This is further complicated by the fact that the behaviour of family members is affected by each others' patterns of care, especially if compensation occurs (Arruda *et al.* 1986; Ingram 1977; Yamamoto *et al.* 1987; Yamamoto 1990).

Other factors may affect the amount of care that parents offer to infants, such as previous experience (Epple 1978*a*; Hoage 1978; Ingram 1978*b*) and group size (Ingram 1977). The presence of older siblings is an important factor in the expression of care by parents. Most authors suggest that it is the mother who is mainly influenced by group size, decreasing the amount of care when there are other caregivers available (Ingram 1977), and increasing it in their absence or in the case of decreased care by the father (Arruda *et al.* 1986; Yamamoto *et al.* 1987; Yamamoto 1990). McGrew (1988) also observed the influence of group size in the amount of care offered by parents, but reports that the increase in group size had a greater decreasing effect for the father, not the mother. McGrew's study, however, presents a much broader definition of care than those previously used, which could account for the difference. Ferrari (in press) and Price (1990*b*) suggest that male infant care may be a form of courting and gaining access to the breeding female. Given the smaller degree of kinship and greater intragroup sexual competition in *Saguinus* than in *Callithrix* (Ferrari and Lopes Ferrari 1989), Ferrari (in press) expects that male care should be more related to breeding status in *Saguinus* than *Callithrix* species. It should be expected, then, that male care would be restricted mainly to the breeding male in *Saguinus* species, and be more widely distributed among all the males in *Callithrix* groups. More detailed studies are necessary to confirm this hypothesis.

The role of older siblings in infant care was quite small in the *Callithrix jacchus* families studied here. There are reports of much more active participation of older siblings in infant care in this and other callitrichid species (Epple 1975*a*; Ingram 1977; Hoage 1982; Cleveland and Snowdon 1984), with a greater involvement of adults and sub-adults. There is good evidence that parental skills are learned, and that laboratory-born animals must have direct experience with infants to become competent parents (Epple 1978; Hoage 1978; Tardif *et al.* 1984). Cleveland and Snowdon (1984) suggest a mechanism for the acquisition of such skills: juveniles learn to tolerate infants on their backs through play, so that by the time they are subadults they are ready to carry them. As a result, as adults, they are prepared to care for their own infants. In this study, neither juveniles nor adult siblings participated actively in infant carrying. As focal animals of families with older siblings in this study were all from the second litter, no sibling had previous experience with infants. This lack of experience, together with observations in our colony of experienced siblings that showed greater participation in infant care (Yamamoto 1990), suggest that the role of older siblings is related not only to age, but also to previous opportunities to interact with younger siblings. However, as in parental care, there is also great variability in sibling care, which, together with our small sample, make any conclusions premature.

Infant care brings costs to the animals involved, as shown by Goldizen (1987*a*). Price (in press *a*) investigated these costs in more detail in captivity, and found that carriers spend less time in feeding, locomotion, and social activities. Ximenes and Sousa (in press) report that in families without helpers the mother decreased the number of feeding bouts, as compared with families with one or two helpers, up to the week 8 of the infants' life. The effects of these costs on various group members should be better investigated, and also which behaviours, besides carrying, alleviate those costs on father and mother, usually the main caregivers, for a better understanding of the role of helpers.

The critical period for independence in our infant *Callithrix jacchus* was weeks 5–10, and especially the first two weeks. At this time infants sought caregivers most, and were highly rejected, marking the end of the carrying and nursing period. At this critical and difficult moment infants find support from older siblings, which engage in social play with them, and from the twin, which provides physical contact. Twin contact may act as an important compensation for decreased care. Increased twin contact has been reported in highly rejected infants (Yamamoto *et al.* 1987), or, as in this study, at periods of increased rejection or at weaning (Locke-Haydon and Chalmers 1983; Yamamoto 1990). Besides, there is evidence that singleton infants are more dependent, and that they maintain their dependence on parents longer (Silva *et al.* 1991), suggesting that the twin is important in the promotion of independence.

The overall similarity of the observed ontogeny of callitrichids suggests a behavioural unity, although it is difficult to confirm, given the high variability, even at the individual level. Some important differences are present, however, such as the exclusive period of carrying by the mother in *Leontopithecus* and active food-sharing in some species. But the paper by Ferrari (1987) suggests that some behaviours may be very rare in occurrence, or restricted to specific conditions, making them difficult to observe in captivity. This is the case for active food-sharing in *Callithrix flaviceps*, seen rarely, and only with animal prey, suggesting that some behaviours not observed in some genera and/or species may simply be infrequent, or that animals in captivity do not meet the conditions stimulating their occurrence.

Some other observed differences may be attributed to the high flexibility of callitrichid behaviour (Goldizen 1988), and interpreted as real. To cite just one example: the early sexual maturation of pre-pubertal females exposed to older animals of the opposite sex, such as reported by Epple (1978*a*).

Nonetheless, true specific differences have been reported lately, such as those related to suppression of ovulation in subordinate females, with consequences for the expression of aggression to conspecifics (French and Inglett 1991; Snowdon 1990).

In conclusion, more studies are necessary on the behavioural ontogeny of Callitrichidae. An emphasis on comparative aspects rather than processes, is required especially directed to those genera and species still poorly known, and to the dissimilarities between them. We should explore the differences and take a closer look at the similarities.

# Part III

## *Ecology*

# 11

# The ecology of the pygmy marmoset, *Cebuella pygmaea*: some comparisons with two sympatric tamarins

*Pekka Soini*

## Introduction

The pygmy marmoset, *Cebuella pygmaea*, an inhabitant of the upper Amazon lowland forests, is the smallest of the callitrichids: the average body mass for wild, non-pregnant, adult animals is 119 g (range 85–140 g, $n = 63$). The ecology and behaviour of *Cebuella* was studied in several areas in northeastern Peru, principally in the Manití, Tahuayo, and Tapiche river basins, from September 1976 to February 1979 (Soini 1982*b*, 1987*a*, 1988). Besides field observations, 115 animals, including over 20 complete troops, were captured live and examined for more detailed data on age/sex composition of the populations. A summary of the principal ecological and demographic data gathered is presented here.

## Troop size and composition

The natural populations are composed of stable, heterosexual troops, transient or incipient pairs or groups, and at least temporarily solitary animals of either sex. Troops range in size from 2–9 independently moving animals, with a modal size of 6–7, and mean size of 5.1 ($n$ = 80 troops). Besides, many troops contain one or two carried infants. Most troops are composed of one reproductive female, her mate, and her offspring of 1–4 successive litters. Some troops also contain an additional adult male or female that do not belong to the reproductive female's progeny. But even in the troops of two adult females, only one of them appears to be reproductively active. In one study troop containing two adult males, both attempted to mate with the female during her post-partum oestrus, but one of the males maintained exclusive copulatory access to the female through intensive guarding of the female and aggression towards the other male. However, whether or not such functional monogamy represents the prevalent mating pattern in wild *Cebuella* remains to be verified.

The sizes of the troops undergo temporal fluctuations; at the Manití study site about 62 per cent of this fluctuation was due to subadult/adult migrations out of and into the troops, and the remaining 38 per cent was due to recruitments through births and losses through juvenile mortality.

## Population structure

At Manití, about 47 per cent of the members of stable troops were adults. However, if we consider the entire population, which includes incipient pairs and solitary animals, adults comprised about 55 per cent. The sex ratio was about even, both amongst adults and non-adults.

## Habitat preference and population density

*Cebuella* is a habitat specialist, inhabiting the edges and interiors of seasonally inundated mature floodplain forests, where the general inundation level does not surpass 2–3 m in depth. Pygmy marmosets are not normally found in the high ground primary forest, except occasionally in the riparian vegetation of small forest streams. However, in areas altered by human activities, they often colonize edges of clearings, fields, orchards and secondary growth, whether in riverine habitat or on higher ground beyond the floodplain.

In the floodplain forest, the highest population densities always occur at the edges of rivers and their backwaters. For example, at the Manití site where the total *Cebuella* populations of a 3 km$^2$ rectangular plot of forest and river were censused in two successive years, the overall ecological densities were found to be 51.5 and 59.0 individuals/km$^2$, and the river edge density 210 and 227 individuals/km$^2$, respectively.

## Use of space

*Cebuella* troops occupy small, exclusive home ranges. The mean home range size for the seven, best-studied troops was 0.3 ha (range 0.1–0.5 ha). Observations on several additional troops suggest that 0.5 ha approximates the upper limit of a circumscribed area over which a stable *Cebuella* troop will normally range during its daily activities. Within these limits, the actual size of a home range area seems to be largely determined by the spatial distribution of the exudate source trees and vines the troop uses for its sustenance. When the exudate yield of a troop's feeding trees and vines drops below a certain critical level, the troop emigrates temporarily or permanently to a new site: alternatively, it may dissipate. For example, of 14 troops monitored at the Manití site, only seven were still resident in their original home range after one year of observation, and only four after two years. The 'lifetime range' of a *Cebuella* troop thus often consists of a series of successive small home ranges, occupied for time periods that vary from a few months to several years.

## Feeding ecology

Pygmy marmosets feed principally on arthropods and exudates of trees and vines. Fruits and other plant parts form only a minor portion of the diet. Exudate is fed on throughout the year. It is obtained principally by persistant gouging of holes through the bark of the boles, stems, and major branches of certain feeding trees and vines. An extensive survey in the Iquitos region has shown that *Cebuella* feeds on the exudate of at least 57 plant species, including 39 trees, 19 vines, and one species of hemiepiphyte. The great majority of the exudate sources belonged to the following four plant families: Vochysiaceae, Leguminosae, Anacardiaceae, and Meliaceae.

A troop usually has one principal feeding tree, on and around which most of the troop's social activities occur. This is the main exudate source for the dominant adults and the younger offspring, and one or several additional trees and vines may be used by them as secondary sources, and as principal sources for other troop members.

## Reproduction

In natural *Cebuella* populations, births occur throughout the year. In the region of this study, however, they tended to cluster around May–June and October–January, with only very few births occurring in August and September. Females often undergo a post-partum oestrus in the third week following delivery, and the interbirth intervals between successive litters of the females in the Manití population were normally 5–7 months. Nearly all the litters consist originally of twins, and at Manití the infant survival rate was estimated to be 67 per cent or slightly more.

## Discussion

In all the study areas, *Cebuella* shared its habitat with two other mutually sympatric callitrichids: the saddle-back tamarin, *Saguinus fuscicollis*, and the

**Table 11.1** Selected population parameters for *Cebuella pygmaea*, *Saguinus fuscicollis*, and *S. mystax* in the area of this study

| | *C. pygmaea* | *S. fuscicollis* | *S. mystax* |
|---|---|---|---|
| Body mass ($\tilde{x}$) | 119 g | 330 g | 515 g |
| Troop size ($\tilde{x}$) | 5.1 ($n$ = 80) | 5.8 ($n$ = 62) | 5.3 ($n$ = 133) |
| Adults/troop ($\tilde{x}$) | 2.8 | 3.2 | 3.7 |
| Adult sex ratio (males to each female) | 1.0 | 1.2 | 1.3 |
| proportion of adults in each troop | 0.47 | 0.61 | 0.65 |
| Average interbirth interval | 6 mo | 8 mo | 11–20 mo |
| Home range size | 0.1–0.5 ha ($n$ = 7) | 16 ha ($n$ = 1) | 25–35 ha ($n$ = 2)[a] |

[a] From Ramirez (1986)

moustached tamarin, *Saguinus mystax*. The ecology and population dynamics of these species were investigated concomittantly with the *Cebuella* study (Soini and Cóppula 1981; Soini and Soini 1982; Snowdon and Soini 1988), and a more detailed study of *S. fuscicollis* was carried out at an adjacent site (Soini 1987*b*). Of the two tamarins, *S. fuscicollis* has a body mass of nearly three times and *S. mystax* four and a half times that of the diminutive *Cebuella*. The lifestyle of *Cebuella* is also markedly different from that of its sympatric relatives. While *Cebuella* is a habitat specialist, almost entirely confined to the seasonally inundated floodplains, both tamarin species are habitat generalists, occupying both floodplains and upland forests. Besides feeding on arthropods, *Cebuella* is a specialized exudate feeder, whereas the tamarins are unspecialized insectivore–frugivores, eating exudate only to supplement their diet at times of low fruit availability. *Cebuella* troops occupy very small home ranges, over which the troop members are often widely scattered during their daily activities, whereas a tamarin troop occupies a relatively large home range, and travels as a single, cohesive, well-coordinated unit. The home range of tamarin troops tends to be temporally stable, lasting for the entire lifetime of the troop; in contrast, *Cebuella* home ranges are more ephemeral, and a troop typically changes its home range site several times in its lifetime.

These, and a few other subsidiary differences in the patterns of use of time and space between *Cebuella* and the tamarins, evidently result from the former's specialization on exudates. In contrast to the fruit resources exploited by the tamarins, which are relatively thinly and evenly distributed, the exudate sources occur in more concentrated and spatially widely scattered packages. *Cebuella* also differs from the sympatric tamarins in terms of troop composition. In the *Cebuella* troops the adult sex ratio is about even, whereas in the tamarins the sex ratio is skewed towards males, with an average of 1.2 or 1.3 males for each female (Table 11.1). Moreover, *Cebuella* troops average a lower number of adult animals but a higher number of immatures than the tamarin troops. This seems to be in part due to a higher reproductive rate in *Cebuella* compared to the tamarins. While wild *Cebuella* troops regularly produce two litters per year, *S. fuscicollis* do so less commonly and *S. mystax* only infrequently (Table 11.1). The higher reproductive success of *Cebuella* may also at least partly be the result of their dietary specialization, since exudate seems to be a temporally more stable resource than fruit, which shows marked seasonal fluctuations in abundance.

## Acknowledgements

The field work was supported primarily by the Peruvian Ministry of Agriculture (Iquitos Office), and partly by the Peruvian National Primate Project, Proyecto Primates (now Proyecto Peruano de Primatologia). Field work at Cahuana, Río Pacaya, received additional support from the Swiss Government through the programme 'Cooperación Técnica Suiza' in Peru.

# 12

# Habitats, feeding ecology, and home range size in the genus *Callithrix*

*Anthony B. Rylands and Doris S. de Faria*

## Introduction

The most recent taxonomic revisions of the marmoset genus *Callithrix* indicate more than 14 species and subspecies, distributed through the southern and eastern part of the Amazon basin (south of the Rio Amazonas and east of its southern tributary, the Rio Madeira) extending into eastern Bolivia and the northeastern tip of Paraguay (eight taxa), and throughout most of central, northeastern and southeastern Brazil (six taxa) (Mittermeier *et al.* 1988*b*; Ferrari and Lopes 1992*b*; Coimbra-Filho 1990; for review see Rylands *et al.* this volume). Although still lacking sufficient extensive field studies, the wide distribution of this genus testifies to the wide variety of habitats that marmosets occupy, and argues for significant ecological and behavioural differences between the various forms. In this paper, we review the current knowledge regarding their feeding and ranging behaviour with the aim of identifying key features which unite the genus and the differences which separate them, and which are related to their use of contrasting habitats. Aspects of their social behaviour have been reviewed by Sussman and Kinzey (1984), Sussman and Garber (1987), Stevenson and Rylands (1988), and Ferrari and Lopes Ferrari (1989) (see also Rothe and Darms, this volume).

## Field studies

The genus was, until recently, very poorly studied in the wild, although since 1978, five species have been subjects of more or less extensive studies of their ecology and behaviour (Table 12.1). The first long-term study was of *Callithrix humeralifer* (we argue that this form is more closely aligned to the bare-ear marmosets, *C. argentata*, see Rylands *et al.* this volume) in the north of the state of Mato Grosso (Rylands 1981, 1982, 1984, 1986*b*; Stevenson and Rylands 1988). Since then, *C. jacchus* has been studied at

**Table 12.1** Field studies of *Callithrix*

| Species | Reference | Duration | Location |
|---|---|---|---|
| *C. humeralifer* | Rylands (1982, 1984, 1986*b*) | 12 mo | Mato Grosso, Brazil |
| *C. kuhli* | Rylands (1982, 1984, 1989*b*) | 6 mo | Bahia, Brazil |
| *C. argentata melanura* | Stallings and Mittermeier (1983) Stallings (1985) Stallings *et al.* (1989) | | N.E. Paraguay |
| *C. flaviceps* | Ferrari (1988*b*) | 12 mo | Minas Gerais, Brazil |
| *C. aurita* | Torres de Assumpção (1983*b*) | 13 mo | São Paulo, Brazil |
| *C. aurita* | Muskin (1984*a*,*b*) | 5 mo | Minas Gerais, Brazil |
| *C. aurita* | Stallings (1988) Stallings and Robinson (1991) | 12 mo | Minas Gerais, Brazil |
| *C. jacchus* | Maier *et al.* (1982) Alonso and Langguth (1989) | 14 mo | Paraiba, Brazil |
| *C. jacchus* | Stevenson and Rylands (1988), Hubrecht (1984, 1985) Scanlon *et al.* (1988, 1989) | 6 mo 7 mo 18 mo | Pernambuco, Brazil |
| *C. penicillata* | Lacher *et al.* (1984) Fonseca and Lacher (1984) | 4 mo | Distrito Federal, Brazil |
| *C. penicillata* | Faria (1984*a*,*b* 1986, 1989) | 12 mo | Distrito Federal, Brazil |

two sites in the northeast of Brazil, at João Pessoa, Paraíba (Maier *et al.* 1982; Alonso and Langguth 1989) and Tapacurá, Pernambuco (Hubrecht 1984, 1985; Stevenson and Rylands 1988, Scanlon *et al.* 1988, 1989); *C. kuhli* was the subject of a short comparative study with the golden-headed lion tamarin, *Leontopithecus chrysomelas*, at Una, Bahia, in 1980 (Rylands 1982, 1984, 1989*b*; Stevenson and Rylands 1988); *C. aurita* was studied as part of a synecological study over 12 months in the state of São Paulo (Torres de Assumpção 1983*b*) and for five months in the south-east of the state of Minas Gerais by Muskin (1984*a*,*b*), and Stallings (1988; Stallings and Robinson 1991) also studied its habitats and densities in the Rio Dôce State Park; and finally *C. flaviceps* was studied during 12 months by Ferrari (1988*b*) in the Caratinga Biological Station in the state of Minas Gerais during 1984–1986. Information on *C. penicillata*, occurring in the savanna region of central Brazil, is unfortunately still rather patchy, but studies have been made on its gum-feeding behaviour by Fonseca and Lacher (1984; Fonseca *et al.* 1980; Lacher *et al.* 1984), and on its feeding, ranging, and social behaviour by Faria (1984*a*,*b*, 1986, 1989), all in central Brazil, near Brasília. Stallings (Stallings and Mittermeier 1983; Stallings 1985; Stallings *et al.* 1989) studied the habitats and densities of *C. argentata melanura* in Paraguay.

## Habitats

The widespread distribution of the genus implies that they are able to occupy a wide range of habitats. The Amazonian *Callithrix* occur in *terra firme* forests but are generally absent from seasonally or permanently flooded forest. *C. argentata* also occupies semi-deciduous forest patches in Amazonian-type savannas to the south of the lower Rio Amazonas in the region of the Rio Tapajós, and *C. humeralifer* has been observed in white sand forest patches in the Rio Aripuanã basin (Rylands 1981, 1982). Both of these vegetation types contain distinct physiognomies and floristic communities. One of the Amazonian subspecies, *C. argentata melanura*, extends far south, occurring in gallery forests and forest patches in the Pantanal of the Mato Grosso region of Brazil and into the *chaco* regions of Bolivia and northeastern Paraguay where it occurs in relatively high subhumid forests (in the far east) and in tall forests with dense undergrowth along ephemeral waterways (*cauces*) (Stallings and Mittermeier 1983; Stallings 1985, Stallings *et al.* 1989). *C. penicillata* occurs in more seasonal environments in the gallery forests, forest patches and savanna forest (*cerradão*) in Central Brazil (Faria 1986, 1989; Fonseca and Lacher 1984; Lacher *et al.* 1984). The common marmoset, *C. jacchus*, occurs in the northern extreme of the Atlantic coastal forest and forest patches and riverine forest in the highly dry thorn scrub or *caatinga* of northeast Brazil. The remaining species, *aurita*, *flaviceps*, *geoffroyi*, and *kuhli*, occur in the Atlantic forest region in the state of Bahia (*kuhli*) and east and south-east Brazil. There are two main types of Atlantic forest: the first is evergreen with a high annual rainfall exceeding 2000 mm and lacking in seasonality; and the second, in more inland areas, is a drier, seasonal semi-deciduous forest. The four species occur in both types, but whereas *C. kuhli* and *C. geoffroyi* are lowland species, *C. aurita* and *C. flaviceps* occur at higher elevations, in upland forest above 400–500 m.

Despite this wide range of habitats, a unifying feature for all the species studied so far is their use of edge formations and secondary growth forest. Irrespective of the forest types they occupy, marmoset groups are always associated with disturbed forest, edge, or secondary growth patches

(Rylands 1986*b*, 1987; Faria 1986; Stallings 1988; Ferrari and Mendes 1991). The same has been found to be true of the tamarins, *Saguinus* (Terborgh 1985; Rylands 1987; Garber this volume), and is related to the fruit types, fruiting patterns, and clumped distribution of pioneer plants. Important too may be a relative abundance of their insect prey in secondary forests and the advantage of dense vegetation for shelter and protection from predators (Terborgh 1983; Rylands 1986*b*, 1987; Snowdon and Soini 1988).

## Feeding ecology

The *Callithrix* species have been variously described as omnivores, frugivore–insectivores, and exudativore or gummivore–insectivores (Sussman and Kinzey 1984; Ferrari and Lopes Ferrari 1989). The significant feature of their feeding habits is their regular use of plant exudates, mostly gums, which they obtain by gouging tree trunks, branches, and vines of certain species, most particularly of the families Anacardiaceae and Leguminosae and, in the case of *C. penicillata*, the Vochysiaceae (Coimbra-Filho 1971; Coimbra-Filho and Mittermeier 1976, 1978; Rylands 1984; Stevenson and Rylands 1988). Coimbra-Filho (1971) reported that *C. jacchus* also feeds on the resinous exudate of *Protium* (Burseraceae), and *C. penicillata* has been observed to gouge and eat the red latex of *Hancornia speciosa* (Apocynaceae) (Rizzini and Coimbra-Filho 1981).

Although a number of primate species regularly include gums in their diets (see Nash 1986), *Callithrix* and the pygmy marmoset, *Cebuella*, are the only genera showing the gouging behaviour specifically to elicit exudate flow. The shape of the lower mandibles, the relatively large lower incisors and the intestinal morphology are adapted for this tree-gouging and gum-feeding behaviour (Hershkovitz 1977; Coimbra-Filho *et al.* 1980; Sussman and Kinzey 1984; Ferrari and Lopes Ferrari 1989). In addition, Rosenberger (1978; see also Nogami and Natori 1986) reported that the lower incisors show a lack of lingual enamel and a thickening of the buccal enamel which results in the lower incisors having a chisel-like structure. Neither *Saguinus* nor *Leontopithecus* possess these adaptations and although they may exploit this resource extensively, their use of gums is opportunistic and restricted to that which is readily available as a result of insect attack or damage (see Peres 1989; Passos and Carvalho 1991; Garber 1984*b*, this volume), or as reported by Soini (1987*b*), *S. fuscicollis* may on occasion pirate the exudate sources of *Cebuella* groups.

Whereas *C. humeralifer* and *C. kuhli* may be considered highly frugivorous (Rylands 1984, 1986*b*, 1989*b*), *C. jacchus*, *C. penicillata*, and *C. flaviceps* groups studied to date are highly exudativorous (Tables 12.2 and 12.3) and include little fruit in their diet (Maier *et al.* 1982; Alonso and Langguth

**Table 12.2** Percentage of daily activity spent in exudate feeding in *Cebuella* and *Callithrix*

| Species | Percentage time | Range[a] | Duration of study | Reference |
|---|---|---|---|---|
| *Cebuella pygmaea* | 32.0 | — | 5 months | Ramirez *et al.* (1978) |
| *C. jacchus* | >29.0 | — | 3 weeks | Maier *et al.* (1982) |
| *C. jacchus* | 15.0 | — | 12 months | Alonso and Langguth (1989) |
| *C. kuhli* | 7.0 | 6.6–7.1 | 3 months | Rylands (1982, 1989*b*) |
| *C. humeralifer* | 3.0 | 0.8–9.8 | 12 months | Rylands (1982) |
| *C. flaviceps* | 8.8 | 6.8–10.1 | 12 months | Ferrari (1988*b*) |

[a] Range of monthly estimates during the study.

**Table 12.3** Percentage composition of the plant part of the diet in *Cebuella* and *Callithrix*

| Species | Exudate (per cent) | Fruit, nectar, flowers, seeds (per cent) | Reference |
|---|---|---|---|
| *Cebuella pygmaea* | 67.0 | not est. | Ramirez *et al.* (1978) |
| *C. humeralifer* | 17.2 | 82.5 | Rylands (1982) |
| *C. flaviceps* | 72.5 | 15.9 | Ferrari (1988*b*) |
| *C. kuhli* | 32.6 | 67.4 | Rylands (1982, 1989*b*) |
| *C. penicillata* | est. >70 | not est. | Fonseca and Lacher (1984) |

1989; Ferrari 1988*b*; Lacher *et al.* 1984). Alonso and Langguth (1989) recorded eight species of fruits eaten by *C. jacchus* over 14 months. Similarly Ferrari (1988*b*) recorded only three species for *C. flaviceps* during 12 months. Faria's (1986, 1989) study of *C. penicillata* demonstrated their use of fruits of nine species during five months (five others were also probably consumed). *C. humeralifer*, on the other hand, was observed consuming fruits of 52 plant species of 24 families over a 12-month period, with one group using between three and 16 species in any one month (Rylands 1982; Stevenson and Rylands 1988). *C. kuhli* groups were observed eating fruits of 21 species over four months (Rylands 1982; Stevenson and Rylands 1988). The relative contribution of fruit to the diet of *C. humeralifer*, *C. kuhli*, and *C. flaviceps* (Table 12.3) reflects the number of species they use. *C. flaviceps* spent little time eating exudates (Table 12.2), although this resource contributed 72.5 per cent of its diet (Table 12.3). This is because exudate-feeding estimates include time spent gouging, which in *flaviceps* was very reduced because the majority of gum feeding, especially for *Anadenanthera* and *Acacia*, was from numerous sites with copious quantities, pre-empting the need to gnaw at holes. Other groups observed near Ferrari's study group were evidently rather more active in gouging, judging from the much greater numbers of gouge marks on trunks of plant species common to the home ranges of the groups.

Brief observations of *C. argentata* in the more seasonal, Amazonian-type savanna, near to Santarém, indicate that they include more gums in the diet than was normally true for *humeralifer*. Although very little is known of the *chaco* populations of *C. argentata melanura*, in eastern Bolivia and northeastern Paraguay they have not to date been observed gum-feeding (Stallings and Mittermeier 1983). The only observations of gum-feeding for *C. aurita* were made by Torres de Assumpção (1983*b*), who observed them feeding on *Astronium graveolens*. Gum-feeding by *C. aurita* was not observed by either Muskin (1984*a*,*b*) or Stallings (1988). *C. geoffroyi* has not been studied but has been observed tree-gouging and gum-feeding in a number of localities (Rylands personal observation) and we would predict that they are similar in the extent of their use of this resource to *kuhli*.

In conclusion, therefore, *C. kuhli* is intermediate, spending less of its feeding time consuming gums than *jacchus* or *penicillata*, but more than the Amazonian *humeralifer*. The study of *kuhli* was carried out in the evergreen humid coastal forest and we would predict that populations inland, in the drier semi-deciduous forest type, include a higher percentage of gums in the diet and that a similar pattern of gum use will be typical of at least *geoffroyi*. *C. aurita* and *C. flaviceps* probably tree-gouge less than the other Atlantic forest species (see below).

In all long-term studies to date, the use of gums has been shown to be more or less seasonal and negatively related to the availability of fruits. In *C. penicillata* gum-feeding was more conspicuous during the dry season (Faria 1986, 1989). A *C. humeralifer* group increased its gum-feeding at the end of the wet season when it was clearly lacking any major fruit sources (Rylands 1986*b*). Similarly *C. jacchus* was also observed to spend more time eating gums in the second half of the wet season, also when fruits were scarce (Alonso and Langguth 1989). It is evident that the exploitation of gums is frequently an energetically costly activity and, as both are basically energy sources, it is to be expected that fruits will be preferred when they are available.

Regarding their categorization as insectivorous, marmosets spend a considerable part of their day in foraging for animal prey. Comparable data are available only for three species, but they demonstrate a remarkable consistency; estimates varying from 24 to 30 per cent of their daily activities (Ferrari 1988*b*; Stevenson and Rylands 1988). Animal prey items eaten include insects (principally orthopterans), spiders, snails, frogs, lizards, and rarely small birds and nestlings (Stevenson and Rylands 1988). Methods of

**Table 12.4** Home range size estimates for *Cebuella* and some *Callithrix* species

| Species | Home range (ha) | No. of groups | Duration of study | Reference |
|---|---|---|---|---|
| *Cebuella* | 0.1–0.5 | 7 | 24 months | Soini (1988, this volume) |
| *C. jacchus* | 0.5 | 1 | 6 months | Stevenson and Rylands (1988) |
| *C. jacchus* | 2.5 | 1 | 3 weeks | Maier *et al.* (1982) |
| *C. jacchus* | 5.0 | 1 | 14 months | Alonso and Langguth (1989) |
| *C. jacchus* | 0.72–1.62 | 5 | 6 months | Hubrecht (1985) |
| *C. jacchus* | 2.5–6.5 | 3 | 18 months | Scanlon *et al.* (1989) |
| *C. penicillata* | 1.25–4.5 | 2 | 4 months | Fonseca and Lacher (1984) |
| *C. penicillata* | 3.5 | 5 | 12 months | Faria (1986, 1989) |
| *C. kuhli* | 10–12 | 8 | 3 months | Rylands (1982, 1984, 1989*b*) |
| *C. aurita* | 11, 16 | 2 | 13 months | Torres de Assumpção (1983*b*) |
| *C. aurita* | 11.5 | 8 | 5 months | Muskin (1984*a*,*b*) |
| *C. humeralifer* | 28.25 | 1 | 12 months | Rylands (1982, 1986*b*) |
| *C. flaviceps* | 35.5 | 1 | 12 months | Ferrari (1988*b*) |

foraging for animal prey are essentially the same for all the *Callithrix* species. They use predominantly a stealthy, stalk and pounce, foliage-gleaning method in the understorey and middle layers of the forest, which has also been described for the pygmy marmoset, *Cebuella* (see Soini 1988), and the moustached tamarins, *S. labiatus*, *S. imperator*, and *S. mystax* (see Garber this volume). This contrasts with the predominantly manipulative, specific-site foraging of the lion tamarins, *Leontopithecus*, and the saddle-back tamarin, *S. fuscicollis*. This difference, implying the exploitation of very different prey items, is perhaps the key feature permitting sympatry between *C. kuhli* and *L. chrysomelas* in southern Bahia (Rylands 1982, 1989*b*), and between *S. fuscicollis* and *Callithrix*, which form mixed groups to the east of the Rio Madeira in the state of Rondônia.

An interesting foraging mode, following army ant swarms (*Labidus praedator* and principally *Eciton burchelli*), has been observed for *C. humeralifer*, *C. kuhli*, *C. flaviceps*, and *C. geoffroyi* (Rylands *et al.* 1989). The flushing of the swarm evidently increases the availability of forest litter fauna, and it is somewhat surprising that this behaviour has not been reported for either *Saguinus* or *Leontopithecus*.

## Range size

The relative proportions of gum in the diet are evidently related to the home range size (Rylands 1984). The highly gummiverous *C. jacchus*, *C. penicillata*, and *Cebuella* have very small home ranges compared to the more frugivorous *humeralifer* (Table 12.4). *C. kuhli* and *C. aurita* are intermediate in their range sizes. Gums are available year round, and just a few good sources are evidently sufficient for a group's needs (Soini 1988; Alonso and Langguth 1989; Faria 1986, 1989; Rylands 1984). Fruits on the other hand are seasonal and sources tend to be more widely dispersed, requiring larger home ranges.

Ferrari's (1988*b*) *flaviceps* provide an exception, being highly gummivorous and yet having the largest home range recorded for any of the *Callithrix* species. I believe that at least one reason for this is their lack of gouging and their exploitation of small gum sources which were widely dispersed, although Ferrari indicates that animal prey availability may be the key factor. Terborgh (1983) argues against insect dispersion and abundance as a factor influencing home range size in *Saguinus fuscicollis* and *S. imperator*, and there is no evidence to suggest that insect abundance is greater, for example, in the very small ranges of *jacchus* and *penicillata*, also in highly seasonal habitats.

## Discussion

The common features of marmoset habitat and diet include (1) their dependence on secondary growth or disturbed forest or forest edge, (2) their ability to exploit gums when faced with permanent or seasonal shortage of fruits, and (3) their methods of foraging for animal prey.

The importance of secondary growth and edge, common to both *Callithrix* and *Saguinus*, has been ascribed largely to the distribution, abundance, fruit types, and fruiting patterns of pioneer trees (Terborgh 1983; Rylands 1986*b*, 1987) but also to the relative abundance of animal prey and the dense vegetation serving as protection from predators. In addition, marmosets and tamarins generally occupy the middle layers and understorey and are, to a greater or lesser extent depending on the species, vertical clingers and leapers, a type of locomotion which for a small animal evidently requires densely packed supports (see Garber 1980*a*,*b*; Ford 1986*b*). Of interest is that the abundance of fruits did not apply to the semi-deciduous, secondary forest occupied by the *C. flaviceps* studied by Ferrari (1988*b*). The pioneer species important in this case were gum sources. Ferrari indicates that his study site, however, was probably not typical for the species as a whole. *C. flaviceps* shows an odd distribution, overlapping to a large extent with *geoffroyi*, but at least in the east of its small range, in the more humid easterly part of the Atlantic forest (see above), being restricted to altitudes above 400–800 m where forests evidently have higher numbers of edible fruits (Ferrari 1988*b*). This dependence on edge and second growth contrasts with the other two callitrichid genera, *Cebuella* and *Leontopithecus*. *Cebuella* is an extreme specialist in gum-feeding and its habitat requirements are determined by the availability of heavily exploited sources of exudate (Soini 1988, this volume). *Leontopithecus*, like *Saguinus*, is unable to exploit gums unless they are readily available. They use second growth forest but require primary forest and swamp forest principally because of their animal-prey foraging method (especially their use of vascular epiphytes) and their dependence on tree holes for sleeping sites. Both epiphytes and sleeping holes are rare or absent in secondary growth (Coimbra-Filho and Mittermeier 1973*a*; Peres, 1986*a*,*b*; Rylands 1982, 1989*a*, this volume).

The extent of the tree-gouging/gum-feeding specialization is not equal amongst the species, with *C. jacchus* and *C. penicillata* excelling and thereby being able to occupy the most unfavourable and seasonal habitats, and frequently in very high densities. The Amazonian marmosets are ecologically very similar to the moustached tamarins except in gum-feeding. Although numerous *Saguinus* species do eat gums, they cannot depend on them and they also exploit nectar sources at times of fruit shortage (Terborgh 1983, 1985; Janson *et al.* 1981; Garber this volume). This is probably also true of *Leontopithecus* (Rylands 1982, 1989*b*). This specialization also means that *Callithrix* generally have smaller home ranges than *Saguinus* (up to 120 ha) and occur in higher densities (Ferrari and Lopes Ferrari 1989). The implications of these differences for the social organization and mating patterns of *Callithrix* on the one hand and *Saguinus* on the other are discussed by Ferrari and Lopes Ferrari (1989; see also Ferrari this volume). Although exploiting a relatively large number of species of fruits, *C. humeralifer*, *Leontopithecus*, and the Amazonian tamarins tend to concentrate on only a few fruiting species at a particular time (Rylands 1982, 1986*b*, 1989*b*; Peres 1986*a*,*b*; Terborgh 1983, 1985). One important aspect of the larger home ranges of *Saguinus* and *Leontopithecus* is that they can make use of a larger number of individuals of a particular fruiting species. Having access to a number of trees, which fruit earlier or later than the norm, enables them to exploit more fully the fruiting season of the species (Rylands 1989*b*).

Narrow hybrid zones occur in the state of Minas Gerais (*C. penicillata* × *C. geoffroyi*; Rylands, personal observation), on the border between the states of Minas Gerais and Espírito Santo (*C. geoffroyi* × *C. flaviceps*; Mendes 1989*a*; Ferrari and Mendes 1991), and in the north of Bahia (*C. jacchus* × *C. penicillata*; Coimbra-Filho 1971, Alonso *et al.* 1987, see Coimbra-Filho *et al.* this volume). In this last case, Coimbra-Filho (1971) argues that *C. jacchus* was introduced, being formerly limited to the north of the Rio São Francisco. Sympatry with mixed-species groups, however, has not been found amongst this genus, as it has for *Saguinus*, and the reason for this probably lies in the lack of any identifiable difference in the animal-prey foraging methods they use, as well as their use of exudate sources to overcome seasonal or permanent fruit shortage.

**Table 12.5** Ecological groupings for the genus *Callithrix*

1. *C. jacchus, C. penicillata*
   Northeast and central Brazil, very seasonal habitat, highly exudativorous, small home ranges
2. *C. kuhli, C. geoffroyi*
   Atlantic coastal, lowland, forest of south-east Brazil, less exudativorous than Group 1 but better adapted for tree-gouging than Groups 3 and 4, home ranges larger than Group 1 but smaller than Group 4.
3. *C. aurita, C. flaviceps*
   Southernmost forms in seasonal, high altitude Atlantic coastal forest in south-east Brazil, relatively poor adaptation for tree-gouging, proportion of exudates in the diet dependent on availability, home ranges larger than Group 1 but smaller than Group 4. (*C. argentata melanura* in highly seasonal extra-Amazonian habitats (e.g. *chaco* of Paraguay).
4. *C. humeralifer, C. argentata*
   Amazonia, highly frugivorous (seasonally exudativorous), relatively poor adaptation for tree-gouging, large home ranges.

The extent of specialization on gums is reflected in the dentition of these marmosets. The lower incisors (used for gouging) of *C. argentata* and *C. humeralifer* are shorter than those of *C. jacchus*, and are transitional in size and shape between *C. jacchus* and the non-gouging tamarins, *Saguinus* (Hershkovitz 1977, p.575). Similarly, the lower canine teeth of *jacchus* are more incisoriform than those of the '*Argentata* group', which again show an intermediate stage to the truly caniniform canines of *Saguinus* (Hershkovitz 1977, p.576). Natori (1986) made a detailed examination of the dental morphology of seven of the *Callithrix* species and constructed a cladogram of phylogenetic relationships, based on the distribution of derived, intermediate-derived, and primitive character states. He maintained the '*Argentata*' and '*Jacchus* groups' identified by Hershkovitz (1977) as separating early in marmoset evolutionary history and, like Hershkovitz (1968), argued that *C. aurita* and *C. flaviceps* differentiated soon after. *C. geoffroyi*, *C. penicillata*, and *C. jacchus* accumulated further derived traits in that order. *C. kuhli* was found to be intermediate but distinct from *C. geoffroyi* and *C. penicillata* (Natori 1990). Natori's (1986) sequence corresponds to the ecological specialization of these marmosets. The results of Natori (1986) and our conclusions regarding the degree of specialization in terms of habitat and diet disagree with Hershkovitz's (1968) appraisal, based on coat colour and hair patterns, only regarding *geoffroyi*, which he places as a derived form of *penicillata*.

In conclusion, a broad characterization of the habitats, feeding ecology and range size of the *Callithrix* species separates them into four main groupings (Table 12.5). The frugivorous forms of *C. argentata* and *C. humeralifer* (group 4), use relatively large home ranges in the humid tall tropical Amazon forests. At the other extreme are the highly exudativorous forms,

*C. jacchus* and *C. penicillata* (group 1), with small home ranges, which have been able to occupy highly seasonal gallery forest, scrubby forests, and forest patches in the northern extreme of the Atlantic forest, semi-arid scrub of the *caatinga* of north-east of Brazil, and *cerrado* (bush savanna) of central Brazil. The Atlantic forest forms, *C. kuhli* and *C. geoffroyi* (group 2), we predict are ecologically similar and intermediate, being less frugivorous with smaller home ranges than the Amazonian marmosets, and less exudativorous with larger home ranges than either *jacchus* or *penicillata*. The most southerly forms, *C. argentata melanura* in the *chaco* and *C. flaviceps* and *C. aurita* (group 3) in the southern subtropical parts of the Atlantic forest, may also be considered separately as a fourth group. The distributions of *C. geoffroyi*, *C. flaviceps*, and *C. aurita* are somewhat confused, forming a still undeciphered mosaic in the south-east of Brazil. *C. aurita* is the southernmost species, occurring in the southern parts of the states of Rio de Janeiro, São Paulo, and Minas Gerais, but extending north up the Rio Dôce basin where it is surrounded by a mosaic of *flaviceps* and *geoffroyi* populations. *C. flaviceps* populations are intermixed, but not generally syntopic (there is one recorded hybrid zone, see below), over a large part of their range with *geoffroyi*. *C. flaviceps* has the most restricted range of the south-east Brazilian marmosets, being limited to the south-east of the states of Minas Gerais and Espírito Santo and possibly the north of the state of Rio de Janeiro. Coimbra-Filho (1971) pointed out that *C. flaviceps* occurs in upland forests in mountainous regions at altitudes above 400 m above sea level, whereas *C. geoffroyi* occupies forests of lower altitudes. This pattern has been confirmed by Mendes (1989*a*) who reported that in Espírito Santo and the east of Minas Gerais, *flaviceps* occurs above 850 m, whereas *geoffroyi* occurs below 500 m. Between these altitudes occur groups of both species along with groups of each with hybrids. Group 3 (Table 12.5) occupies relatively harsh, cool, and highly seasonal subtropical forests at the southernmost limits of the family's geographical distribution and yet have evidently not evolved the degree of specialization for exudate-eating which is the key to the success of *C. jacchus* and *C. penicillata* in occupying the *caatinga* and *cerrado* of north-east and central Brazil.

# 13

# Feeding ecology and behaviour of the genus *Saguinus*

*Paul A. Garber*

## Introduction

Tamarins of the genus *Saguinus* represent a successful radiation of New World primates that exploit a wide range of primary, secondary, and forest edge habitats from western Panama throughout much of the Amazon Basin of Colombia, Ecuador, Peru, Bolivia, and northern Brazil (Hershkovitz 1977; Mittermeier and Coimbra-Filho 1981, Snowdon and Soini 1988). In terms of species number and subspecific differentiation, this genus is the largest and most diverse platyrrhine lineage with three primary groupings (*hairy-face taxa*, *mottled-face taxa*, and *bare-face taxa*), 11 species, and 29 recognized subspecies (Hershkovitz 1977; Mittermeier and Coimbra-Filho 1981; Mittermeier *et al.* 1988*b*) (Table 13.1). Although there is little fossil evidence from which to reconstruct the evolutionary history of this group, differentiation of many of the current species is likely to have occurred sometime during the Pleistocene (Hershkovitz 1977; Kinzey 1982; Ford 1986*a*).

A major controversy in the evolution of New World primates has centred on the question of the derived or primitive nature of the callitrichine radiation. Although researchers such as Beattie (1927), Eisenberg (1978), and Hershkovitz (1977) have argued that tamarins and marmosets are among the most primitive of living primates and have diverged little from an ancient anthropoid stock, there is a growing consensus that numerous features of callitrichine biology and behaviour represent a set of derived and highly specialized character traits (Table 13.2). Given that each of these traits is present in all living callitrichines, and virtually absent in other New World taxa, they are likely to reflect a related adapted pattern inherited from a common ancestor (Leutenegger 1973, 1980; Rosenberger

Although the taxonomic scheme proposed by Rosenberger (1981) places the genera *Saguinus*, *Leontopithecus*, *Callithrix*, *Cebuella*, and *Callimico* in the subfamily Callitrichinae, in this chapter the term callitrichine is restricted to tamarins and marmosets and does not include Goeldi's monkey (*Callimico*).

**Table 13.1** Systematics of the callitrichine genus *Saguinus* (Adapted from Hershkovitz 1977, Mittermeier and Coimbra-Filho 1981, and Mittermeier *et al.* 1988*b*)

**Hairy-face group**
*S. fuscicollis*
*S. imperator*
*S. labiatus*
*S. midas*
*S. mystax*
*S. nigricollis*
Total: 6 species and 26 subspecies

**Mottled-face group**
*S. inustus*
Total: 1 species

**Bare-face group**
*S. bicolor*
*S. geoffroyi*[a]
*S. leucopus*
*S. oedipus*
Total: 4 species and 3 subspecies

Total for all 3 groupings: 11 species and 29 subspecies

[a] Hershkovitz (1977) does not assign full species status to the rufous-naped tamarin of Panama and northeastern Colombia. He considers it to be a subspecies of *Saguinus oedipus* (*S. oedipus geoffroyi*).

**Table 13.2** Biological and behavioural traits characteristic of the primate subfamily Callitrichinae. For a more detailed description of callitrichine mating patterns and social organization see Goldizen (1987*a*,*b*, 1990), Sussman and Garber (1987), Snowdon and Soini (1988), and Stevenson and Rylands (1988)

Small body size
Claw-like nails on all digits excluding the hallux
Ability to adopt a clinging posture on large vertical supports
Loss of third maxillary and mandibular molars
Upper molars tritubercular and lacking a hypocone
Procumbent lower incisors with thick labial enamel and an absence of enamel on the lingual aspect[a]
Reproductive twinning
High litter weight
Ability to produce young twice during the year
Evidence of suppressed ovulation among subordinate adult female group members
Extensive male assistance in infant care
Helpers and communal care of the young

[a] Found only in the marmoset genera *Cebuella* and *Callithrix*.

1977, 1981; Ford 1980*b*, 1986*a*; Garber 1980*a,b*, in press; Byrd 1981; Sussman and Kinzey 1984).

Evidence presented by Leutenegger (1973, 1980), Rosenberger (1977, 1981) and Ford (1980*b*, 1986*a*) suggests that many callitrichine adaptations are a direct result of phyletic dwarfism and the constraints that reduced body size have imposed on diet, foraging patterns, and reproductive success. Although dwarfing is likely to have been an important factor in callitrichine evolution, it provides only a partial explanation for the existence of these traits. Despite the fact that certain aspects of tamarin and marmoset biology parallel those found in other lineages of small (but not necessarily dwarfed) primates, the combination of features exhibited by callitrichines appears to reflect a series of specific adaptations to a particular ecological niche (Garber 1980*a,b*, 1984*a*, in press; Sussman and Kinzey 1984). The purpose of this paper is to examine questions regarding behavioural variability and adaptive unity in tamarin feeding ecology, and identify relationships between habitat use, ranging, and dietary patterns which characterize species in this taxonomically diverse primate genus.

## Field studies

Prior to 1975 our knowledge of the behaviour and ecology of tamarins in the wild was based solely on anecdotal accounts, censuses, and brief surveys. In more recent years however, detailed and systematic field investigations of five months or more have been conducted on seven species (Table 13.3). *Saguinus fuscicollis* is the most commonly studied species. This is related to its widespread geographical distribution and tendency to form stable longlasting associations with other callitrichine taxa such as *S. mystax*, *S. labiatus*, *S. imperator*, and *S. nigricollis*, as well as *Callithrix argentata*. Although the ecological constraints associated with single and multispecies tamarin troops are likely to differ (i.e. predation pressure, feeding competition, etc.), studies on both types of social groupings provide an expanded data base from which to identify and compare species-specific patterns of behaviour and ecology.

## Feeding behaviour

Tamarins are monomorphic primates exhibiting only minor differences in body and canine size between the sexes. In some species males are heavier than females, whereas in others females are heavier than males. These differences rarely exceed 5 per cent (Dawson 1976; Hershkovitz 1977; Soini and Cóppula 1981; Leutenegger and Larson 1985; Garber and Teaford 1986).

**Table 13.3** Long-term field studies of *Saguinus*

| Species | Duration | Location | Reference |
|---|---|---|---|
| *S. geoffroyi* | 19 mo | Panama | Dawson (1976, 1978, 1979) |
| *S. geoffroyi* | 8 mo | Panama | Garber (1980*a*,*b*, 1984*a*,*b*) |
| *S. oedipus* | 24 mo | Colombia | Neyman (1978) |
| *S. imperator*[a] | 24+ mo | Peru | Terborgh (1983, 1985, 1986) |
| *S. labiatus*[a] *labiatus* | 6 mo | Bolivia | Izawa and Yoneda (1981) Yoneda (1981, 1984a) |
| *S. labiatus*[a] *labiatus* | 5 mo | Bolivia | Pook and Pook (1982) |
| *S. labiatus*[a] *labiatus* | 5 mo | Bolivia | Buchanan-Smith (1990) |
| *S. mystax*[a] *mystax* | 12 mo | Peru | Garber (1986, Garber 1988*a*,*b*) Garber and Teaford (1986) |
| *S. mystax mystax* | 5 mo | Peru | Pruetz and Garber (1991) |
| *S. mystax*[a] *mystax* | 10 mo | Peru | Norconk (1986) |
| *S. mystax*[a] *mystax* | 15 mo | Peru | Ramirez (1989) |
| *S. nigricollis* | 7 mo | Colombia | Izawa (1978) |
| *S. fuscicollis illigeri* | 15 mo | Peru | Soini (1987*b*) |
| *S. fuscicollis weddelli* | 15 mo | Peru | Crandlemire-Sacco (1986) |
| *S. fuscicollis*[a] *weddelli* | 5 mo | Bolivia | Pook and Pook (1981, 1982) |
| *S. fuscicollis*[a] *weddelli* | 6 mo | Bolivia | Izawa and Yoneda (1981) Yoneda (1981, 1984*a*,*b*) |
| *S. fuscicollis*[a] *weddelli* | 5 mo | Bolivia | Buchanan-Smith (1990) |
| *S. fuscicollis*[a] *weddelli* | 24+ mo | Peru | Terborgh (1983, 1985, 1986) |
| *S. fuscicollis*[a] *weddelli* | 24+ mo | Peru | Goldizen (1987*a*,*b*, 1990) |
| *S. fuscicollis*[a] *nigrifrons* | 12 mo | Peru | Garber (1986, 1988*a*,*b*) Garber and Teaford (1986) |
| *S. fuscicollis*[a] *nigrifrons* | 10 mo | Peru | Norconk (1986) |

[a] Formed a mixed species troop with another tamarin species.

**Table 13.4** Body weight of adult wild-caught tamarins

| Species | Mean adult body weight (males) | Sample size | Reference |
|---|---|---|---|
| *S. mystax* | 564 g | 16 | Garber and Teaford (1986) |
| *S. mystax* | 505 g | 161 | Soini (1982*c*) |
| *S. midas* | 533 g | 3 | Fleagle and Mittermeier (1980) |
| *S. labiatus* | 491 g | 17 | Yoneda (1981) |
| *S. geoffroyi* | 486 g | 53 | Dawson (1976) |
| *S. oedipus* | 432 g | 25 | Hershkovitz (1977) |
| *S. oedipus* | 406 g | 6 | Neyman (1978) |
| *S. fuscicollis nigrifrons* | 413 g | 33 | Garber and Teaford (1986) |
| *S. fuscicollis weddelli* | 405 g | 4 | Yoneda (1981) |
| *S. fuscicollis nigrifrons* | 354 g | 39 | Soini and Cóppula (1981) |
| *S. fuscicollis illigeri* | 320 g | ca. 13 | Soini and Cóppula (1981) |

Information on the mean body weight of wild-trapped adult male tamarins is presented in Table 13.4. These data indicate a range of values from approximately 320 g in populations of the smallest species, *S. fuscicollis*, to 564 g in *S. mystax*. Values for females were excluded in order to avoid problems associated with changes in weight during reproduction (in the wild, saddle-back tamarin females may lose as much as 22 per cent of their body weight during lactation; Garber and Teaford 1986).

Due to their small body size, limited gut volume, and rapid rate of food passage (Crandlemire-Sacco 1986; Garber 1986) tamarins require a diet high in nutrient quality and available energy. In all species studied, insects, ripe fruits, plant exudates, and nectar are the primary components of the diet. Fundamental differences in the nutritional content, seasonal availability, distribution, and habitat location of these resources have a major impact on tamarin feeding and ranging patterns.

## Insectivory

Insects provide tamarins with a dependable source of proteins and lipids throughout the entire year. These small-bodied primates expend a significant amount of their daily activity searching for invertebrate prey. Field studies indicate that arthropods account for 30–77 per cent of total feeding and foraging time (Garber 1980*a*,*b*, 1984*a*, 1988*a*,*b*, in press; Mittermeier and Roosmalen 1981; Soini 1987*b*; Soini and Cóppula 1981; Terborgh 1983; Snowdon and Soini 1988). Hunting techniques include stealth, turning over leaves, exploring crevices and knotholes, pouncing, rummaging through palm fronds, and jumping rapidly to the ground to seize cryptic prey. Group members forage for insects independently and are often 1–10 m from their nearest neighbour (Garber 1980*b*; Yoneda 1984*b*; Soini 1987*b*). In those cases when individuals are found in close proximity, there is no evidence of a co-ordinated or co-operative effort. Rather, foraging success appears to be dependent on selecting appropriate areas of the forest and times of the day when the opportunity for prey detection and capture are high. A comparison of species-specific differences within the genus indicates at least three distinct insect foraging patterns. These are characterized by differences in hunting techniques, substrates exploited, vertical stratification, and modes of positional behaviour.

### *Pattern 1*

In the case of the Panamanian tamarin (Garber 1980*a*,*b*, 1984*a*) insects are hunted on thin flexible branches in the low shrub layer of the forest understorey. Foraging substrates are small (66 per cent were ≤5 cm in diameter) and located some 1–5 m above the ground (Fig. 13.1A). *S. geoffroyi* actively engages in a series of energetically costly locomotor and postural activities during insect foraging, including climbing, grasping, and jumping. These account for 81 per cent of the positional repertoire. More stationary postures like sitting, standing, and walking account for only 18 per cent. Foraging success appears to be related to the ability of this primate to move cryptically and with minimal disturbance in the understorey. These microhabitats include areas of low shrubs and vine tangles adjacent to tree-fall gaps and the forest margin. Prey are gleaned from leafy substrates by striking rapidly with the forelimbs while the hindlimbs maintain a firm grasp on the supporting vegetation.

### *Pattern 2*

A second pattern of insect foraging characterizes species such as *S. mystax*, *S. labiatus*, and *S. imperator*, and possibly *S. midas*. These primates exploit insects on leaves and branches in the lower and middle levels of the forest canopy (Mittermeier and Roosmalen 1981; Yoneda 1981, 1984*a*; Terborgh 1983, 1985; Garber 1988*b*). Detailed observations of insect foraging in *S. imperator* (Terborgh 1983, 1985) indicate that visual scanning plays a primary role in the detection of exposed and mobile prey. The emperor tamarin captures insects on the leaves of trees and vines using '. . . a stealthy advance and lightning attacks . . .' (Terborgh 1983, p.107).

The pattern of foraging for canopy insects described for *S. imperator* is extremely similar to that observed in *S. mystax*. In this latter species, relatively sedentary postures such as quadrupedal walking, standing, and

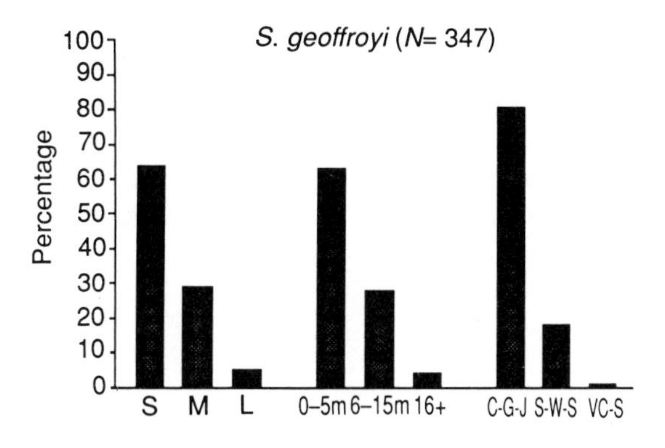


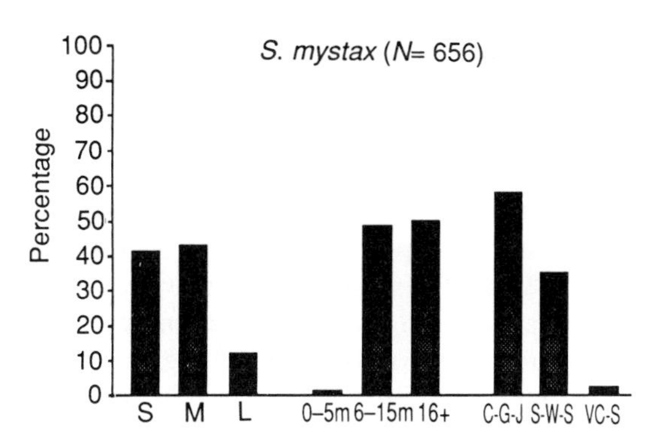


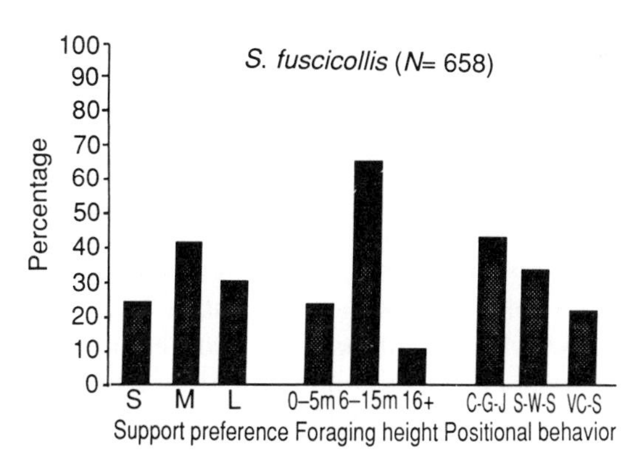


**Fig. 13.1** Support preference during insectivory: S, small (supports of approximately ≤5 cm in diameter, which tamarins can grasp with their hands and feet); M, medium (supports of approximately 5–10 cm in diameter, which tamarins can grasp with their arms and legs); L, large (supports in excess of 10 cm in diameter, too large to be grasped by tamarins). Positional behaviour: C-G-J, climbing, grasping, and jumping; S-W-S, standing, walking, and sitting; VC-S, vertical clinging and scansorial locomotion.

sitting account for 35 per cent of positional behaviour (Fig. 13.1B). Although the moustached tamarin was frequently observed to exploit branches of small diameter, 45.5 per cent of insect foraging and feeding occurred on supports of moderate size (5–10 cm in diameter). These activities were concentrated in the lower and middle layers of the crown, with 49 per cent of all prey captured above a height of 15 m. In contrast to Pattern 1, less than 1 per cent were captured below 6 m.

*Pattern 3*

The third and perhaps most distinctive foraging pattern is reported for *S. fuscicollis* and possibly *S. nigricollis* and *S. bicolor*. In all areas where they have been studied, saddle-back tamarins concentrate their feeding efforts on relatively large and cryptic invertebrate prey (Izawa 1978; Terborgh 1983; Yoneda 1984*b*; Crandlemire-Sacco 1986; Soini 1987*b*; Snowdon and Soini 1988). For example, Terborgh (1983) found that 42 per cent of the invertebrates eaten by *S. f. weddelli* in southeastern Peru were 1 cm or more in length. Similarly, Yoneda (1981) reports that many of the grasshoppers commonly captured by saddle-back tamarins in Bolivia approach a length of 8 cm. These insects are procured in all levels of the forest canopy and on a wide range of substrates includes leaves, branches, palm crowns, bark surfaces, knotholes, and the ground. When exploiting canopy insects, foraging techniques in *S. fuscicollis* closely resemble those described in Pattern 2 and involve sitting, standing, and quadrupedal walking (Fig. 13.1C).

However, in contrast to other tamarin species the majority of insect feeding in *S. fuscicollis* occurs on supports of moderate (42.5 per cent) and large (31.6 per cent) size, principally trunks, in the forest understorey (Fig. 13.1C). Saddle-back tamarins exploit these trunks as a primary foraging platform using a combination of vertical clinging postures (79 per cent) and scansorial locomotion (21 per cent) (Garber, unpublished data). By embedding its elongated and laterally compressed claw-like nails into the bark, *S. fuscicollis* is able to adopt a stable posture from which to manually explore knotholes, crevices and other regions of the trunk. Prey are seized in one or both hands and killed with a rapid bite to the head. Vertical trunks also serve as a perch from which to locate terrestrial prey. 25–75 per cent of insect foraging in this species is reported to occur at a height of less than 6 m above the ground (Terborgh 1983; Yoneda 1984*a*,*b*; Norconk 1986; Soini 1987*b*; Garber 1988*b*, in press).

Thus, although *S. fuscicollis* resembles *S. geoffroyi* in exploiting insects in the forest understorey and resembles *S. mystax* in exploiting insects in the canopy, 'by virtue of its great agility on vertical surfaces, the saddle-back tamarin has a foraging niche all to itself' (Terborgh 1985, p.299).

## Prey choice

Field observations indicate that tamarins are highly selective in their choice of arthropod prey. An analysis of the stomach contents of 129 wild-shot *S. geoffroyi* (Dawson 1976) indicates that irrespective of the time of year orthopteran insects contribute 65.9–77.3 per cent of the volume of the stomach ingesta. Although many varieties of orthopterans were present, large-bodied long-horned grasshoppers of the family Tettigoniidae were the most frequently consumed prey. Similarly, large-bodied orthopterans are the most commonly observed insects captured by *S. fuscicollis* (Soini and Cóppula 1981; Yoneda 1981, 1984*a*,*b*; Terborgh 1983; Crandlemire-Sacco 1986; Soini 1987*b*; Garber, personal observation), and *S. nigricollis* (Izawa 1978). Estimates of the percentage of orthopteran insects in the diet of *S. fuscicollis* range from 61 to 67 per cent (Terborgh 1983; Crandlemire-Sacco 1986). Tamarin species such as *S. mystax*, *S. imperator*, and *S. labiatus* were also observed to consume significant quantities of orthopteran prey, but these appear to be somewhat smaller in size and more mobile (Terborgh 1983, 1985; Yoneda 1984*a*; Garber, personal observation) than those exploited by *S. fuscicollis*. A preliminary study of feeding behaviour of *S. midas* (Mittermeier and Roosmalen 1981) suggests a similar pattern.

## Plant resources

### *Frugivory*

Ripe fruits and nectar are a high-energy resource rich in non-structural carbohydrates and simple sugars. Data collected on *S. midas* (Fleagle and Mittermeier 1980), *S. fuscicollis* (Crandlemire-Sacco 1986; Garber 1986, 1988*a*,*b*), *S. mystax* (Garber 1986, 1988*a*,*b*), *S. geoffroyi* (Garber 1980*a*, 1984*a*,*b*), *S. oedipus* (Neyman 1978), *S. labiatus* (Yoneda 1981, 1984*a*) and *S. imperator* (Terborgh 1983) indicate that the majority of fruit and nectar feeding takes place on small to moderate-sized branches in the periphery of the tree crown. In all species studied, ripe fruits account for 20–65 per cent of total feeding time. Nectar is consumed principally in the dry season, and during certain years, appears to be an important dietary staple.

A more detailed and quantitative comparison of substrate preferences, vertical ranging patterns, and positional behaviour during frugivory is presented in Fig. 13.2. The data indicate that in contrast to insect hunting, *S. mystax*, *S. fuscicollis*, and *S. geoffroyi* all exhibit an extremely similar pattern. Each species avoids exploiting the lowest layers of the forest and feeding on large branches and trunks. The preferred positional behaviours are sitting and grasping postures.

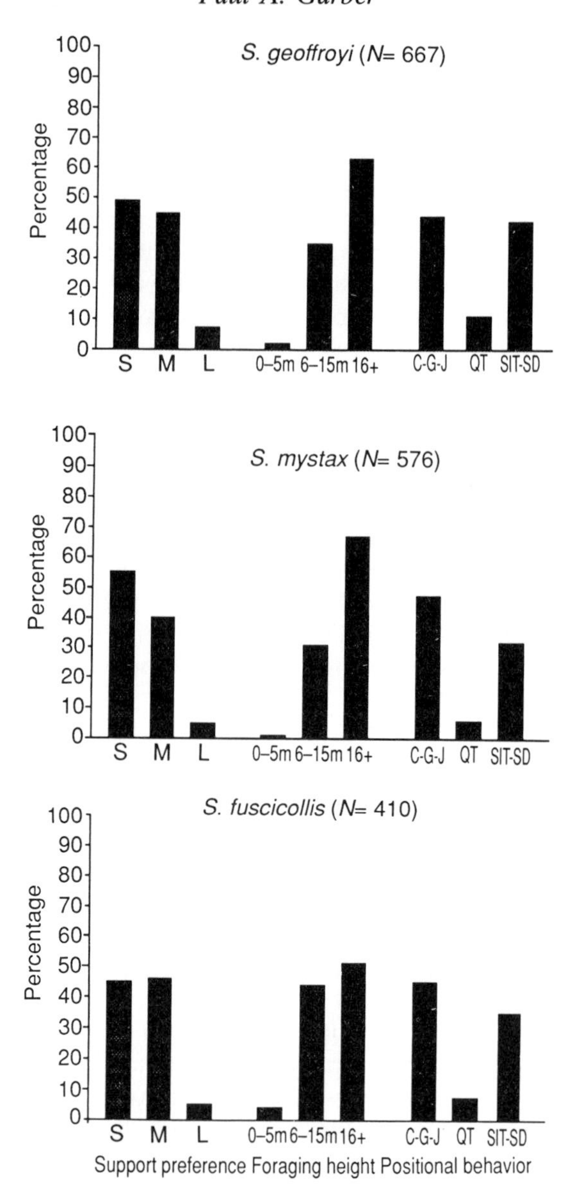


**Fig. 13.2** Support preference during frugivory: S, small (supports of approximately ≤ 5 cm in diameter, which tamarins can grasp with their hands and feet); M, medium (supports of approximately 5–10 cm in diameter, which tamarins can grasp with their arms and legs); L, large (supports in excess of 10 cm in diameter, too large to be grasped by tamarins). Positional behaviour: C-G-J, climbing, grasping, and jumping; QT, quadrupedal walking and running; SIT-SD, sitting and standing.

**Table 13.5** Characteristics of seeds swallowed and passed through the tamarin digestive tract. Seed volume was calculated using the formula $V = R^2 (L\text{-}2/3R)$, where $V$ is Volume, $R$ is (width + height)/4, and $L$ is length.

| Species | Mean seed dimensions | | | | |
|---|---|---|---|---|---|
| | Weight (g) | Length (cm) | Width (cm) | Volume ($cm^3$) | Sample size |
| *S. geoffroyi* | 0.27 ±.02 | 1.14 ±.40 | 0.63 ±.20 | 0.29 ±.23 | 129 |
| *S. mystax* | 0.66 ±.31 | 1.37 ±.30 | 0.87 ±.15 | 0.57 ±.29 | 122 |
| *S. fuscicollis* | 0.64 ±.30 | 1.47 ±.40 | 0.94 ±.16 | 0.69 ±.34 | 35 |

Many of the fruiting species eaten by *Saguinus* are drupes or arillate fruits which contain a single, or small number of, large seeds (Terborgh 1983; Crandlemire-Sacco 1986; Garber 1986; Soini 1987*b*; Snowdon and Soini 1988). Generally, the sticky pulp or aril is swallowed while still attached to the seed. The seed is undigested and generally passes unharmed through the digestive tract in 1–3 h (Crandlemire-Sacco 1986; Garber 1986). Although many species of primates play a role in seed dispersal, tamarins are unusual in that seeds are voided singly and are extremely large in comparison to the size of the animal. Data collected by Garber (1986, unpublished data) indicate that the mean length of seeds voided by *S. geoffroyi*, *S. fuscicollis*, and *S. mystax* exceeds 1.1 cm (Table 13.5). A single individual may have as may as 13 seeds, each in excess of 1.2 cm in length in its digestive tract at one time. Large undigested seeds have also been reported in the stomachs or faeces of *S. oedipus* (Hampton *et al.* 1966), *S. nigricollis* (Izawa 1978), *S. imperator* (Terborgh 1983), and *S. midas* (Hershkovitz 1977; Mittermeier and Roosmalen 1981). Although the frequency with which seeds are ingested appears to vary between species (Garber 1986), voided seeds exhibit high germination success (Hladik and Hladik 1969; Crandlemire-Sacco 1986; Garber 1986, unpublished data). The physiological costs and benefits of passing large seeds and the effectiveness of tamarins as agents of seed dispersal require further study.

## Exudate feeding

Many if not all species of tamarins are reported to consume plant exudates (Nash 1986; Garber in press). These are principally gums (although marmosets may consume saps and latex; Ramirez *et al.* 1978; Stevenson and Rylands 1988) from a limited number of plant families such as the

Anacardiaceae, Leguminosae, Combretaceae, and Vochysiaceae (Izawa 1975, 1978; Garber 1980*a*,*b*, 1984*b*; Crandlemire-Sacco 1986; Soini 1987*b*). Unlike marmosets, tamarins do not possess the specialized incisor morphology for gouging holes in tree bark and directly stimulating the flow of gum. Rather, the availability of gum sites is dependent upon natural damage to the bark and the activity of wood-boring insects. It is likely, however, that by revisiting the site and removing the exudate seal, the tamarins are able to exert some influence on gum production.

For most tamarin species, exudates represent a highly seasonal food resource. Virtually all observations of exudate feeding in the Panamanian tamarin have been reported for the wet season months of May, June, and July (Garber 1984*b*). In saddle-back tamarins, Soini (1987*b*) reports that exudates account for 58 per cent of feeding time during the dry season. In this latter species however, gums may be consumed throughout the year (Garber, personal observation).

Little is known concerning the chemical composition of most plant exudates (see Nash 1986 for a comprehensive review of gum feeding in primates). However, it has been proposed (Garber 1984*b*) that in addition to providing a source of non-structural complex polysaccharides, certain exudates are rich in calcium. This mineral is often lacking in other parts of the diet and may be particularly important for reproductive females during periods of pregnancy and lactation. In the case of *S. geoffroyi*, exudate feeding appears to coincide temporally with the terminal phase of gestation and lactation. Calcium levels in the *Anacardium excelsum* gum consumed by the Panamanian tamarin ranges from 310 to 1110 mg/100 g (Table 13.6).

In a related study, Garber (in preparation) examined the nutritional content of exudates consumed by moustached and saddle-back tamarins in northeastern Peru. The most commonly eaten exudates (*Inga* sp. and *Parkia oppositifolia*) exhibited high levels of calcium and crude protein (Table 13.6). These gums were consumed principally during the late dry and early wet season months of September to December. Alpha females in both study groups gave birth to twin infants during November.

At least in the case of *S. fuscicollis*, dominant females control access to gum licks which are typically located on small (10–20 cm in diameter) defensible regions of the trunk. In a marked group composed of six adults and one juvenile, the individual most frequently consuming exudates was the breeding female (35 per cent of all cases in which the individual feeding could be identified; Garber, unpublished data). Although physical aggression was rare at these feeding sites, the dominant female was the only group member observed to scent-mark gum licks. Moreover since there is usually one or at most two gum licks on a given tree, only a single or small number of group members can exploit this resource at a given time. Detailed

**Table 13.6** Nutritional composition of exudates exploited by tamarins. Crude protein analysis was performed on the sample on an as-received basis, without drying and based on a protein conversion factor of 6.25. Mineral content is expressed as mg/100 g dry weight.

| | Crude protein (per cent) | Moisture (per cent) | Calcium | Phosphorus | Ca/PC ratio |
|---|---|---|---|---|---|
| *Anacardium excelsum* tree 01 (December) | 1.8 | 55.5 | 710 | 05 | 142:1 |
| *Anacardium excelsum* tree 03 (May) | 9.3 | 45.2 | 620 | 10 | 62:1 |
| *Anacardium excelsum* tree 04 (May) | 9.9 | 40.3 | 310 | 10 | 31:1 |
| *Anacardium excelsum* tree 07 (January) | 3.5 | 30.1 | 1110 | 10 | 110:1 |
| *Inga sp.* tree 236 (August) | 20.9 | 88.5 | 610 | 90 | 6:1 |
| *Inga sp.* tree 348 (September) | 37.5 | 93.4 | 830 | 10 | 83:1 |
| *Parkia oppositifolia* tree 02 (August) | 32.2 | 93.6 | 940 | 160 | 6:1 |
| *Parkia oppositifolia* tree 08 (August) | 13.7 | 9.1 | 220 | 20 | 11:1 |
| *Parkia oppositifolia* tree 02 (September) | 39.7 | 96.7 | 610 | 300 | 2:1 |

studies on inter- and intrasexual exudate feeding patterns are needed in order to examine more fully the nutritional importance of this resource in the callitrichine diet.

A comparison of substrate preference and positional behaviour during exudate feeding is presented in Fig. 13.3. Except for those occasions when tamarins exploit the syrupy liquid produced in the elongate and fibrous pods of *Parkia oppositifolia*, exudates are procured principally on large-diameter trunks and vines using a vertical clinging posture. Claw-like nails enable these small-bodied primates to exploit a food resource and set of arboreal substrates that would otherwise be inaccessible (Garber 1980*b*, 1984*a*, in press). In the three tamarin species examined, 85–99 per cent of exudate feeding and foraging involved the use of large vertical supports.

*P. oppositifolia* pods are located on small flexible supports high in the forest canopy. In exploiting this resource the animal is frequently suspended by hindlimbs only, while the forelimbs and mouth aid in extracting the exudate.

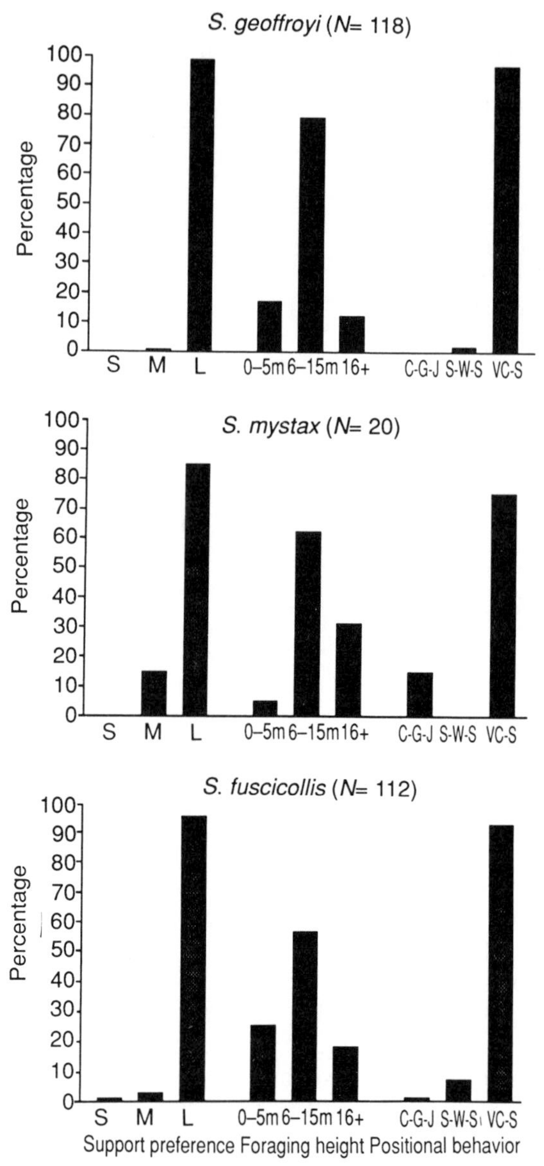


**Fig. 13.3** Support preference during exudate-feeding: S, small (supports approximately ≤5 cm in diameter, which tamarins can grasp with their hands and feet); M, medium (supports of approximately 5–10 cm in diameter, which tamarins can grasp with their arms and legs); L, large (supports in excess of 10 cm in diameter, too large to be grasped by tamarins). Positional behaviour: C-G-J, climbing, grasping, and jumping; S-W-S, standing, walking, and sitting; VC-S, vertical clinging and scansorial locomotion.

## Resource exploitation

When feeding on plant foods tamarins exhibit two distinct patterns of resource exploitation (Pook and Pook 1982; Soini 1987*b*; Garber 1988*b*, 1989). One involves the use of very small-crowned feeding trees that are exploited by one or only a few group members. These trees appear to be fed in opportunistically and are rarely revisited. In the case of mixed species troops of *S. mystax* and *S. fuscicollis*, 55 per cent of all trees/lianas fed in during a three-month period were visited on only one occasion (Garber 1988*b*). Single-species groups of *S. fuscicollis* exhibit a similar pattern (Crandlemire-Sacco 1986; Soini 1987*b*). For example, Soini (1987*b*) reports that one-third of the fruiting species eaten by *S. f. weddelli* are exploited in a highly opportunistic fashion. These resources are rarely defended from other groups, and their overall contribution to the diet is small.

The second group of plant species, however, appears to be the primary focus of feeding and ranging activities. During the course of each day, tamarins concentrate their feeding efforts on a large number of trees/lianas from a small number of species. These trees are exploited in a coordinated manner, with all or most group members feeding in the same tree at the same time. Aggression between group members is minimal, and trees of the same species are often visited during successive feeding bouts. This pattern has been reported in *S. mystax*, *S. fuscicollis*, *S. imperator*, and *S. geoffroyi* (Janson *et al.* 1981; Terborgh 1983, 1985; Yoneda 1984*b*; Crandlemire-Sacco 1986; Garber 1986, 1988*a*,*b*; Soini 1987*b*).

Tree species important in the tamarin diet are generally characterized by small to moderate-sized crowns (Terborgh 1983, 1986), a high degree of intraspecific fruiting synchrony, the production of small amounts of ripe fruit each day, and a scattered and patchy distribution (Terborgh 1983, 1986; Crandlemire-Sacco 1986; Garber 1986, 1988*b*; Soini 1987*b*). Garber (1988*b*) reports, for example, that during the wet season 20 important fruiting species exploited by moustached and saddle-back tamarins in northeastern Peru exhibited individual densities of less than 0.5 tree/ha in the 40 ha home range of the study troop. Mean nearest-neighbour distances between trees of these species averaged 140 m, although Terborgh (1983) reports higher densities for tree species visited by tamarins in southeastern Peru. The exploitation of a specific set of temporally predictable tree species is an important factor in tamarin feeding ecology.

## Ranging patterns and territorial behaviour

Compared to primates of similar body weight such as *Aotus* (Wright 1984, 1986), *Callicebus* (Kinzey 1977, 1981 Wright 1986), and various prosimian

**Table 13.7** Home range, day range, and body weight in small-bodied ceboid primates

| Species | Body weight (g) | Home range (ha) | Mean day range (m) | Reference |
|---|---|---|---|---|
| *Callicebus moloch* | 800 | 8 | 670 | Wright (1986) |
| *Callicebus personatus* | — | 5 | 695 | Kinzey and Becker (1983) |
| *Aotus trivirgatus* | 800 | 10 | 710 | Wright (1986) |
| *Callicebus torquatus* | 900 | 18 | 820 | Kinzey (1981) |
| *Saimiri sciureus* | 900 | 250 | 2100 | Wright (1986) |
| *Saimiri oerstedi* | 750 | 24–40 | 3350 | Baldwin and Baldwin (1981) |
| *Callimico goeldii* | 650 | 30–60 | 2000 | Pook and Pook (1981) |

**Table 13.8** Home range, day range, and home range overlap in *Saguinus*

| Species | Home range (ha) | Mean day range (m) | Home range overlap (per cent) |
|---|---|---|---|
| *S. geoffroyi* | 10–43 | 2061 | 13–83 |
| *S. oedipus* | 8–10 | 1750 | 20–27 |
| *S. mystax* | 30–40 | 1946 | 23–30 |
| *S. labiatus* | 23–41 | — | 40 |
| *S. imperator* | 30–120 | 1420 | ca. 10 |
| *S. nigricollis* | 30–50 | 1000 | extensive |
| *S. fuscicollis nigrifrons* | 30–40 | 1849 | 23–30 |
| *S. fuscicollis illigeri* | 16–17 | 1405 | 21 |
| *S. fuscicollis weddelli* | 26–40 | 1285 | 79 |
| *S. fuscicollis weddelli* | 33+ | — | extensive |
| *S. fuscicollis weddelli* | 20–35 | — | 40 |
| *S. fuscicollis weddelli* | 30–120 | 1220 | ca. 10 |

taxa (Charles-Dominique 1977; Bearder 1987), tamarins exploit moderate to large home ranges and extremely large day ranges. These data are reviewed in Tables 13.7 and 13.8. For most tamarin species, home ranges of 20–40 ha are common. Home ranges of over 100 ha, however, have been reported in mixed-species troops of *S. imperator* and *S. fuscicollis* in Peru. In all species for which quantitative data are available, mean day range exceeds 1200 m. An analysis of ranging patterns among six tamarin species using a Spearman's rank correlation indicates no direct relationship between home range and day range sizes ($R_s = -0.16$, $P > 0.34$). However, time spent feeding on plant resources ($R_s = 0.88$, $P < 0.01$) and adult male

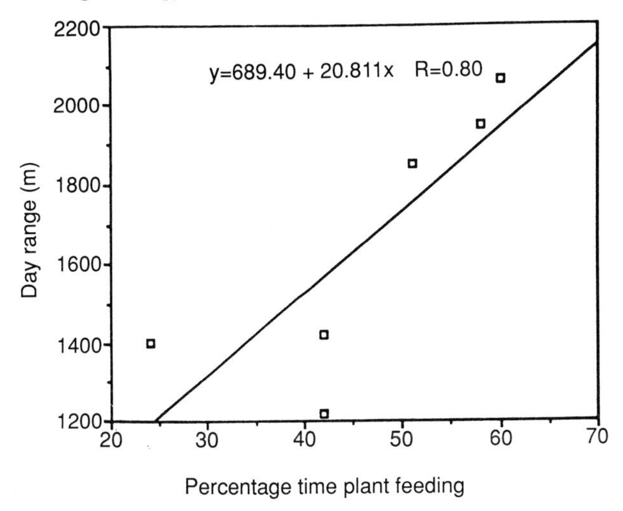


**Fig. 13.4** Mean daily path length plotted against the percentage of feeding time that individual tamarin species devote to consuming plant resources. The regression equation and correlation coefficient are given in the figure.

body weight ($R_s = 0.86$; $P < 0.02$) were positively correlated with day range (Figs 13.4 and 13.5). Although these results must be viewed with caution because of differences in data collecting techniques employed by various field workers, many researchers have suggested that the distribution of available plant resources is a primary determinant of primate daily ranging patterns (Clutton-Brock and Harvey 1977*b*; Oates 1987).

## Resource defence

There appears to be considerable variability in the expression, extent, and function of intergroup aggressive behaviour in the genus *Saguinus*. Territorial battles among tamarin troops can be highly aggressive, involving both vocal challenges and physical combat. These interactions often occur on the periphery of the troop's home range, and have generally been interpreted as representing strict territorial defence (Thorington 1968; Yoneda 1981; Terborgh 1983, Goldizen 1987*b*). However, with increasing field information it has become apparent that tamarin home ranges are not areas of exclusive use, and that overlap between neighbouring groups can be quite extensive. Home range overlap of 13–83 per cent has been reported for *S. geoffroyi*, 20–27 per cent for *S. oedipus*, 40 per cent for *S. labiatus*, 23 per cent for *S. mystax*, and 10–79 per cent *S. fuscicollis* (Table 13.8). The degree of home range overlap may depend on the locations and

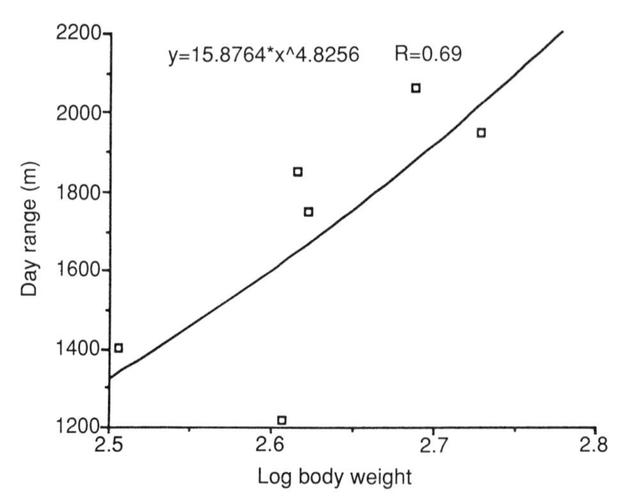


**Fig. 13.5** Mean daily path length plotted against mean adult male body weight in individual tamarin species. The regression equation and correlation coefficient are given in the figure.

distribution patterns of major food patches. When resources are clumped or restricted in their distribution, the mutual attraction of neighbouring groups to the same feeding sites can play an important role in the evolution of resource defence (Altmann 1974; Dawson 1979; Garber 1988*b*; Pruetz and Garber 1991).

Although few researchers have carefully mapped the spatial relationship between intertroop encounters and food resources, there is much qualitative evidence to suggest that among tamarins areas of home range overlap are frequently associated with major feeding trees (Dawson 1976, 1979; Pook and Pook 1982; Crandlemire-Sacco 1986; Garber 1988*b*; Pruetz and Garber 1991). For example, Dawson (1978) reports that in the Panamanian tamarin two major tree species, *Spondias mombin* and *Inga* sp., were commonly found in areas of home range overlap. Similarly, in the case of the saddle-back tamarin, Crandlemire-Sacco (1986) has suggested that rather than territorial boundaries, fruit trees are defended from neighbouring groups.

Garber (1988*b*) has conducted a quantitative study of feeding behaviour and resource defence in a mixed species troop of moustached and saddle-back tamarins. He reports a significant relationship between the location of important feeding sites and the location of intertroop conflicts. Seventy-four per cent of all major feeding trees (individual trees which account for the top 50 per cent of feeding time) were located within 75 m of an intertroop conflict. Less than 35 per cent of non-major feeding trees were located in this same area. Boundary areas not associated with major feeding

**Table 13.9** Taxonomic list of some of the major plant taxa exploited by tamarins

| | *Inga* | *Parkia* | *Cecropia* | *Pourouma* | *Spondias* | *Symphonia* | *Combretum* |
|---|---|---|---|---|---|---|---|
| *S. geoffroyi* | + | | + | + | + | | |
| *S. mystax* | + | + | + | + | + | + | |
| *S. labiatus* | | + | + | + | | + | |
| *S. midas* | | + | | | | + | |
| *S. imperator* | + | | | + | | | + |
| *S. nigricollis* | | + | | + | | | |
| *S. fuscicollis weddelli* | + | + | + | + | | + | + |
| *S. fuscicollis illigeri* | + | + | | + | + | | + |
| *S. fuscicollis nigrifrons* | + | + | + | + | | + | |

trees were rarely defended. Although the locations of these battles changed seasonally, the spatial relationship between major feeding sites and areas of intertroop conflict remained constant. Given that these feeding sites generally produced only a small amount of ripe fruit each day, and were widely scattered throughout the troop's home range, first or priority access is likely to be a critical factor in foraging success. On days in which the troop was unsuccessful in resource defence, the distances travelled to the next major feeding sites were considerably greater than that recorded for days of successful defence. The ability to monitor, defend, and exploit productive feeding trees from neighbouring troops is an important part of the feeding ecology of both single and mixed species tamarin troops (Pruetz and Garber 1991).

Table 13.9 provides a taxonomic list of some of the major fruiting and flowering species exploited by various *Saguinus* species. Despite the fact that these data were collected in diverse forest types in Panama, Peru, Colombia, and Bolivia there is marked similarity in feeding preferences. For example, the arillate seed covering and exudate of *Inga* and *Parkia*, and fruits of the genus *Cecropia* and *Pourouma*, are important food resources in the diet of *S. geoffroyi* (Dawson 1976; Garber 1980*a*), *S. fuscicollis nigrifrons* (Garber 1986, 1988*b*), *S. f. weddeli* (Izawa 1975; Yoneda 1984*a*,*b*; Crandlemire-Sacco 1986), *S. f. illigeri* (Soini 1987*b*), *S. mystax* (Garber 1986, 1988*b*), *S. midas* (Sussman and Kinzey 1984), *S. nigricollis* (Izawa 1978), *S. labiatus* (Yoneda 1984*a*; Buchanan-Smith 1990), *S. imperator* (Terborgh 1983), as well as Amazonian *Callithrix humeralifer* (Rylands 1986*b*). These same resources, however, are rarely among the primary food plants exploited by other New World primates such as *Saimiri sciureus* (Terborgh 1983), *Callicebus moloch* (Kinzey 1981; Terborgh 1983; Crandlemire-Sacco 1988), *C. torquatus* (Kinzey 1977; Easley 1982), *Cebus olivaceus* (Robinson 1986), *C. capucinus* (Oppenheimer 1982), *C. apella* (Terborgh 1983; Janson

1985), *C. albifrons* (Terborgh 1983), and *Aotus trivirgatus* (Wright 1984, 1986).

Similarly, during the dry season when fruit production in the forest is generally low, many tamarin species are reported to exploit nectar from the flowers of *Combretum* or *Symphonia* as an alternative or keystone resource (Janson *et al.* 1981; Yoneda 1984*a,b*; Terborgh 1986; Soini 1987*b*; Garber 1988*b*; Buchanan-Smith 1990; E. Pages, and personal communication). Thus, it appears that many tamarin species exhibit a relatively common feeding plan in terms of both the particular plant species exploited and the phenological characteristics of these plants. It is likely that, due to their relatively limited daily fruit production and scattered distribution, resources exploited by tamarins are of only limited profitability to larger-bodied cebids (Terborgh 1983, 1985).

## Discussion

Numerous researchers have commented on the fact that tamarin home ranges are composed of a variety of habitat types including primary and secondary forest, edge vegetation, swamps, inundated areas, and tree fall gaps (Moynihan 1970; Dawson 1976; Garber 1980*a*, 1984*a*, 1986; Mittermeier and Roosmalen 1981; Sussman and Kinzey 1984; Soini 1987*b*). Although a preference for areas with a high ratio of edge to non-edge vegetation has been noted for many species (*S. geoffroyi*, *S. midas*, *S. fuscicollis*, *S. labiatus*), a mix of forest types appears to be an essential requirement of tamarin ecology (Izawa and Bejarano 1981, Izawa and Yoneda 1981; Mittermeier and Roosmalen 1981; Snowdon and Soini 1988). The need for microhabitat diversity is likely to reflect differences in the temporal and spatial distribution of food types such as insects and ripe fruits, as well as the particular manner in which tamarins exploit their environment.

Data presented in this review indicate that there is evidence for at least three distinct insect foraging patterns in the genus *Saguinus*. Although these patterns are characterized by differences in foraging technique, positional behaviour, and substrate preference, they are all associated with the capture of large orthopteran prey. The ability to exploit orthopteran (tettigoniids, acridids) insects appears to be of critical importance in tamarin feeding behaviour. Unlike certain larger and more powerful primate species (i.e. *Cebus*), tamarins are not destructive foragers and do not break open branches or strip wood with their incisors and canines to extract concealed prey or larvae. Rather, they satisfy their daily nutritional requirements by relying on stealth in order to locate a small number of large prey (Terborgh 1983). Although the data are complex and often contradictory, there is evidence that, in comparison to other arthropod forms, orthopterans such as grasshoppers, crickets, and katydids represent a relatively stable food resource, and especially during the dry season comprise a substantial proportion of the insect biomass (Janzen and Schoener 1968; Janzen 1973; Buskirk and Buskirk 1976; Tanaka and Tanaka 1982; Boinski and Fowler 1989).

The ability of large orthopterans to survive the dry periods in an adult stage of development provides a high-quality food source during what may otherwise be a food-limited time of the year. In the case of the Panamanian tamarin, the size of orthopterans harvested in the dry season and late wet season 'was larger than that observed during the rest of the year' (Dawson 1976, p.57). Similarly, Izawa (1978) and Yoneda (1984*b*) report that large orthopterans are a common food item consumed by *S. nigricollis* and *S. fuscicollis* during the dry season. In addition, Garber (1991*a*) found only minor differences in the rate of capture of insects in the dry and wet season in moustached and saddle-back tamarins. Although the tamarins spent more time insect foraging during the dry season months of mid-July, August, and September than at other times of the year, the number of insects consumed was consistent from season to season. Thus, given their small group size, limited individual feeding requirements, and selection of relatively large prey, tamarins are able to exploit invertebrates without major shifts in ranging and patterns of habitat use which characterize another insectivorous New World primate, the squirrel monkey (*Saimiri oerstedi*) (Boinski 1987*a*).

One of the most distinctive features of callitrichine primates is the presence of claw-like tegulae on all manual and pedal digits excepting the hallux. Numerous researchers have commented on the importance of this trait and its influence on the manner in which tamarins exploit their environment (Cartmill 1974; Rosenberger 1977; Garber 1980*a*, 1980*b*, 1984, in press; Sussman and Kinzey 1984; Ford 1986*b*). Specifically, it has been suggested that claw-like digits permit a small-bodied primate to use large vertical or sharply inclined supports that are otherwise too wide to be spanned by their tiny hands and feet. Given the growing body of information on tamarin positional behaviour in the wild, there is no evidence that claw-like nails provide any significant disadvantages for movement on small flexible supports. This trait does, however, enable tamarins to use a series of resources such as plant exudates and insects which are harvested on the trunks of large trees. Although there is some disagreement on whether the evolution of claw-like nails is best understood in terms of an adaptation for foraging (Garber 1980*b*, 1984*a*; 1991*b*; in press) or travelling (Ford 1986*b*), these specialized digits do serve a critical function in feeding behaviour.

## Conclusion

Field and laboratory studies conducted over the past 10 years have provided a firm basis from which to begin to understand the behaviour and ecology of callitrichine primates. Conclusions based on these data must be viewed with caution however, and are best regarded as a set of working hypotheses requiring rigorous testing. In the case of the genus *Saguinus*, only one species, *S. fuscicollis*, has been studied in a variety of habitats and over a period exceeding two years. A review of the present information on seven tamarin species indicates that despite certain important differences, there exists marked similarity or unity in diet, foraging behaviour, and habitat utilization. The expression of these patterns is probably constrained by a set of shared derived traits (Table 13.2) which characterize all tamarin species. In identifying common behavioural patterns, I do not seek to minimize the extent of variability and flexibility in tamarin feeding ecology, rather my goal is to define the underlying structure upon which such variation is built. In summary:

1. Insects, ripe fruits, exudates, and nectar are the primary components of the tamarin diet. These resources differ in nutritional content and are located in different parts of the canopy and in different microhabitats within the forest.

2. Large orthopterans are the most important invertebrate prey. Compared to other insects they may represent a relatively stable food resource.

3. There is evidence for at least three distinct insect foraging patterns or techniques. These differ with respect to substrates exploited, vertical ranging patterns, and positional behaviour.

4. The pattern exhibited by *S. fuscicollis* and *S. nigricollis* is the most distinctive and involves the exploitation of trunks and other large vertical substrates in search of cryptic and hidden prey. Insects captured by *S. fuscicollis* are generally larger than those captured by other tamarin species.

5. During fruit and exudate feeding differences in foraging patterns, substrate preference, and positional behaviour between species are minimized.

6. Tamarins concentrate their daily feeding efforts on many individual trees from a small number of tree species. These trees represent small and widely scattered feeding sites.

7. Compared with other small-bodied primates (excluding *Saimiri*), tamarins have large day and home ranges. The distance travelled each day appears

to be more closely related to the distribution of plant resources than invertebrate prey. Foraging time, however, is constrained principally by the rate at which invertebrates are encountered and captured.

8. The ranges of neighbouring tamarin troops overlap, and there is evidence that the ability to defend major feeding sites is an important aspect of tamarin feeding ecology.
9. *S. fuscicollis* is the smallest and most behaviourally flexible tamarin species. It has the greatest geographical distribution and forms mixed species troops with *S. mystax*, *S. labiatus*, *S. imperator*, *S. nigricollis*, and *Callithrix argentata*. The ability of saddle-back tamarins to exploit insects in all layers of the canopy and on a variety of large and small surfaces is probably an important factor in its ecological success.

## Acknowledgements

I thank Anthony Rylands for inviting me to participate in the symposium on callitrichine behavior and ecology. Critical comments on earlier drafts of this manuscript were provided by Dr Lynette Norr and Dr Anthony Rylands. I wish to thank Sara, Jenni, and Lynette for their love and patience, and for allowing me to gain greater insight into the benefits that callitrichine males must receive when caring for two offspring.

A draft of this paper was presented in a symposium on adaptive unity in the callitrichidae: a systematic comparison of species differences in the family at the XIIth Congress of the International Primatological Society, Brasília, 24–29 July 1988. Funds to attend this Congress were provided by the Scholars Travel Fund, the Center for Latin American and Caribbean Studies, and the Department of Anthropology, University of Illinois. This research was supported by funds from the National Science Foundation (BNS 8310480), the Research Board of the University of Illinois, and the Center for Field Research.

# 14

# The ecology of the lion tamarins, *Leontopithecus*: some intrageneric differences and comparisons with other callitrichids

*Anthony B. Rylands*

## Introduction

The four lion tamarin species, *Leontopithecus rosalia*, *L. chrysomelas*, *L. chrysopygus*, and *L. caissara* (following Rosenberger and Coimbra-Filho 1984), occur exclusively in the Atlantic coastal forest of Brazil. Unlike the other callitrichid genera, their distributions are disjunct, and there is no evidence for them having been connected even in historic times (Coimbra-Filho and Mittermeier 1973*a*; Coimbra-Filho 1976*a*; Hershkovitz 1977). They are the largest of the callitrichids, adults weighing more than 500–550 g. Carvalho and Carvalho (1989) obtained a mean adult weight of 572.5 g (range 516–647 g) for *L. chrysopygus*. Kleiman *et al.* (1986) give an adult weight for *L. rosalia* as 500 g or more.

Despite the limited information available on the ecology of the lion tamarin species, it is evident that they may differ in certain key aspects, contrasting particularly the black lion tamarin with the remaining three coastal species. In addition, the lion tamarins show a number of traits regarding their habitat preferences, diet, ranging, and reproduction which set them apart from other callitrichids, and which undoubtedly contribute to their highly endangered status.

## Habitats

The lion tamarins occur in the tropical and subtropical forests of the Atlantic coast of Brazil. All but *L. chrysopygus* occur in the coastal lowlands, generally below 300 m above sea level. *L. chrysopygus* occurs in the inland (westward) extension of the Atlantic forest in the state of São Paulo, and has been recorded at altitudes of up to 700 m (Coimbra-Filho 1970*a*,*c*, 1976*b*).

The northernmost species, *L. chrysomelas*, occurs in two types of forest in southern Bahia: tall evergreen broadleaf tropical forest, lacking in seasonality, with the heaviest rains generally from March to June and November to December, but distributed throughout the year and exceeding 2000 mm; and in drier semideciduous forests inland, which have a marked seasonality with annual rainfall around 1000 mm. Both forest types are characterized by an abundance of bromeliad epiphytes (Coimbra-Filho 1970*a*, 1976*a*; Rylands 1982). *L. rosalia* is confined to seasonal tropical forest of part of the coastal lowlands of the state of Rio de Janeiro. Annual rainfall is around 1500 mm, with a short but pronounced dry season from May to July/August (Coimbra-Filho and Mittermeier 1973*a*; Peres 1986*a*). *L.caissara* occurs in the coastal lowlands of the northern tip of the state of Paraná and part of São Paulo. This region is considered to be transitional tropical to subtropical, although lacking a dry season and with annual rainfall being higher than in Rio de Janeiro; around 2000 mm. It has been recorded in evergreen submontane forest (below 100 m altitude), but is found mainly in the coastal lowland pioneer formations, including sub-xeromorphic *restinga* (a low coastal forest of 2–10 m height); taller (15–25 m) hygrophyllic *restinga* forest, abundant in palms, epiphytes, and lianes; and *caxetais*, which are low inundated forests (8–10 m), characterized by *caxeta, Tabebuia cassinoides* (Persson and Lorini in press). All of these forest types are abundant in bromeliad epiphytes. Lastly, *L. chrysopygus* occurs in the inland semideciduous forests, considered riparian by Rizzini (1963), of the state of São Paulo, with a pronounced dry season from April to September, and annual rainfall around 1130 mm. *L. chrysopygus* has been observed using three types of vegetation: tall semideciduous forest, swamp forest, and *macega*, a low bushy forest up to 5 m in height (Carvalho *et al.* 1989). There is a notable lack of epiphytic bromeliads in the forests occupied by this species.

To summarize, high rainfall and a lack of seasonality is characteristic of the range of *L. caissara* and a large part of that of *L. chrysomelas*. Seasonality and a reduced annual precipitation characterizes the range of *L. rosalia* and even more so the inland *L. chrysopygus*. The forests occupied by all the species, except for *L. chrysopygus*, are abundant in epihytic bromeliads, a highly characteristic foraging site.

## Group size and structure

Four *L. chrysomelas* groups observed by Rylands (1982) had a mean group size of 6.7 individuals (range 5–8). Kleiman *et al.* (1986) recorded a mean group size of 7.2 (range 4–11, $n$ = 10), although Dietz and Kleiman (1986) provide a smaller average group size of 5.8 (range 3–11) for 21 groups of

*L. rosalia*, both for the Poço das Antas Reserve. For *L. chrysopygus*, Carvalho and Carvalho (1989) obtained a mean group size of 3.6 individuals (range 2–7, $n = 9$) in the Morro do Diabo State Park. The evidence to date suggests, therefore, that group sizes are similar for the two coastal species, but rather smaller for *L.chrysopygus* (see also Coimbra-Filho 1976*a*).

A comparison with the group sizes recorded for other callitrichids indicates that lion tamarins are more similar to *Saguinus* than to *Callithrix*. Ferrari and Lopes Ferrari (1989) showed that *Callithrix* tend to live in larger groups, the smallest mean obtained for five species studied being that of *C. kuhli* (mean group size 6.6, $n = 8$, Rylands 1982, 1989*b*), the remainder having mean group sizes between 8.9 and 11.5. Mean group sizes for *Saguinus*, on the other hand, range from 3.4 to 6.9. Ferrari and Lopes Ferrari (1989) calculated a mean of 6.4 individuals/group for 76 *Cebuella* groups.

Kleiman *et al.* (1986) recorded the structure of 10 groups of *L. rosalia* in the Poço das Antas Reserve. In the nine groups in which the sexes of the adults were identified, all but one contained at least two adult males, two contained just one adult female, four contained two adult females, and three contained three adult females. None of the nine groups had just one adult female and male. Three *L. chrysomelas* groups studied by Rylands (1982) also contained supernumerary adults (four, four, and at least three). Very little information is available for *L. chrysopygus*, but the group observed by Keuroghlian (1990) contained, at least ephemerally, three adults, and Carvalho and Carvalho (1989) recorded that groups in the Morro do Diabo State Park are normally composed of two to three adults. Considering the rather smaller group sizes recorded for *L. chrysopygus*, the occurrence of groups with more than a single pair of adults may be rather less common than in the coastal species. However, it is evident that, as with all other callitrichids, groups of the three species of lion tamarins may frequently contain adult females in addition to the breeding female, as well as supernumerary adult males (see below).

## Population densities

Rylands (1982, 1989*b*) provided density estimates of between 5 and 17 individuals/km$^2$, or 0.9 to 3 groups/km$^2$ for *L. chrysomelas* in the Lemos Maia Experimental Station, at Una in southern Bahia. Kleiman *et al.* (1986) estimated a population of 153 *L. rosalia* in the Poço das Antas Biological Reserve of 5200 ha. About 60 per cent (3120 ha) of the reserve contains habitat suitable for lion tamarins (Green 1980), which therefore indicates a population density of 4.9 individuals/km$^2$. This is at the lower end of the estimates for *L. chrysomelas*, and considering that the two species have similar home ranges with a similar overlap of about 10 per cent (Rylands 1982, 1989*b*; Peres 1986*a,b*), it may be expected that the Poço das Antas population can still increase, to at least 8–9 individuals/km$^2$, taking the middle of the range of estimates for *L. chrysomelas*. Population density estimates of *L. chrysopygus* are not yet available. With the smaller group sizes and larger home ranges (see below) observed in the Morro do Diabo and Caetetus reserves, they will evidently be considerably lower. Padua *et al.* (1990) indicated a minimum of 80 and a maximum of 450 *L. chrysopygus* in the Morro do Diabo State Park, which would indicate a density of 0.2–1.3 individuals/km$^2$ for the Park (unsuitable habitat not taken into account).

## Home range

The lion tamarins have the largest home ranges of the callitrichids. No data are available for *L. caissara*, but whereas *L. chrysomelas* and *L. rosalia* groups have home ranges of the order of 40 ha (Kleiman *et al.* 1986; Peres 1986*a,b*; Rylands 1989*b*), *L. chrysopygus* groups have been observed using areas of over 100 ha. Carvalho and Carvalho (1989) estimated home range sizes of more than 66 ha for one group and 133 ha for a second in the Morro do Diabo State Park. Keuroghlian (1990) observed a group in the Caetetus State Reserve using more than 118 ha. An unqualified home range size of 200 ha was given for *L. chrysopygus* by Padua *et al.* (1990).

The difference in home range size, between *L. chrysomelas* and *L. rosalia* on the one hand and *L. chrysopygus* on the other, is also reflected in the day ranges. Rylands (1982, 1989*b*) observed monthly mean day ranges of between 1552 m and 1954 m (a minimum of 1410 m and maximum of 2175 m) during three months for *L. chrysomelas*. Peres (1986*a*) obtained mean path lengths of between 1339 m (15 days in 1984) and 1533 m (40 days in 1985) for *L. rosalia*. Keuroghlian (1990) estimated day range lengths to average 2289 m, with a minimum of 2061 and a maximum of 2611 m, during seven complete observation days of *L. chrysopygus* in the Caetetus State Reserve.

Day range length and home range size data available for callitrichids are summarized by Ferrari and Lopes Ferrari (1989). The day range lengths for *L. chrysomelas* and *L. rosalia* are larger than any reported for *Callithrix* (0.5–1.5 km) and at the upper end of the range reported for *Saguinus* (1.2–2.1 km). Home range size shows a similar pattern, with the largest reported for *Callithrix* being 35.5 ha (*C. flaviceps* by Ferrari 1988*b*), and that for *Saguinus* being 120 ha (Terborgh and Stern 1987), although *Saguinus* home ranges in six of the ten studies listed by Ferrari and Lopes Ferrari (1989) are between 20 and 50 ha, the other exceptions being the smaller ranges observed during studies of *S. fuscicollis illigeri* with 15.7–16.5 ha (Soini 1987*b*), *S. geoffroyi* with 9.4 ha (Garber 1984*a*) and *S. oedipus* with 7.8–10 ha (Neyman 1978).

## Diet and feeding behaviour

As with the majority of other callitrichids, the lion tamarins may be classified as frugivore-insectivores. The fruits used by each of the species will obviously vary considering the distances between the forests where they occur, although little information is available. Peres (1986*a*) listed 43 plant species of 17 different families used by *L. rosalia* in the Poço das Antas Biological Reserve during five months: ripe, and in some cases green, fruits were eaten of 38 species; nectar and flowers of three species; and fluids or exudates of three species (Table 14.1). In this study, feeding data were obtained during the period April–August in two consecutive years (1984 and 1985). In the first year fruits comprised 78 per cent of the diet, whereas in 1985, nectar of *Symphonia globulifera* (Guttiferae), which failed to flower in 1984, was the most important item, comprising 44 per cent of the feeding records (including animal prey). *Symphonia globulifera* nectar was also found to be a significant item in the diet of *L. chrysomelas* during four (August–November) of five months (July–November) of observations in 1980 by Rylands (1982). During this time, two *L. chrysomelas* groups were observed eating 16 species of fruits, and nectar from two species (Table 14.1). Quantitative data on feeding behaviour were obtained from September to November, during which time fruit comprised 74–89 per cent of the diet, flowers and nectar 2–20 per cent, and exudates 3–11 per cent of the plant part of the diet. Exudate feeding was restricted to the readily available gum in the pods of *Parkia pendula* (Leguminosae).

Although providing no quantitative data, Carvalho and Carvalho (1989) observed two groups of *L. chrysopygus* during 10 months (December–September) in the Morro do Diabo State Park, and recorded the consumption of fruits of 16 species and gums of a further four (see Table 14.1). Keuroghlian (1990) observed one group of *L. chrysopygus* in the Caetetus State Reserve during two months (November and December). During this time fruits comprised 90 percent or more of the diet, and exudates 4–6 per cent. The species eaten are listed in Table 14.1, although the exudate species involved were not identified. Passos and Carvalho (1991) reported on the ingestion of exudates by a group of *L. chrysopygus*, also at the Caetetus State Reserve, during the period May–July 1990. At the height of the dry season, and lacking fruits, exudate consumption was much higher

**Table 14.1** Plant species in the diet of *L. rosalia* (GLT), *L. chrysomelas* (GHLT) and *L. chrysopygus* (BLT). Fr, fruit; Fl, flowers; Ne, nectar; Ex, exudate.

| Family, species | Plant part | GLT | GHLT | BLT |
|---|---|---|---|---|
| Annonaceae | | | | |
| *Guatteria* sp. | Fr | | Rylands (1982) | |
| Apocynaceae | | | | |
| *Couma rigida* | Fr | | Rylands (1982) | |
| Unidentified sp. | Fr | | Rylands (1982) | |
| Araceae | | | | |
| Unidentified sp. | Fr | Coimbra-Filho (1976) | | Keuroghlian (1990) |
| Anacardiaceae | | | | |
| *Tapirira guianensis* | Ex | Coimbra-Filho (1976) | | Passos and Carvalho (1991) |
| Boraginaceae | | | | |
| *Cordia sellowiana* | Fr | Peres (1986) | Rylands (1982) | |
| *Cordia ecalyculata* | Fr | | | Carvalho and Carvalho (1990) |
| *Cordia sagotii* | Fr | | Rylands (1982) | |
| *Cordia* sp. | Fr | Peres (1986) | | |
| Burseraceae | | | | |
| *Protium* sp. | Fr | Rylands (1982) | | |
| Cactaceae | | | | |
| *Rhipsalis* spp.[a] | Fr | | | Carvalho and Carvalho (1990); Keuroghlian (1991) |
| *Zygocatus* sp. | Fr | | | Carvalho and Carvalho (1990) |
| Unidentified sp.1 | Fr | | | Keuroghlian (1990) |
| Unidentified sp.2 | Fr | | | Keuroghlian (1990) |
| Unidentified sp.3 | Fr | | | Keuroghlian (1990) |
| Unidentified sp.4 | Fr | | | Keuroghlian (1990) |
| Combretaceae | | | | |
| *Combretum fruticosum* | Ne | Peres (1986) | | |
| *Combretum* sp. | Ex | | | Carvalho and Carvalho (1991) |

**Table 14.1** (*cont.*)

| Family, species | Plant part | GLT | GHLT | BLT |
|---|---|---|---|---|
| Guttiferae | | | | |
| *Symphonia globulifera* | Ne | Peres (1986) | Rylands (1982) | |
| Hippocrataceae | | | | |
| Unidentified sp. | Fr | | Rylands (1982) | |
| Leguminosae, Caesalpinoidae | | | | |
| *Dialium guianense* | Fr | | Rylands (1982) | |
| Leguminosae, Mimosoideae | | | | |
| *Enterolobium contorsiliguum* | Fr, Ex | | | Carvalho and Carvalho (1990) |
| *Inga edulis* | Fr | Peres (1986) | | |
| *Inga leptantha* | Fr | Peres (1986) | | |
| *Inga fagifolia* | Fr | Peres (1986) | | |
| *Inga* sp. | Fr | Coimbra-Filho (1976) | | Carvalho and Carvalho (1990), Keuroghlian (1990) |
| *Inga* sp. | Ex | | | Keuroghlian (1990), |
| *Parkia pendula* | Ex | | Rylands (1982) | Passos and Carvalho (1991) |
| *Pithecolobium edwallii* | Fr | | | Carvalho and Carvalho (1990) |
| Unidentified sp. | Fr | | | Keuroghlian (1990) |
| Liane sp.1 | Ex | Peres (1986) | | |
| Legumionosae, Papilionaceae | | | | |
| *Machaerium* sp. | Ex | Peres (1986) | | |
| Loganiaceae | | | | |
| *Strychnos* sp. | Fr | | Rylands (1982) | |
| Loranthaceae | | | | |
| *Struthanthus vulgaris* | Fr | | | Keuroghlian (1990) |

**Table 14.1** (*cont.*)

| Family, species | Plant part | GLT | GHLT | BLT |
|---|---|---|---|---|
| Rubiaceae | | | | |
| *Coussarea* sp. | Fr | Peres (1986) | | |
| *Genipa americana* | Fr | Peres (1986) | | |
| *Posoqueria longiflora* | Fr | | Rylands (1982) | |
| *Randia spinosa* | Fr | Peres (1986) | | |
| Unidentified sp.1 | Fr | Peres (1986) | | |
| Rutaceae | | | | |
| *Pilocarpus pauciflorus* | Ex | | | Passos and Carvalho (1991) |
| Sapindaceae | | | | |
| *Paullinia* sp. | Fr | Peres (1986) | | |
| Sapotaceae | | | | |
| *Chrysophyllum gonocarpum* | Fr | | | Carvalho and Carvalho (1990) |
| *Manilkara* sp.1 | Fr | | Rylands (1982) | |
| *Manilkara* sp.2 | Ne | | Rylands (1982) | |
| *Mimusops* sp. | Fr | Peres (1986) | | |
| *Pouteria parviflora* | Fr | Peres (1986) | | |
| *Pouteria* sp. | Fr | Peres (1986) | | |
| Unidentified sp. | Fr | | Rylands (1982) | |
| Ulmaceae | | | | |
| *Celtis* sp. | Fr | Peres (1986) | | |
| Verbenaceae | | | | |
| *Vitex polygama* | Fr | Peres (1986) | | |
| Vitaceae | | | | |
| *Cissus* sp. | Fr | Peres (1986) | | |

[a] Fruits of four unidentified species of *Rhipsalis* were recorded by Keuroghlian (1990).

**Table 14.1** (*cont.*)

| Family, species | Plant part | GLT | GHLT | BLT |
|---|---|---|---|---|
| Myrtaceae | | | | |
| *Campomanesia* cf. *meschalanthra* | Fr | | | Carvalho and Carvalho (1990) |
| *Eugenia glomerata* | Fr | Peres (1986) | | |
| *Eugenia sulcata* | Fr | | | Coimbra-Filho (1976) |
| *Eugenia* spp. | Fr | | | Coimbra-Filho (1978), Carvalho and Carvalho (1990), Keuroghlian (1990) |
| *Myrceugenia ovata* | Fr | | | Carvalho and Carvalho (1990) |
| *Myrcia* sp. | Fr | Peres (1986) | | |
| *Myrciaria* cf. *cauliflora* | Fr | | | Carvalho and Carvalho (1990) |
| *Psidium guineense* | Fr | Peres (1986) | | |
| *Psidium* sp.1 | Fr | Peres (1986) | | |
| *Psidium* sp.2 | Fr | | | Carvalho and Carvalho (1990) |
| *Syzygium jambos* | Fr | Peres (1986) | | |
| Unidentified sp.1 | Fr | Peres (1986) | | |
| Unidentifiwed sp.2 | Fr | | Rylands (1982) | |
| Nyctaginaceae | | | | |
| *Neea* sp. | Fr | | Rylands (1982) | |
| *Pisonia* sp. | Fr | | | Carvalho and Carvalho (1990) |
| Palmae | | | | |
| *Bactris* sp. | Fr | Peres (1986) | | |
| *Euterpe edulis* | Fr, Ex | Peres (1986) | | Passos and Carvalho (1991) |
| *Syagrus romanzoffiana* | Fr | | | Carvalho and Carvalho (1990) |
| Unidentified sp.1 | Fr | Peres (1986) | | |
| Unidentified sp.2 | Fl | Peres (1986) | | |
| Passifloraceae | | | | |
| *Passiflora* sp. | Fr | | Rylands (1982) | |
| Polygonaceae | | | | |
| *Ruprechtia laxiflora* | Ex | | | Carvalho and Carvalho (1990) |

**Table 14.1** (*cont.*)

| Family, species | Plant part | GLT | GHLT | BLT |
|---|---|---|---|---|
| Melastomataceae | | | | |
| *Clidemia bulbosa* | Fr | Peres (1986) | | |
| *Clidemia biserrata* | Fr, Fl | Peres (1986) | | |
| *Henriettea saldanhai* | Fr | Peres (1986) | | |
| *Henriettea succosa* | Fr | | Rylands (1982) | |
| *Miconia candolleana* | Fr | Peres (1986) | | |
| *Miconia dodecandra* | Fr | | Rylands (1982) | |
| *Miconia hypoleuca* | Fr | Peres (1986) | Rylands (1982) | |
| *Miconia ibaguensis* | Fr | Peres (1986) | | |
| *Miconia* sp.1 | Fr | | Rylands (1982) | |
| *Miconia* sp.2 | Fr | | Rylands (1982) | |
| *Platycentrum clidemioides* | Fr | | Rylands (1982) | |
| Menispermaceae | | | | |
| *Abuta sellowiana* | Fr | Peres (1986) | | |
| *Anemospermum* sp. | Fr | Peres (1986) | | |
| Moraceae | | | | |
| *Brosimum* sp. | Fr | | Rylands (1982) | |
| *Cecropia cinerea* | Fr | Peres (1986) | | Carvalho and Carvalho (1990) |
| *Cecropia* spp. | Fr | Coimbra-Filho (1976) | | |
| *Chlorophora tinctoria* | Fr | | | Carvalho and Carvalho (1990) |
| *Ficus obtusiuscula* | Fr | Peres (1986) | | |
| *Ficus clusiaefolia* | Fr | Coimbra-Filho (1976) | | |
| *Ficus* spp. | Fr | Peres (1986) | | Coimbra-Filho (1976), Keuroghlian (1990) |
| *Helicostylis tomentosa* | Fr | | Rylands (1982) | |
| *Pourouma velutina* | Fr | | Rylands (1982) | |
| Musaceae | | | | |
| *Musa paradisiaca* | Fr | Peres (1986) | | |
| *Musa rosaceae* | Fr | Peres (1986) | | |

than reported by Keuroghlian (1990) and higher than recorded for any of the lion tamarins to date (see below).

Despite the variable and short duration of the studies, some interesting patterns emerge when comparing the fruiting species used by the three lion tamarins (Table 14.1). Notable, comparing the two coastal species, is the importance of *Symphonia* and the Melastomataceae, especially *Miconia* and *Henriettea*, typical of secondary growth and edge habitats. Fruit of only one species of the family Myrtaceae was eaten by *L. chrysomelas*, but the family was very well represented in the Una forest, and it probably provides a number of important fruiting species at other times of the year (Rylands 1982). This family, along with the Leguminosae, is represented in the diets of all three species. Of interest is the lack of Melastomataceae in the diet of *L. chrysopygus*, and the inclusion of cactus fruits, not eaten by the two coastal species. Overlap in the diet recorded for *L. chrysomelas* and *L. rosalia* on the one hand and *L. chrysopygus* on the other, is restricted to seven genera (*Cordia*, *Ficus*, *Combretum*, *Inga*, *Cecropia*, *Eugenia*, and *Psidium*), but not one species, and indicates, for sure, very different floristic communities between the coastal forest and the inland semideciduous forest occupied by the black lion tamarins.

Small but significant amounts of plant exudates are eaten by all three species. Peres (1986*a*, 1989) found that plant exudate consumption increased in the late dry season, when fruit availability was lowest, a pattern also found for *Callithrix* (Rylands 1984). Passos and Carvalho (1991) registered a high proportion of gum in the diet of *L. chrysopygus* in the Caetetus State Reserve in the dry season, also when fruits were scarce: 21 per cent (May), 55 per cent (June) and 37 per cent (July) of the feeding records for the group. The lion tamarins lack the behavioural and morphological adaptations for tree-gouging, typical of the two marmoset genera, *Cebuella* and *Callithrix*, and are typically opportunistic feeders, as with the tamarins, *Saguinus*, (see Garber 1984*b*, this volume; Rylands 1984). Peres (1989), however, recorded adult (but not juvenile) *L. rosalia* biting the base of *Machaerium* lianas specifically to elicit exudate flow. Gouging behaviour has not been observed for *L. chrysopygus* and *L. chrysomelas* (Rylands 1989*b*; Carvalho and Carvalho 1989).

*L. chrysomelas* is syntopic with *Callithrix kuhli* and may even form mixed groups, although considerably more ephemeral than those recorded for the moustached and saddle-back tamarins (see Yoneda 1981; Terborgh 1983). The principal feature which separates the two species is their mode of foraging for animal prey (Rylands 1989*b*). The long slender hands and fingers are used for probing for concealed prey in specific microhabitats, and they may be categorized as specific-site foragers for largely non-mobile prey (Hershkovitz 1977; Coimbra-Filho 1970*a*, 1976*a*, 1981; Peres 1986*a*;

Rylands 1982, 1989*b*). As pointed out by Peres, a considerable proportion of prey items are located by touch rather than sight. Peres (1986*a*) quantified the foraging of *L. rosalia* in different microhabitats and, as in Rylands' (1982, 1986*b*) study of *L. chrysomelas*, found the most important site to be epiphytic bromeliads. This foraging microhabitat was ignored by *C. kuhli*. Other sites foraged by both *L. rosalia* and *L. chrysomelas* included accumulations of leaf litter (in palm crowns and vine tangles for example), under loose and rotting bark, and crevices and holes in trees. Carvalho and Carvalho (1989) and Keuroghlian (1990) recorded very similar specific-site foraging for *L. chrysopygus*, and also observed them foraging in leaf litter, or picking up insects which had fallen to the ground, a behaviour which was not observed for *L. chrysomelas* by Rylands (1982). Interestingly, the forests of the two areas where *L. chrysopygus* has been studied are lacking in the bromeliads which are such important foraging sites for *L. rosalia* and *L. chrysomelas*, and undoubtedly *L. caissara*. The marmosets, and the moustached tamarins show a more predominantly visual, foliage-gleaning, stalk-and-pounce foraging for largely mobile prey (typified by orthopterans). Specific-site, tactile foraging is also typical of the saddle-back tamarin, *S. fuscicollis* (see Pook and Pook 1982; Yoneda 1981, 1984*a*; Terborgh 1983) and, as in *C. kuhli* and *L. chrysomelas*, is undoubtedly the key to permitting sympatry, and mixed-species groups, with the moustached tamarins (*S. mystax*, *S. labiatus*, and *S. imperator*).

The animal prey foraging methods of the lion tamarins indicate that they are exploiting a very different (and diverse) fauna from that available to the marmosets (see Coimbra-Filho 1981). Another indication of this is provided by the difference in forest levels typically used for animal prey foraging by sympatric *C. kuhli* and *L. chrysomelas*; 6–13 m for *C. kuhli*, and 12–20 m for *L. chrysomelas* (see Rylands 1982, 1989*b*). *L. chrysomelas* foraged more in the upper levels of the forest, and *C. kuhli* more in the middle levels and understorey. *L. rosalia* and *L. chrysopygus* spend much more time at lower levels of the forest than was recorded for *L. chrysomelas* (Carvalho and Carvalho 1989; personal observation). Neither of these species has been observed in sympatry with marmosets (although *C. jacchus* has been introduced in various forests within the range of *L. rosalia*), and foraging levels are undoubtedly influenced by the structure and height of the vegetation. Although employing a similar technique to lion tamarins, the saddle-back tamarin, *Saguinus fuscicollis*, spends most of its foraging time in the lower levels of the forest, below 10 m, while the sympatric moustached tamarins, with a foraging technique similar to *Callithrix*, use more the middle layers, between 10–20 m (Yoneda 1981, 1984*a*). Sympatry between these species is enabled by the differences in foraging techniques as well heights in the forest, and the exploitation of different faunas.

## Sleeping sites

A unifying feature of the lion tamarins is their use of holes in tree trunks and branches as sleeping sites (Coimbra-Filho 1970*a*, 1976*a*, 1978*a*; Coimbra-Filho and Mittermeier 1973*a*; Rylands 1982; Peres 1986*a*; Carvalho and Carvalho 1989). This behaviour is only rarely observed for *Callithrix* (see Rylands 1982; Stevenson and Rylands 1988) and *Saguinus* (see Snowdon and Soini 1988). The reason for this behaviour is not known, although Snowdon and Soini (1988) outlined the adaptive feature of sleeping in huddles, typical of all callitrichid species. One effect of the lion tamarin's dependence on tree holes, however, is that they are limited to forests where they are available: an ecological constraint in the use of less disturbed forest (Rylands 1982, 1989*b*; Peres 1986*a*). The *L. chrysomelas* group studied by Rylands (1982, 1989*b*) had access to rather few holes (six identified during four months), and all in the primary forest of their range, no holes being observed at all in the secondary forest. Keuroghlian (1990) recorded 10 sleeping sites during two months in the Caetetus State Reserve. The *L. chrysopygus* population in the Morro do Diabo State Park have access to extremely numerous holes, probably as a result of the extensive selective cutting and degradation of the forest over the last decades which has resulted in extraordinary large numbers of dying or dead trees, at least in parts of the Park (personal observation). Peres (1986*a*) recorded 39 distinct sleeping sites for his study group of *L. rosalia*, only three of which were common to both years of the study.

## Social organization and reproduction

The presence of more than one adult of either sex in lion tamarin groups conforms with the pattern observed for all other callitrichid species studied. This indicates the possibility of a flexible breeding system, which may be monogamous, polyandrous, or even polygynous. Although callitrichid groups normally contain only one breeding female (Sussman and Kinzey 1984; Ferrari and Lopes Ferrari 1989; see also Rothe and Darms, this volume), rare cases of polygyny and polyandry in captive *Callithrix jacchus* and *Saguinus oedipus* have been documented by Rothe and Koenig (1991) and Price and McGrew (1991), respectively. These authors conclude that when these alternative mating strategies do occur they are temporary, and reflect a transitional period between monogamous phases.

Polygyny in wild groups has been recorded for *Saguinus fuscicollis* by Terborgh and Goldizen (1985; also Goldizen 1986) and *L. rosalia* by Dietz and Kleiman (1986) and Dietz and Baker (1991). In *L. rosalia*, females in polygynous groups are normally mother and daughter, with the mother being dominant (Dietz and Baker 1991). These authors found no significant reduction in the survival of the infants of the dominant female in this situation, and although the breeding success of the daughter was found to be approximately one-third that of her mother, it was two times higher than that observed for females which dispersed to other groups. The occurrence of groups with two breeding females was not related to the availability of helpers, nor deviations from a 1:1 sex ratio, nor to limiting resources such as sleeping sites and fruiting trees. A positive correlation was found, however, with the size of the home range (Dietz and Baker 1991). Dietz and Baker (1991) argued that the relaxation of reproductive inhibition by the mother on her daughter is explicable in terms of kin selection in a situation where opportunities for reproduction outside the natal group are very limited, or inducive to very low reproductive success on the part of her daughter. This indicates that polygyny is likely to arise in overcrowded populations where opportunities for dispersal are limited. Rylands (1982, 1989*a*) gave a similar argument to explain the occurrence of polyandrous mating, evidently a possibility in all groups containing more than one adult male. With only one reproductive female per group, breeding opportunities for males are limited, and polyandry may arise either through the need for collaboration between males to defend breeding females, or through straight manipulation by the female to increase the number of male helpers, through uncertainty of paternity (Rylands 1982, 1986*a*, 1989*a*; Terborgh and Goldizen 1985). In both cases, opportunities for dispersal and suitable habitat are considered the principal factors involved (see also Terborgh and Goldizen 1985; Goldizen 1986, 1987*a*; Ferrari and Lopes Ferrari 1989; Caine this volume; Garber this volume).

The occurrence of polygyny is particularly interesting because it indicates a breakdown, at least temporary, of the remarkable phenomenon of reproductive inhibition by a single, dominant, breeding female—a feature typical of the family (Abbott and Hearn 1978; Abbott 1984; Abbott *et al.* this volume). Transient polygyny (or at least two females breeding in the same group perhaps with two breeding males) may occur during a change in the dominance status of two adult females, predictably in many cases a mother giving way to a daughter. However, it is becoming evident that the lion tamarins differ in the mechanism by which inhibition is maintained (French 1987; French *et al.* 1989; Abbott *et al.* this volume). Whereas neither daughters nor unrelated subordinates show ovarian cycles in *Callithrix jacchus*, Tardif (1984) demonstrated that in some cases daughters in *Saguinus oedipus* may cycle, when in the presence of the reproductive dominant female, although this is related to moments of social disruption or change in group composition. French *et al.* (1989; see also French and Stribley 1987) have shown, however, that in golden lion tamarins (*L. rosalia*) older daughters and unrelated subordinate females are capable of expressing a normal ovarian function in the presence of the breeding female (see also Abbott *et al.* this volume). In this case, behavioural rather than physiological inhibition is involved when breeding is restricted to just one female. Supporting this, Kleiman (1979), French and Inglett (1989), and Inglett *et al.* (1989) have reported high levels of aggression between mothers and daughters and between breeding females and females strangers, not typical of other callitrichid species studied (French 1987; French *et al.* 1989) and Baker (1987) reported that almost all emigrations of *L. rosalia* recorded during 10 years in the Poço das Antas Reserve resulted from aggression between same-sex, reproductive age group members.

French *et al.* (1989) and Abbott *et al.* (this volume) conclude that the lion tamarins differ from all other callitrichids with regard to their behavioural (as opposed to physiological/behavioural) mechanism of inhibiting reproduction in subordinate females. This evidently has important ecological implications, permitting a greater flexibility with regard to the dominant female's prohibition, or otherwise, of reproductive activity by other females in the group. In order to understand the reasons for this difference, it will be necessary to understand the adaptive function of reproductive inhibition in callitrichids in the first place. Abbott *et al.* (this volume) discuss this, emphasizing that our knowledge is still limited to suppositions and conjecture. They consider the principal aspect to be related to the advantages of subordinate females staying in their natal groups, (1) because of the lack of opportunities or difficulty of obtaining breeding opportunities elsewhere, (2) because of their contribution to rearing siblings, and (3) because they can gain breeding experience. In summary, Abbott *et al.* argue that it involves making the best of a situation where breeding within their natal group is disadvantageous, but the opportunities for breeding elsewhere are limited. Proof of this requires an understanding of the population dynamics (especially mechanisms, rates and patterns of dispersal, new group formation, and reproductive success), and habitat requirements (Rylands 1982, 1989*a*; Ferrari and Lopes Ferrari 1989).

In the wild, lion tamarins generally breed once a year (from September to February), even though they have the capacity to breed twice (Kleiman *et al.* 1988), and may do so in years of exceptional fruit abundance. The same is true for *Saguinus*, but *Callithrix* and *Cebuella* generally breed twice a year (for review see Ferrari and Lopes Ferrari 1989). Ferrari and Lopes Ferrari (1989) attribute this to the capacity of the marmoset genera to supplement their diet with plant exudates at times of fruit scarcity. This dietary limitation is another factor contributing to the relatively restrictive conditions determining habitat suitability when compared to the marmosets, for example. As pointed out by Ferrari and Lopes Ferrari (1989), the marmosets are able occupy forests which are more seasonal than any occupied by *Saguinus* because of their gum-feeding capacity. *L. chrysopygus* and *L. rosalia*, at least, also occupy highly seasonal habitats, and yet can only exploit gums opportunistically (Peres 1989), also depending, like *Saguinus*, on such alternatives as nectar.

An understanding of dispersal and new group formation in callitrichids is fundamental for an understanding of the social organization of the four genera. Both sexes disperse, and may succeed in occupying positions in nearby groups with breeding vacancies or establish new groups, in some cases through lone individuals joining forces (Baker 1987). Baker (1987) found, however, that immigration was delayed following emigration, and some animals would remain alone for several months. Ferrari and Lopes Ferrari (1989) argue that *Callithrix* groups are more stable, and that dispersal events involving just single animals are rare, when compared to *Saguinus* and *Leontopithecus*. This may be related to the greater stability of the marmoset's food resources (Ferrari and Lopes Ferrari 1989) which also contributes to them having smaller home ranges. Whereas *Callithrix* and *Saguinus* have the options of either dispersing or waiting until a dominant position can be achieved in their natal group, it seems that female lion tamarins have a third option of orchestrating a relaxation of the behavioural inhibition of the dominant female. Whether more than one male breeds in a group is still a moot point, although based on dimorphism in male canine tooth size, Dietz and Kleiman (1987) argued that males in multimale groups 'are father and son(s), and that monogamy and not polyandry is the predominant mating system for golden lion tamarins'.

The relative importance of dispersal with regard to its effects on genetic relatedness, stability, and mating systems in callitrichid groups is somewhat disputed. Whereas Rothe and Darms (this volume; see also Koenig and Rothe 1991*a*) argue that migration events are too rare to have any significant effect on group structure and composition (compared to births and deaths), Ferrari and Lopes Ferrari (1989) argue to the contrary, and indicate that *Callithrix* groups are more stable than *Saguinus*, and the degree of relatedness within marmoset groups consequently much higher. Following this reasoning, polyandry in *Callithrix* is more likely to be fraternal, rather than a strategy to increase reproductive success in the short term through recruitment of helpers, which may be more typical for tamarins (Ferrari and Lopes Ferrari 1989). More long-term studies are necessary to clarify the relative importance of migration on the group structure and breeding strategies of marmosets and tamarins.

Even subtle differences in the floristic composition of a forest can have a major influence on its suitability for callitrichid groups. The study of Goldizen *et al.* (1988) provided a good example of few key seasonal resources being fundamental for the survival and breeding success of *S. fuscicollis* groups and, as argued by Rylands (1982, 1986*b*, 1987; Rylands and Faria, this volume) and Terborgh (1983), edge habitat and secondary forest are extremely important for at least *Callithrix* and *Saguinus*. However, although lion tamarins exploit fruiting trees typical of edge and pioneer formations (especially Melastomataceae), they have determinants of habitat suitability not shared by other callitrichid species (Rylands 1982, 1989*a*,*b*). These include adequate foraging microhabitats (especially epiphytic bromeliads in the case of at least *L. rosalia* and *L. chrysomelas*) and tree-holes for sleeping sites. *S. fuscicollis* is the only other callitrichid with a similar animal foraging technique (Yoneda 1981, 1984*a*), but undoubtedly significant is its smaller size; about 70 per cent of the weight of a lion tamarin. The lack of epiphytic bromeliads in the forests occupied by *L. chrysopygus*, along with a more pronounced seasonality which probably implies more severe fruit (and insect) shortages than experienced by the coastal species, are probably the reasons for their extraordinarily large home ranges, and it is significant that black lion tamarins have been observed eating gums (an alternative to fruits) considerably more than the other species (Passos and Carvalho 1991). As argued by Coimbra-Filho (1976*a*) and Rylands (1982, 1989*b*), lion tamarins, when compared to other callitrichids, are unable to adapt to severely degraded forests which lack especially tree-hole shelters and adequate densities of their animal prey foraging sites; a reason for their endangered status. *Callithrix* and *Saguinus* use primary forest, but require areas of second growth and edge, whereas the reverse is true for lion tamarins, which use second growth and edge, but require intact primary forest for the two reasons given above. The lack of a physiological reproductive inhibition in subordinate female lion tamarins undoubtedly implies a more flexible reproductive strategy on the part of the mother in occasionally permitting philopatry and reproduction in a situation where habitat availability is more restricted than is typical for other callitrichids. Again more studies are required on the resources necessary and available for lion tamarins, particularly investigating such aspects as seasonality in animal prey and fruit resources. Because the lion tamarins have a specific site foraging technique, it would also be enlightening (and possible) to obtain an understanding of the rates of renewal of prey in such as bromeliads and palm crowns.

## Conclusions

The admittedly scarce information available on lion tamarins indicates that *L. chrysopygus* is ecologically distinct from the three coastal species in a number of aspects, which include at least a highly seasonal forest, a lack of epiphytic bromeliads, possibly a greater (seasonal) use of plant exudates, very large home ranges, and possibly smaller group sizes. Furthermore, although probably ecologically very similar to *L. chrysomelas*, *L. rosalia* occupies a more seasonal habitat, possibly important in determining reproductive patterns, group size, mating strategies, and dispersal. Little can be said of *L. caissara* except that, like *L. chrysomelas*, it occupies a less seasonal environment, but possibly of a very different floristic composition to the forests found in southern Bahia. The foraging techniques and sleeping sites of lion tamarins differ from all other callitrichids, and are significant determinants of habitat quality and carrying capacity.

Following Ferrari and Lopes Ferrari (1989; also Rylands 1984), the gum-feeding specialization of the marmosets has important implications with regard to group stability (possibly more stable in marmosets), dispersal patterns, home range size, seasonal food shortage, and reproductive strategies, when compared to the tamarins and lion tamarins. Whereas the marmosets breed twice a year in the wild even in very seasonal habitats, the tamarins and lion tamarins tend to breed only once. The large home ranges of lion tamarins in comparison to marmosets is related not only to their larger size, but also possibly to seasonal fruit shortage, their opportunistic use of plant exudates, and the animal prey they exploit. Finally, the links between reproductive strategies, habitat quality, and dispersal in lion tamarins must take into account the lack of physiological inhibition of ovarian function in subordinate females; a characteristic not found in other callitrichids studied to date.

# 15

# Ecological differentiation in the Callitrichidae

*Stephen F. Ferrari*

## Introduction

The most obvious and absolute difference between callitrichids and all other anthropoid primates is their small size, a characteristic which, along with other diagnostic features of their biology, is intimately related to an evolutionary process of specialization for the efficient colonization of marginal and disturbed forest habitats (Eisenberg 1978; Garber 1980*b*; Sussman and Kinzey 1984). All four callitrichid genera also share a range of morphological, physiological, ecological, and behavioural characteristics, many of which are discussed at length in other chapters of this book. The family is less homogeneous than it might seem, however, depending on the parameters chosen and how they are evaluated (Ferrari and Lopes Ferrari 1989). While all callitrichids are small, for example, they are not equally so. Body size, which plays a fundamental role in a primate's ecology (Milton and May 1976; Clutton-Brock and Harvey 1977*a*), increases approximately sixfold between the smallest (*Cebuella pygmaea*) and the largest (*Leontopithecus chrysopygus*) callitrichid species. This is consistent with body size variation within the Cebidae, if the number of genera is taken into account. The present overview of the data on callitrichid ecology reveals that such differences have probably played a fundamentally important role in the evolutionary history of a highly successful primate radiation.

## Gummivory in callitrichids

A basic variable within the Callitrichidae is dental morphology. In the 'short-tusked' marmosets (*Callithrix* and *Cebuella*), the lower anterior dentition is modified for bark-gouging, which permits the systematic

The species classification followed in this chapter is that of Mittermeier *et al.* 1988*b* (see Rylands *et al.*, Chapter 1, this volume), except for the Amazonian marmosets, which follows Vivo (1988) and Ferrari and Lopes (1992*b*).

exploitation of plant exudates (Coimbra-Filho and Mittermeier 1976, 1978; Sussman and Kinzey 1984). There is no such specialization in the tamarins (*Leontopithecus* and *Saguinus*), which, when feeding on exudates, do so opportunistically. Ferrari and Lopes Ferrari (1989) have proposed that this fundamental difference between marmosets and tamarins has far-reaching implications for many aspects of their behaviour, ecology and, ultimately, social organization.

By gouging the bark of gum-producing plants, marmosets are able to ensure a regular supply of nutrients, in particular carbohydrates (Anderson *et al.* 1972; Bearder and Martin 1980; Garber 1984*b*; Nash 1986), throughout the year. Marmosets also appear to be adapted for the efficient digestion of the complex polysaccharides which constitute the carbohydrate component of gums (Coimbra-Filho *et al.* 1980; Power *et al.* 1990; Ferrari and Martins in press). These specializations permit marmosets to subsist on a diet of gum and insects whenever or wherever preferred sources of carbohydrate, such as fruit, are scarce.

Tamarins are unable to exploit plant exudates as systematically as marmosets, although most, if not all species probably feed on gum in the wild (Izawa 1978; Rylands 1982; Terborgh 1983; Garber 1984*b*, this volume; Soini 1987*b*; Peres 1989). Garber (1980*b*) proposes that exudates are a basic component of the diets of all callitrichids, and that the animals' pointed, keeled nails are primarily an adaptation for clinging to the relatively broad trunks and boughs of trees on which gum deposits are typically found. Small body size, a correlate of gummivory in primates (Nash 1986), would be a concomitant of this adaptation (Garber 1980*b*).

While dental specializations are absent in the tamarins, at least two species (*Saguinus fuscicollis*, Soini 1987*b*; *Leontopithecus rosalia*, Peres 1989) are known to damage gum-producing plants during feeding, an activity which stimulates further exudate flow, although probably to a much lesser extent than gouging by marmosets. The extent to which other tamarin species engage in such behaviour is unclear, but the available data indicate that it is probably not significant overall.

*S. fuscicollis* is also known to exploit marmoset gum-feeding holes where it occurs syntopically with *Cebuella pygmaea* (Soini 1987*b*) and *Callithrix emiliae* (Ferrari and Martins in press). Similar behaviour has not been reported for any of the four other *Saguinus* species that are sympatric with *Cebuella*, although the same can also be said for the majority of *S. fuscicollis* studies. Major ecological differences between *Cebuella* and other callitrichids probably reduce significantly the chances of contact between them. *S. fuscicollis* is not only far more similar to *C. emiliae* in body size and habits, but also associates with this species during foraging (Martins *et al.* 1987; Lopes and Ferrari in preparation), a form of behaviour which may facilitate the exploitation of *C. emiliae* gum-feeding sites.

**Table 15.1** Proportion of plant exudates in tamarin diets

| Species | Percentage gum in plant diet | Site | Reference |
|---|---|---|---|
| *S. fuscicollis* | 2.3 | Peru | Fang (1990) |
| *S. mystax* | 0.2 | | |
| *S. geoffroyi* | 24.6 | Panama | Garber (1984*b*) |
| *S. fuscicollis* | 7.6 | Peru | Garber (1988*b*) |
| *S. mystax* | 1.5 | | |
| *S. midas* | 0 | Surinam | Mittermeier (1977) |
| *L. rosalia* | 1.5 | Brazil | Peres (1989) |
| *L. chrysomelas* | 3–11[a] | Brazil | Rylands (1989*b*) |
| *S. fuscicollis* | 1–42[a] | Peru | Soini (1987*b*) |
| *S. fuscicollis* | 3–9[a] | Peru | Terborgh (1983) |
| *S. imperator* | 1–2[a] | | |

[a] Range of values recorded during partial sample periods.

*S. fuscicollis* is the only tamarin species known to occur in sympatry with both *Cebuella* and *Callithrix*, although at least two other tamarin species are sympatric with *Callithrix*. Rylands (1982, 1989*b*) carried out a detailed three-month study of syntopic *Leontopithecus chrysomelas* and *Callithrix kuhli* at Una in the Brazilian state of Bahia. While the former species was observed feeding on gum and the latter gouging gum-producing plants in typical marmoset fashion, Rylands did not observe the exploitation of *C. kuhli* feeding holes by *L. chrysomelas*. The species generally had little contact with one another except at fruit-feeding sites.

Syntopy between *Callithrix argentata* and *Saguinus midas* has recently been confirmed in the flood-plain of the Tocantins–Xingu interfluvium (Ferrari and Lopes Ferrari 1990*a*). The evidence from two sites in this region (Ferrari and Lopes Ferrari 1990*a*, personal observation) indicates a similar situation to that at Una, although it is probably too early to confirm that *S. midas* does not exploit *C. argentata* gum-feeding holes. There is no evidence of mixed-species associations from any of these sites, however, and Ferrari and Lopes Ferrari (1990*a*) have proposed that the formation of such associations is a behavioural-ecological specialization of *S. fuscicollis*.

The available feeding data (Table 15.1), especially those from studies of syntopic tamarins, indicate that *S. fuscicollis* is more gummivorous than most other tamarins, a characteristic which is possibly related to the relatively small size of this species. Body size varies considerably among the 13 *S. fuscicollis* subspecies recognized by Hershkovitz (1977), but weights are generally very similar to, or even smaller than those of *Callithrix* marmosets (Table 15.2).

Small size may be favourable for the acquisition of exudates, but whether the difference in size between *S. fuscicollis* and other tamarins would confer a significant advantage is not clear. Similarly, while data on gut morphology

**Table 15.2** Adult body weights for *Callithrix* species and *S. fuscicollis* subspecies (wild-caught animals)

| Taxon | Mean body weight (gn) | *N* | Reference |
|---|---|---|---|
| *C. argentata* | 355.6[a] | 14 | Ferrari and Lopes (1992*b*) |
| *C. emiliae* | 313.3[a] | 12 | Ferrari and Lopes (1992*b*) |
| *C. emiliae* | 326.9 | 8 | Ferrari and Martins (in press) |
| *C. jacchus* | 332.9[a] | 10 | L.J. Digby (personal communication) |
| *C. nigriceps* | 370.0[a] | 3 | Ferrari and Lopes (1992*b*) |
| *S. f. fuscicollis* | 328.0[a] | 9 | Snowdon and Soini (1988) |
| *S. f. illigeri* | 292.0[a] | 9 | Snowdon and Soini (1988) |
| *S. f. nigrifrons* | 412.8[a] | 33 | Garber and Teaford (1986) |
| *S. f. nigrifrons* | 354.0 | 51 | Snowdon and Soini (1988) |
| *S. f. weddelli* | 315.0 | 8 | Ferrari and Martins (in press) |

[a] Adult males only.

(Ferrari and Martins in press), and digestive efficiency (Power *et al.* 1990) indicate that *S. fuscicollis* is less specialized for gummivory than marmosets, there is no conclusive evidence to support the hypothesis that it is more specialized phylogenetically than other tamarins. This may not necessarily be essential for the digestion of gum in the quantities normally consumed by *S. fuscicollis*, however, especially as some adaptation of the digestive tract may take place in response to the inclusion of gum in the diet (e.g. Johnson and Gee 1986).

The exploitation of plant exudates by *S. fuscicollis* is little more systematic than that of other tamarin species. This is exemplified most clearly through a comparison of the *S. fuscicollis* studies carried out at two Peruvian sites: Río Pacaya (Soini 1987*b*; Snowdon and Soini 1988) and Cocha Cashu (Terborgh 1983; Terborgh and Stern 1987; Goldizen *et al.* 1988). The major contrast between the two studies, in the context of the present discussion, is the difference in the availability/use of plant exudates, and concomitant variations in factors such as home range size and breeding patterns, as proposed by Ferrari and Lopes Ferrari (1989).

Plant exudates were apparently abundant at Río Pacaya, being made available predominantly through wind damage and insect boring. In the dry season at this site, gum constituted 42 per cent of plant-feeding records, and the nectar of *Combretum fruticosum* 4 per cent (Soini 1987*b*). During the same period at Cocha Cashu, *S. fuscicollis* spent only 9 per cent of plant-feeding time eating ‘sap’, but 75 per cent consuming the nectar of *Combretum assimile* and *Quararibea cordata* (Terborgh 1983). Reliance on nectar during the dry season at Cocha Cashu apparently has deleterious consequences for the *S. fuscicollis* population (Terborgh and Stern 1987; Goldizen *et al.* 1988), and may account for the large home ranges (up to

120 ha) of the species at this site. While the Río Pacaya study group was larger (6–10 members), its home range was only 16 ha, half that of the *S. fuscicollis* group at Cocha Cashu (3–5 members).

Differences in the availability of resources at the two sites are also reflected in breeding patterns. Breeding was almost continuous over a four-year period at Río Pacaya (Soini 1987*b*, Table III), in a pattern similar to that recorded for marmosets such as *Callithrix flaviceps* (Ferrari and Diego in press) and *Callithrix humeralifer* (Rylands 1986*b*). On the other hand, breeding at Cocha Cashu is generally far more seasonal (Goldizen *et al.* 1988).

## Foraging strategies and interspecific competition

Marmosets and tamarins can be characterized ecologically by a number of common features (Sussman and Kinzey 1984; Chapters 11 to 14 in this volume). In very general terms, callitrichids prefer the dense vegetation of the lower forest strata in marginal or disturbed habitats, where they typically prey on large-bodied mobile insects, in particular orthopterans. Whether feeding predominantly on reproductive plant parts or exudates, in addition, callitrichids typically concentrate on a relatively small selection of the plant species available to them at any one time (e.g. Rylands 1982; Terborgh 1983; Soini 1987*b*; Ferrari 1988*b*). Body size plays an important role in these and many other aspects of callitrichid ecology and, indirectly, in the avoidance of competition with larger primates.

Small body size enables callitrichids to trapline plant resources, for example, whose characteristics make them unattractive for systematic exploitation by larger-bodied platyrrhines. Many of the most frequently exploited fruit sources are small trees or lianes which produce relatively small fruits in 'piecemeal' crops (Opler *et al.* 1980), that is in small quantities over relatively long periods. Gum sources, especially those established in response to marmoset gouging, present similar characteristics. In both cases, the distribution of the resource in time and space renders it inappropriate for larger primates, reducing potential competition.

Similarly, the small size of the callitrichids enables them to prey more efficiently on relatively large, mobile insects than the larger insectivorous platyrrhines such as *Cebus* and *Saimiri* (Terborgh 1983). In the mobile environment of the forest canopy, callitrichids are able to forage for insects efficiently by stealth, in contrast with the larger cebids which generally forage manipulatively for immobile prey such as immature lepidopterans and hymenopterans. Capturing relatively large prey enables callitrichids to maintain a larger proportion of animal material in the diet while spending less time foraging, and thus effectively follow a time-minimizing foraging strategy (Schoener 1971; Terborgh 1983), appropriate for animals of small body size which are relatively vulnerable to predation.

In general, marmosets and tamarins forage for insects in the foliage of tree crowns, following either 'scan-and-pounce' or leaf-gleaning techniques. *C. flaviceps*, for example, spent 89.2 per cent of its foraging time scanning for signs of prey and only 7.1 per cent in manipulative behaviour (Ferrari 1988*b*). However, two forms, the lion tamarins (*Leontopithecus* spp.) and the saddle-back tamarin (*S. fuscicollis*), exhibit more manipulative foraging behaviour, predominantly at specific sites such as holes in branches and trunks. A specialization for foraging in holes, and in particular the leaf axils of bromeliads, has been proposed for *Leontopithecus* on the basis of both morphological (elongated cheiridia: Coimbra-Filho 1970*a*; Hershkovitz 1977) and ecological (Rylands 1989*b*) data. Rylands (1989*b*) proposes that this foraging technique may enable *L. chrysomelas* to capture larger prey than *C. kuhli*.

*S. fuscicollis* also appears to capture larger prey than its syntopic congeners by following a similar strategy of foraging at specific sites such as holes and fissures in bark, and leaf litter accumulations (Yoneda 1981, 1984a; Terborgh 1983), but does not investigate bromeliads in *Leontopithecus* fashion. Like *Leontopithecus*, however, *S. fuscicollis* appears to be morphologically specialized for manipulative foraging, although to a lesser degree (Ferrari in preparation).

The two forms exhibit important contrasts in a number of other characteristics, including body size. Lion tamarins are the largest callitrichids, while saddle-backs are the smallest tamarins. Behaviourally, the two forms contrast most significantly in the use of forest strata. At Una, *Leontopithecus* is active at higher levels in the forest than *C. kuhli* (Rylands 1989*b*) whereas all studies of syntopic tamarins have recorded *S. fuscicollis* using lower levels than its congeners (Pook and Pook 1982; Terborgh 1983; Yoneda 1984*a*; Buchanan-Smith 1990; Fang 1990; Heymann 1990*a*; Norconk 1990). In all cases, the smaller species (*C. kuhli* or *S. fuscicollis*) uses the lower forest strata, a pattern also observed in sympatric lorisids (Bearder 1987).

This contrast reflects other ecological differences between the two forms. The apparent importance of foraging in bromeliads for *Leontopithecus* may be a fundamental factor determining the preference not only for relatively high levels, but also for primary and certain specific forest habitats in this genus (Coimbra-Filho 1970*a*; Rylands 1989*b*). Coimbra-Filho (1978*a*) has proposed that the use of tree-holes as sleeping sites may also underlie the preference of this genus for relatively undisturbed forest.

In broad terms, then, *Leontopithecus* appears to be ecologically inflexible in comparison with other callitrichids, a characteristic which may be reflected in its restricted geographical distribution, especially in relation to that of eastern Brazilian *Callithrix*. *S. fuscicollis*, by contrast, is one of the most widespread of all callitrichids and apparently one of the most adaptable. Given the adaptive complex of the Callitrichidae, body size differences may be fundamentally important here.

In addition to body size, the adaptability and apparent success of *S. fuscicollis* in comparison with *Leontopithecus* may be linked to a number of factors, in particular its propensity for the formation of mixed-species associations. Such associations probably bring a number of benefits for *S. fuscicollis* in, for example, predator defence, resource monitoring, and territorial defence (see for example Terborgh 1983; Garber 1988*b*), and there is some evidence that this species achieves significantly higher population densities in the presence of a congener in comparison with sites at which other tamarins are absent (Norconk 1990).

In addition to benefits with regard to predator defence (Sussman and Kinzey 1984; Ferrari and Lopes Ferrari 1990*b*; Heymann 1990*a*), the use of relatively low levels in the forest by *S. fuscicollis* when associating with congeners may also entail significant advantages when foraging for insect prey. From observations of *C. flaviceps*, Ferrari (1988*b*) proposed that the disturbance of prey may be an important component of callitrichid foraging behaviour, given the typical predator avoidance strategies followed by the animals most commonly captured by these monkeys. Orthopterans, caterpillars, coleopterans, spiders, snails, tree-frogs, and lizards constituted 94.1 per cent of identified *C. flaviceps* prey ($n$ = 1146) and this appears to be typical of other callitrichids (e.g. Coimbra-Filho 1981; Terborgh 1983; Stevenson and Rylands 1988).

In the vast majority of cases, these animals rely on camouflage, rather than agility or flight, to avoid predation, with a last line of defence that involves leaping or falling away rapidly to a new position when disturbed. Stick insects exemplify this strategy, the efficiency of which is greatly reduced at low levels in the forest, and especially once the animal comes to the ground, from which further escape is normally impossible.

Ferrari (1988*b*) noted that the disturbance of prey by group members other than that capturing the item was an important component of *C. flaviceps* foraging behaviour. The systematic exploitation of prey disturbed by army ant columns has also been recorded for this and other marmoset species (Rylands *et al.* 1989), and the capture of flying invertebrates by associating birds has been recorded for a number of insectivorous platyrrhines (e.g. Terborgh 1983; Boinski and Scott 1988; Siegel *et al.* 1989; Ferrari 1990; Egler 1991). While there is no specific evidence on this aspect of their foraging behaviour, it seems likely that the disturbance of prey by a group of congeners foraging at higher levels in the forest may be an additional significant benefit of interspecific association for *S. fuscicollis* groups.

## Dwarfism, ecology, and body size constraints

Similar in size (Table 15.2) and many aspects of their ecology (e.g. relative degree of gummivory, foraging levels), *Callithrix* and *S. fuscicollis* can be seen as the two most successful callitrichid radiations in terms of variables such as geographical distribution (Hershkovitz 1977), population density and habitat exploitation (Ferrari and Lopes Ferrari 1989; Rylands 1989*b*; Norconk 1990). Within the marmosets, there is much evidence to suggest that *Callithrix jacchus* and *Callithrix penicillata* are also the ecologically most successful species with regard to these same variables (Stevenson and Rylands 1988). At least one important factor here is the relatively high degree of morphological specialization of these two species for gummivory (Hershkovitz 1977; Rylands and Faria, this volume; Ferrari and Martins in press), which may include relatively small body size (Table 15.2).

Together, *Callithrix* and *S. fuscicollis* are distributed over approximately two-thirds of the total geographic range of the family (Hershkovitz 1977), encountering almost the full range of ecological variables experienced by callitrichids in the wild. Only *Leontopithecus caissara* (Lorini and Persson 1990) occurs at higher latitudes, for example. *C. jacchus* group (*sensu* Hershkovitz 1977) marmosets occur at altitudes varying from sea level to 2000 m (Ferrari 1988*b*), and in a wider range of habitat types and climates than those in which other callitrichids are found. In some areas species of this group are the only endemic primates, whereas some *S. fuscicollis* populations are sympatric with 12 other platyrrhines, in the largest known Neotropical primate communities (Terborgh 1983).

Nowhere, however, does the body weight of *Callithrix* species or *S. fuscicollis* subspecies fall below 250 g (Table 15.2). While small body size may play a fundamental role in callitrichid ecology, a body weight of 250–400 g possibly represents a threshold below which a number of disadvantages may begin to outweigh its benefits. At an adult body weight of 100–120 g *Cebuella* represents the extreme of the dwarfing continuum, and also something of an exception to the general pattern outlined here, given that its small size and adaptations for gummivory might be expected to have resulted in a species capable of achieving exceptionally high population densities in most if not all habitats.

Studies of the pygmy marmoset (Soini 1988, this volume) have shown, on the contrary, that it is not only highly habitat-specific, but also forms small social groups, in comparison with other marmosets (Ferrari and Lopes Ferrari 1989), and populations of relatively low density/biomass. This is emphasized by the exceptional ranging behaviour of *Cebuella* groups, which leave much of the available space unoccupied over the short/medium term, even in apparently ideal habitats (Soini 1988).

Given its extremely small size, factors such as predation pressure, reproductive parameters, and energetic constraints may have played a far more important role in the evolution of *Cebuella* behavioural ecology than they have for either *Callithrix* or *S. fuscicollis*. The influence of predation pressure can be clearly seen in the highly cryptic nature of *Cebuella* colouration and habits (Soini 1988). While this can be seen as an extreme form of the time-minimizing foraging strategy common to other callitrichids, this type of predator avoidance behaviour may in itself have placed restrictions on the foraging and ranging behaviour of the species, and possibly also on group size (Ferrari and Lopes Ferrari 1990*b*). Given the relationship of metabolic expenditure to body size (Kleiber 1961), this type of restriction would be reinforced by the relatively high energetic demands of any activity for a primate of this size.

Reproduction may be an additionally critical energetic burden for *Cebuella*, given that neonatal litter weight may reach as much as 24 per cent of female body weight (Leutenegger 1973). As in other callitrichids, dependent infants are carried at all times during the daily activity period rather than being left in nests, although the burden of infant care is distributed among group members (e.g. Goldizen 1987*b*; Ferrari in press). Even so, the demands of infant care may also place a number of important restrictions on the behaviour and ecology of the species.

The importance of such body size constraints for the evolution of *Cebuella* behavioural ecology is supported by a comparison with other primate species exhibiting body sizes as small as or smaller than that of the pygmy marmoset. All these species (*Galago demidovii*, *Galago zanzibaricus*, *Microcebus murinus*, and *Tarsius* spp.) are nocturnal prosimians which are predominantly solitary in their behaviour (Martin 1972; Bearder 1987). Both kinds of behaviour probably greatly increase the crypticity of these prosimians in comparison with *Cebuella*. In addition, diurnal raptors, the primary predators of callitrichids (Ferrari and Lopes Ferrari 1990*b*), are relatively far more abundant than nocturnal species in tropical forests (e.g. Terborgh *et al.* 1990). Predation pressure is nevertheless a fundamentally difficult aspect of primate ecology to evaluate, and the evidence from the two nocturnal primates for which data are available is inconclusive (Cheney and Wrangham 1987).

A further important difference between *Cebuella* and the small-bodied prosimians is that all species, with the exception of *Tarsius bancanus*, leave their young in nests (Martin 1972; Bearder 1987). The avoidance of infant-carrying has obvious benefits in energetic terms, and in the specific case of co-operatively breeding callitrichids such as *Cebuella*, may also reduce the influence of a possibly significant determinant of social behaviour. Without the need to carry infants, *Cebuella* might be able to form smaller or less cohesive social units which would be more cryptic and less vulnerable to predation.

Many other factors may be important in defining ecological differences between *Cebuella* and the small-bodied prosimians, and it is impossible to make more than very broad comparisons between them. However, these prosimians appear to be far less specialized in terms of habitat preference and normally occur at higher densities in the wild (Bearder 1987; Soini 1988). Overall, this evidence appears to suggest that a body size as small as that of *Cebuella* may be sub-optimal for a morphologically or behaviourally unspecialized callitrichid primate.

Following the present discussion of ecological differentiation in the Callitrichidae, interspecific competition may have been a key factor in *Cebuella* evolutionary history. *Cebuella* is at least potentially sympatric with two tamarins throughout its geographical range (Hershkovitz 1977), whereas no more than two callitrichid species are present at any other site within the distribution of the family. Assuming, as argued below, that the radiation of the two tamarins preceded that of *Cebuella*, and given the apparent importance of body size in callitrichid niche partitioning, the small size of *Cebuella* can be seen as an extension of the dwarfing pattern under exceptional conditions. In this case, the prior occupation of a marmoset-like niche (with regards to body size, gum feeding, foraging levels, etc.) by the highly successful *S. fuscicollis* may have been the most important factor in the evolution of *Cebuella* ecology.

## Zoogeography and evolutionary history of the Callitrichidae

At the present time, there is little general consensus on either the taxonomy of the Callitrichidae (Hershkovitz 1977; Vivo 1988; Mittermeier *et al.* 1988*b*; Rylands *et al.* this volume) or the evolutionary relationships within the family (Hershkovitz 1977; Cronin and Sarich 1978; Byrd 1981; Rosenberger 1984; Rosenberger and Coimbra-Filho 1984; Ford 1986*a*; Kay 1990; Rosenberger *et al.* 1990). However, the present discussion of ecological differentiation does appear to support the widely accepted hypothesis that the evolutionary history of the family is characterized by a process of phyletic dwarfism (Leutenegger 1973, 1979; Rosenberger 1977, 1981, 1984; Maier 1978; Ford 1980*b*, 1986*a*; Garber 1980*b*, this volume; Byrd 1981; Sussman and Kinzey 1984; Martin 1990). Following the same reasoning, the evolutionary history of the family can also be discussed, in very broad terms, on the basis of the present-day distribution of callitrichid taxa (Fig. 15.1), given the prior assumption that the common ancestor of all callitrichids (possibly *Branisella boliviana*: Rosenberger *et al.* 1990) would have been similar in size to the largest present-day tamarins.

Callitrichid zoogeography is characterized by a relative lack of interspecific sympatry, especially in comparison with most other polyspecific primate families (or even subfamilies), and despite the fact that the Callitrichidae

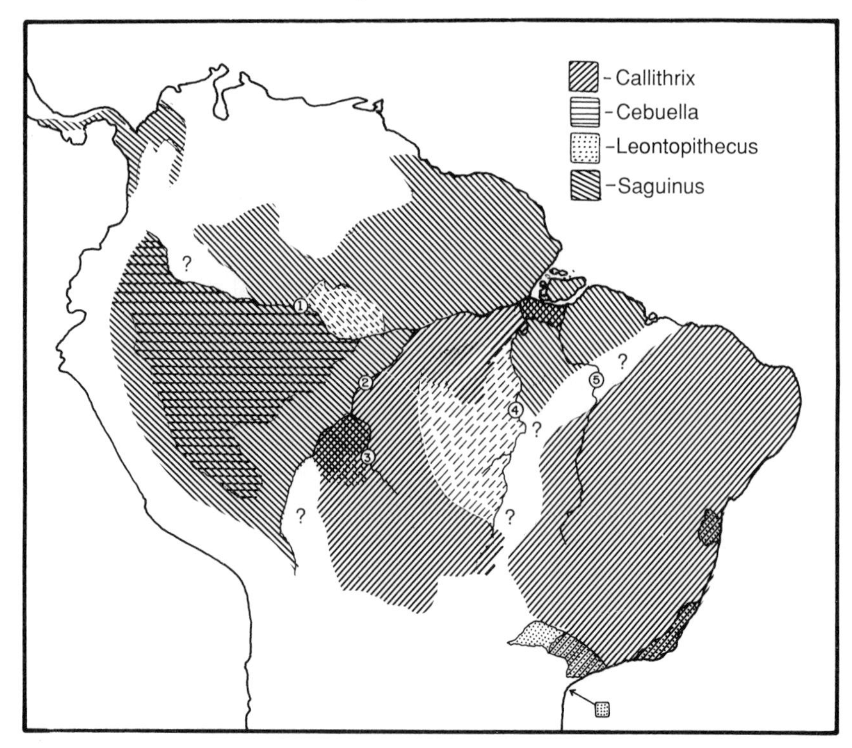


**Fig. 15.1** Geographical distribution of the four callitrichid genera, following Hershkovitz (1977), Stallings and Mittermeier (1983), Terborgh (1983), Vivo (1988), Ferrari and Lopes Ferrari (1990*a*), Lorini and Persson (1990), Ferrari and Lopes (1992*a*) personal obervation. Key to rivers: ① Amazonas/Solimões/Japurá; ② Madeira; ③ Jiparaná; ④ Xingu; ⑤ Tocantins.

constitutes a group of at least 15 (Hershkovitz 1977) and possibly many more than 20 species (Mittermeier *et al.* 1988*b*; Vivo 1988: Lorini and Persson 1990; Ferrari and Lopes 1992*b*). Where sympatry does occur, as described above, niche partitioning appears generally to have exerted a strong pressure on the selection of morphological and behavioural characteristics especially if, as suggested by Pook (1978*a*), the callitrichid radiation has been relatively rapid, at least in comparison with that of the cebids.

An additional characteristic of callitrichid zoogeography is the restriction of all genera but *Saguinus* to the south of the Amazonas/Solimões/Japurá river system (Fig. 15.1), which also implies a relatively recent radiation for some taxa. All proven cases of sympatry between callitrichid species, including those of *S. fuscicollis* and its congeners, are also restricted to the south of this river system.

On the basis of shared derived traits, Rosenberger (1984) and Ford (1986*a*) have proposed that the marmosets are more closely related to *Leontopithecus* than to *Saguinus*. Given the present-day distributions of the latter two genera and the relatively weak ties between them (Ford 1986*a*), it seems reasonable to assume that their recent evolution has taken place in isolation in the pre-cursors of the Amazonian and Atlantic forest ecosystems, respectively. Following the same reasoning, *Callithrix* would also have appeared originally in the Atlantic Forest of eastern Brazil, where it would have either been derived from, or shared a common ancestor with *Leontopithecus*.

As argued above, competition between *Callithrix* and *Leontopithecus* prototypes would have been a fundamental factor determining morphological, behavioural and ecological differences between these forms. The restricted and disjunct distribution of the present-day lion tamarins, especially in comparison with that of *Callithrix*, is probably the result of a combination of the formation of forest refuges during the Pleistocene (Kinzey 1982) and the competitive advantages of marmoset body size and gum-feeding specializations (Ferrari and Lopes Ferrari 1989). These same specializations would also have been a fundamental variable in the expansion of the marmosets westwards and northwards into the woodland and savanna habitats which separate the Amazonian and Atlantic forest ecosystems and are probably inappropriate for colonization by most tamarins.

The subsequent radiation of the marmosets in the forested Amazon basin would have been influenced by two principal factors, the presence of tamarin populations, and the existence of major geographical barriers, such as the Amazon Lake (Frailey *et al.* 1988) and in the more recent past, the Amazonas/Solimões/Japurá river system. Given their present distribution (Fig. 15.1) and in the absence of evidence to the contrary, it seems reasonable to assume that tamarins would have been distributed throughout the Amazon basin prior to the arrival of marmosets in this region. In this case, the absence of *Saguinus* between the lower Madeira and Xingu rivers (Fig. 15.1) may be at least partly due to competitive exclusion by the smaller-bodied, gum-feeding marmosets.

The limited degree of overlap between the distributions of *Callithrix* and *Saguinus* (Ferrari and Lopes Ferrari 1990*a*; Ferrari and Lopes 1992*a*; Fig. 15.1) would appear to support this hypothesis, although a number of important questions remain. A fuller understanding of callitrichid ecology and zoogeography in southwestern Amazonia is critical, especially given the apparent discontinuity between the present-day ranges of *Callithrix* and *Cebuella* (cf. Hershkovitz 1977). However, Terborgh (1983) records *Cebuella* on the upper Río Madre de Dios and, given the incomplete nature of our knowledge of primate zoogeography in this region (e.g. Vivo 1985; Ferrari and Lopes 1992*a*), further study may reveal an area of at least parapatry between these two genera. The Madre de Dios, for example, is a tributary of the Río Guaporé, which implies potential parapatry between *Cebuella* and *C. emiliae* in northeastern Rondônia. If, in addition, the range of *Saguinus imperator* is restricted to the north/west bank of the Madre de Dios (Hershkovitz 1977), the distribution of *C. melanura* in Bolivia may feasibly extend as far north as this river, again suggesting possible contact with the range of *Cebuella*.

The most relevant recent finding in this region was the extension of the range of *S. fuscicollis* to the right or east bank of the Rio Madeira in the Brazilian state of Rondônia (Vivo 1985; Martins *et al.* 1987; Ferrari and Lopes 1992*a*). *S. fuscicollis* had previously been restricted to the west of the Rio Madeira and the Río Mamoré in Bolivia (Hershkovitz 1977), eliminating the possibility of sympatry between this species and the westernmost *Callithrix* forms, *C. emiliae*, and *C. melanura*. It is now known that *S. fuscicollis* is found on the right bank of the Madeira as far east as the Rio Jiparaná (Ferrari and Lopes 1992*a*), a region in which it is sympatric with *C. emiliae*.

The southern limits of the distribution of *S. fuscicollis* in this region, the presence of the species in the Mamoré–Guaporé interfluvium and contact with *C. melanura* remain unknown. Given the paucity of the information available, it does not seem unreasonable to suppose that the ranges of these two species may overlap on one or even both margins of the Guaporé, possibly extending as far north as the Madre de Dios (as suggested above).

Given the present discussion of ecological differentiation in the Callitrichidae, further information on the nature of sympatry between *S. fuscicollis* and *Callithrix* marmosets and the ecological relationships between these forms would provide a number of insights into the evolutionary history of the family. Preliminary observations of syntopic *C. emiliae* and *S. fuscicollis* populations at Samuel in Rondônia (Martins *et al.* 1987; Lopes and Ferrari in preparation have revealed a number of parallels with the mixed-species tamarin associations observed at other sites in western Amazonia. *C. emiliae* are slightly larger-bodied than *S. fuscicollis* at Samuel and assume an ecological role similar to that filled by the larger tamarins with which *S. fuscicollis* associates at other sites, in terms of variables such as foraging height and technique.

In addition to the usual benefits of association for *S. fuscicollis*, however, co-existence with *Callithrix* marmosets also implies access to supplies of plant exudates (Ferrari and Martins in press) that are more abundant and readily available than those normally encountered by tamarin populations in other areas. Given evidence (e.g. Soini 1987*b*) that *S. fuscicollis* can include large quantities of plant exudate in its diet, the ability to exploit *Callithrix* gum-feeding holes probably confers an important competitive advantage on this species. Given this, and assuming, as proposed above, that marmosets have colonized southern Amazonia through the competitive exclusion of tamarins, present-day overlap between *S. fuscicollis* and *Callithrix* suggests the recent colonization of the range of the latter by the former. Information on the factors which determine the southern limits of *S. fuscicollis* distribution in this region would thus be of considerable value for the understanding of callitrichid ecology.

Sympatry between *C. argentata* and *S. midas* in eastern Amazonia appears, by contrast, to represent an ongoing expansion of the range of the former species which is currently delimited by the Tocantins river to the east and upland forest habitats associated with the Precambrian Brazilian Shield to the south (Ferrari and Lopes Ferrari 1990*a*). Assuming that this expansion has been relatively recent, this situation may be partly related to differences in the topography of the Tocantins and Xingu rivers, specifically, the complex of relatively narrow channels and islands associated with the major S-bend of the Xingu in the vicinity of Altamira which may have permitted the eastward migration of *C. argentata* at some time in the recent geological past. Similar narrowing of the Tocantins occurs much further up-river, far to the south of the present range of *C. argentata* on the west bank of this river.

The apparent inability of *C. argentata* to colonize the southern uplands occupied by *S. midas* suggests that the competitive advantages of marmoset gum-feeding specializations (including small body size) may be outweighed by other factors, at least under existing conditions. Evidence on morphology (Hershkovitz 1977; Ferrari and Martins in press) and feeding behaviour (Rylands 1984) indicate that exudates are probably a less important resource for Amazonian marmosets than for those of the *C. jacchus* group (e.g. Lacher *et al.* 1984; Rylands 1984; Ferrari 1988*b*; Alonso and Langguth 1989), which generally experience more pronounced seasonal fluctuations in the availability of plant resources. As suggested above, however, advantages associated with the ability to feed systematically on gum may come into play on a broader time scale, that of the formation of Pleistocene refugia (Kinzey 1982) and periods during which the expansion of drier woodland habitats would have favoured marmosets. The expansion of *C. argentata* to the east of the Xingu may thus have been no more than simultaneous with the most recent of these periods.

## Conclusion

To summarize, following the processes of ecological differentiation discussed here and the present-day distribution of the family, the principal events in the evolutionary history of the Callitrichidae would be the following:

1. The initial radiation of a tamarin-like ancestor throughout the forested Neotropics, giving rise to two different forms following isolation of the Amazonian and Atlantic forest ecosystems.

2. The appearance of a marmoset prototype in the Atlantic forest and its subsequent expansion westwards and northwards to the Amazon basin.
3. The appearance and radiation of *S. fuscicollis* in the western Amazon basin.
4. The appearance of *Cebuella* in the western Amazon basin.

This hypothetical sequence of events is broadly consistent with some interpretations of evolutionary relationships within the Callitrichidae based on morphological and physiological characteristics (e.g. Rosenberger 1984; Ford 1986*a*), although not with others, above all Hershkovitz (1977). In spite of recent finds (Rosenberger *et al.* 1990), an adequate fossil record on which the confirmation of either hypothesis might be based is unavailable at the present time and all such interpretations, like that presented here, must include a degree of speculation.

## Acknowledgements

I am grateful to the National Research Council (CNPq) of the Brazilian government, Anthony Rylands, Cida Lopes, Leslie Digby, and Mário de Vivo.

# References

Abbott, D.H. (1978). The physical, hormonal and behavioural development of the common marmoset, *Callithrix jacchus jacchus*. In *Biology and behaviour of marmosets* (ed. H. Rothe, H.-J. Wolters and J.P. Hearn), pp.99–106. Eigenverlag H. Rothe, Göttingen. (Chapters 7, 10)

Abbott, D.H. (1984). Behavioural and physiological suppression of fertility in subordinate marmoset monkeys. *Am. J. Primatol.*, 6, 169–86. (Chapters 4, 5, 7, 10, 14)

Abbott, D.H. (1986). Social suppression of reproduction in subordinate marmoset monkeys (*Callithrix jacchus jacchus*). In *A primatologia no Brasil—2* (ed. M.T. de Mello), pp.16–31. Sociedade Brasileira de Primatologia, Brasília. (Chapters 5, 7)

Abbott, D.H. (1987). Behaviourally mediated suppression of reproduction in female primates. *J. Zool., Lond.*, 213, 455–70. (Chapter 5)

Abbott, D.H. (1988). Natural suppression of fertility. In *Reproduction and disease in captive and wild animals* (ed. J.P. Hearn and G.R. Smith), pp. 7–28. Oxford University Press, Oxford. (Chapter 5)

Abbott, D.H. (1989). Social suppression of reproduction in primates. In *Comparative socioecology: the behavioural ecology of humans and other mammals* (ed. V. Standen and R.A. Foley), pp.284–304. Blackwell Scientific Publications, Oxford. (Chapter 5)

Abbott, D.H. (1991). Social control of fertility. In *Primate responses to environmental change* (ed. H.O. Box), pp.75–89. Chapman and Hall Ltd., London. (Chapter 5)

Abbott, D.H. and George, L.M. (1991). Reproductive consequences of changing social status in female marmoset monkeys. In *Primate responses to environmental change* (ed. H.O. Box), pp.295–309. Chapman and Hall Ltd., London. (Chapter 5)

Abbott, D.H. and Hearn, J.P. (1978). Physical, hormonal and behavioural aspects of sexual development in the marmoset monkey (*Callithrix jacchus*). *J. Reprod. Fert.*, 53, 155–66. (Chapters 5, 7, 10, 14)

Abbott, D.H., McNeilly, A.S., Lunn, S.F., Hulme, M.J., and Burden, F.J. (1981). Inhibition in ovarian function in subordinate female marmoset monkeys (*Callithrix jacchus jacchus*). *J. Reprod. Fert.*, 63, 335–45. (Chapters 5, 7)

Abbott, D.H., Hodges, J.K., and George, L.M. (1988). Social status controls LH secretion and ovulation in female marmoset monkeys (*Callithrix jacchus*). *J. Endocr.*, 117, 329–39. (Chapter 5)

Abbott, D.H., O'Byrne, K.T., Sheffield, J.W., Lunn, S.F., and George, L.M. (1989). Neurendocrine suppression of LH secretion in subordinate female marmoset monkeys (*Callithrix jacchus*). In *Comparative reproduction in mammals and man* (ed. R.M. Eley), pp.63–7. National Museums of Kenya, Nairobi. (Chapter 5)

Abbott, D.H., Barrett, J., George, L.M., and Cheesman, D.J. (1991). Olfactory cues and the suppression of ovulation in socially subordinate female marmoset

monkeys. *Abstracts—Symposium on Chemical Signals in Vertebrates VI*, Philadelphia, p.21. (Chapter 4)
Achilles, L. (1991). Experimentelle Untersuchungen zum Einfluss von Verwandtschaft und Vertrautheit auf das Interaktionsverhalten adulter männlicher Weissbüschelaffen (*Callithrix jacchus* Erxleben 1777). Unpublished diploma thesis, University of Göttingen, Göttingen. (Chapter 7)
Addington, R.A., Caine, N.G., and Schaffner, C. (1991). Factors affecting the food calls of red-bellied tamarins. *Am. J. Primatol.*, 24, 85. Abstract. (Chapter 8)
Adler, H.J. (1988). Observation of the birth and postnatal phase in the cotton-top tamarin (*Saguinus oedipus oedipus*) at the Zoologischer Garten Leipzig. *Prim. Rep.*, 18, 45–49. (Chapter 10)
Albone, E.S. (1984). *Mammalian semiochemistry*. John Wiley, New York. (Chapter 4)
Aldrich-Blake, F.P.G. (1968). A fertile hybrid between two *Cercopithecus* spp. in the Budongo forest, Uganda. *Folia primatol.*, 9, 5–21. (Chapter 3)
Almeida, F.B. and Deane, L.M. (1970). *Plasmodium brasilianum* reencontrado em seu hospedeiro original, o macaco uakari branco, *Cacajao calvus*. *Bolt. Inst. Nac. Pesq. Amazônia (INPA)—Patologia Tropical*, 4, 1–9. (Chapter 1)
Alonso, C. (1984). Observações de campo sobre o cuidado a prole e o desenvolvimento dos filhotes de *Callithrix jacchus jacchus*. In *A primatologia no Brasil* (ed. M.T. de Mello), pp.67–78. Sociedade Brasileira de Primatologia, Brasília. (Chapter 10)
Alonso, C. (1986). Fracasso na inibição da reprodução de uma fêmea subordinada e troca de hierarquia em um grupo familiar de *Callithrix jacchus jacchus*. In *A primatologia no Brasil—2* (ed. M.T. de Mello), p.203. Sociedade Brasileira de Primatologia, Brasília. Abstract. (Chapter 7)
Alonso, C. and Langguth, A. (1989). Ecologia e comportamento de *Callithrix jacchus* (Primates: Callitrichidae) numa ilha de floresta Atlântica. *Rev. Nordestina Biol.*, 6, 107–37. (Chapters 7, 12, 15)
Alonso, C. and Porfiro, S. (1989). Comportamento afiliativo entre duas fêmeas de *Callithrix kuhlii* (Primates, Callitrichidae) durante a criação simultânea de seus filhotes. In *Resumos. XVI Congresso Brasileiro de Zoologia*, pp.124–5. Federal University of Paraíba, João Pessoa. (Chapter 7)
Alonso, C., Faria, D.S. de, Langguth, A., and Santee, D.F. (1987). Variação da pelagem na área de intergradação entre *Callithrix jacchus e Callithrix penicillata*. *Rev. Brasil. Biol.*, 47, 465–70. (Chapters 1, 3, 12)
Altmann, S.A. (1974). Baboons, space, time, and energy. *Am. Zool.*, 14, 221–48. (Chapter 3)
Amsler, F. (1982). Familienauflösung bei Weissbüscheläffchen (*Callithrix jacchus*). Eine experimentelle Studie zum Interesse einer Familiengruppe an fremden Artgenossen. Unpublished Master's thesis, University of Zürich, Zürich. (Chapter 7)
Anderson, D.M.W., Hendrie, A., and Munro, A.C. (1972). The amino acid and amino sugar composition of some plant gums. *Phytochem.*, 11, 733–6. (Chapter 15)
Andrew, R.J. and Klopman, R.B. (1974). Urine washing: comparative notes. In *Prosimian biology* (ed. R.D. Martin, G.A. Doyle and A.C. Walker), pp.303–12. Duckworth, London. (Chapter 4)
Anzenberger, A.A. (1983). Bindungsmechanismen in Familiengruppen von Weissbüscheläffchen (*Callithrix jacchus*). Unpublished Ph.D thesis, University of Zürich, Zürich. (Chapter 7)

Anzenberger, A.A. and Simmen, C. (1987). Father-daughter incest in a family of common marmosets (*Callithrix jacchus*). *Int. J. Primatol.*, 8, 524. Abstract. (Chapter 7)

Araujo, A. and Yamamoto, M.E. (In press). Reação a intrusos da mesma espécie em *Callithrix jacchus*: influência do status social. In *A primatologia no Brasil—4* (ed. M.E. Yamamoto and M.B.C. de Sousa). Sociedadade Brasileira de Primatologia and Fundação Biodiversitas, Belo Horizonte. (Chapter 10)

Arruda, M.F., Yamamoto, M.E., and Bueno, O.F.A. (1986). Interactions between parents and infants, and infants-father separation in the common marmoset (*Callithrix jacchus*). *Primates*, 27, 215–28. (Chapter 10)

Ávila-Pires, F.D. de. (1969). Taxonomia e zoogeografia do gênero 'Callithrix' Erxleben, 1777 (Primates, Callithricidae). *Rev. Brasil. Biol.*, 29(1), 49–64. (Chapters 1, 3)

Ávila-Pires, F.D. de. (1974). Caracterização zoogeográfica da Província Amazônica. II. A famìlia Callitrichidae e a zoogeografia Amazônica. *An. Acad. Brasil. Ciênc.*, 46(1), 159–81. (Chapter 1)

Ávila-Pires, F.D. de. (1986). On the validity and geographical distribution of *Callithrix argentata emiliae* Thomas, 1920 (Primates, Callithrichidae). In *A primatologia no Brasil—2* (ed. M.T. de Mello), pp.319–22. Sociedade Brasileira de Primatologia, Brasília. (Chapter 1)

Ávila-Pires, F.D. de and Gouveia, E. (1977). Mamìferos do Parque Nacional do Itatiaia. *Bol. Mus. Nac., Rio de Janeiro, Nova Série Zoologia*, (291), 1–29. (Chapter 1)

Axelrod, R. and Hamilton, W.D. (1981). The evolution of cooperation. *Science*, 211, 1390–6. (Chapter 7)

Ayres, J.M. (1983). Conservation of primates in Brazilian Amazonia: problems and strategies. In *Proceedings of the Symposium on the Conservation of Primates and Their Habitats, Vol. 1, Primate Conservation in the Wild* (ed. D. Harper), pp.2–33. University of Leicester, Leicester. (Chapter 1)

Ayres, J.M. (1986). Uakaris and Amazonian flooded forest. Unpublished doctoral thesis, Sub-department of Veterinary Anatomy, University of Cambridge, Cambridge. (Chapter 1)

Ayres, J.M. and Milton, K. (1981). Levantamento de primatas e habitat no Rio Tapajós. *Bol. Mus. Para. Emílio Goeldi, Nova Série, Zoologia*, (111), 1–11. (Chapter 1)

Ayres, J.M., Mittermeier, R.A., and Constable, I.D. (1980). A distribuição geográfica e situação atual dos saguis-de-cara-nua (*Saguinus bicolor*). *Bol. FBCN*, Rio de Janeiro, 16, 62–8. (Chapter 1)

Ayres, J.M., Mittermeier, R.A., and Constable, I.D. (1982). Brazilian tamarins on the way to extinction? *Oryx*, 16(4), 329–33. (Chapter 1)

Baal, F.L.J., Mittermeier, R.A., and van Roosmalen, M.G.M. (1988). Primates and protected areas in Suriname. *Oryx*, 22(1), 7–14. (Chapter 1)

Baker, A.J. (1987). Emigration in wild groups of golden lion tamarins (*Leontopithecus rosalia*). *Int. J. Primatol.*, 8, 500. Abstract. (Chapters 5, 7, 14)

Baldwin, J.D. and Baldwin, J.I. (1981). The squirrel monkeys, genus *Saimiri*. In *Ecology and behavior of Neotropical primates*, vol. 1 (ed. A.F. Coimbra-Filho and R.A. Mittermeier), pp.277–330. Academia Brasileira de Ciências, Rio de Janeiro. (Chapter 13)

Ballou, J.D. (1989). Emergence of the captive population of golden-headed lion tamarins *Leontopithecus chrysomelas*. *Dodo, J. Jersey Wildl. Preserv. Trust*, 26, 70–7. (Chapter 1)

Barlow, G.W. (1988). Monogamy in relation to resources. In *The ecology of social behavior* (ed. C.N. Slobodchikoff), pp.55–79. Academic Press, New York. (Chapter 7)

Barrett, J. and Abbott, D.H. (1989). Pheromonal cues suppress ovulation in subordinate female marmoset monkeys (*Callithrix jacchus*). *J. Reprod. Fert.*, Abstract Series No. 3, 33. Abstract. (Chapter 5)

Barrett, J., Abbott, D.H., and George, L.M. (1990). Extension of reproductive suppression by pheromonal cues in subordinate female marmoset monkeys, *Callithrix jacchus. J. Reprod. Fert.*, 90, 411–18. (Chapters 4, 5)

Bartecki, U. and Heymann, E.W. (1987). Field observation of snake-mobbing in a group of saddle-back tamarins, *Saguinus fuscicollis nigrifrons. Folia primatol.*, 48, 199–202. (Chapter 8)

Bartecki, U . and Heymann, E.W. (1990). Field observations on scent-marking behaviour in saddle-back tamarins, *Saguinus fuscicollis* (Callitrichidae, Primates). *J. Zool., Lond.*, 220, 87–99. (Chapter 4)

Barton, N.H. and Hewitt, G.M. (1981). Hybrid zones and speciation. In *Evolution and speciation* (ed. W.R. Atchley and D.S. Woodruff), pp.109–45. Cambridge University Press, Cambridge. (Chapter 3)

Barton, N.H. and Hewitt, G.M. (1985). Analysis of hybrid zones. *Ann. Rev. Ecol. Syst.*, 16, 113–48. (Chapter 3)

Beach, F.A. (1976). Sexual attractivity, proceptivity and receptivity in female mammals. *Horm. Behav.*, 7, 105–38. (Chapter 6)

Bearder, S.K. (1987). Lorises, bushbabies, and tarsiers: diverse societies in solitary foragers. In *Primate societies* (ed. B.B. Smuts, D.L. Cheney, R.M. Seyfarth, R.W. Wrangham and T.T. Struhsaker), pp.11–24. Chicago University Press, Chicago. (Chapters 13, 15)

Bearder, S.K. and Martin, R.D. (1980). Acacia gums and its use by bushbabies *Galago senegalensis* (Primates: Lorisidae). *Int. J. Primatol.*, 1, 103–28. (Chapter 15)

Beattie, J. (1927). The anatomy of the common marmoset (*Hapale jacchus* Kuhl). *Proc. zool. Soc. Lond.*, 1927, 593–718. (Chapter 13)

Belcher, A.M., Epple, G., Kostelc, J.G., and Smith, A.B., III (1982). Changes in chemical composition due to aging in the scent marks of a primate, *Saguinus fuscicollis. Abstracts. IVth Annual Meeting of the Association of Chemoreception Sciences*, Sarasota, Florida. Abstract. (Chapter 4)

Belcher, A.M., Smith, A.B., III, Jurs, P.C., Lavine, B., and Epple, G. (1986). Analysis of chemical signals in a primate species (*Saguinus fuscicollis*): use of behavioral, chemical and pattern recognition methods. *J. Chem. Ecol.*, 12, 513–31. (Chapter 4)

Belcher, A.M., Epple, G., Küderling, I., and Smith, A.B., III (1988). The volatile components of scent material from the cotton-top tamarin (*Saguinus o. oedipus*): a chemical and behavioral study. *J. Chem. Ecol.*, 14, 1367–84. (Chapter 4)

Belcher, A.M., Epple, G., Greenfield, K.L., Richards, L.E., Küderling, I., and Smith, A.B., III (1990). Proteins: biologically relevant components of the scent marks of a primate (*Saguinus fuscicollis*). *Chemical Senses*, 15, 431–46. (Chapter 4)

Bender, M.A. and Mettler, L.E. (1960). Chromosome studies of primates. II. *Callithrix*, *Leontocebus*, and *Callimico. Cytologia*, 25, 400–4. (Chapter I.1)

Benirschke, K., Anderson, J.M., and Brownhill, L.E. (1962). Marrow chimerism in marmosets. *Science*, 138, 513–15. (Chapter 6)

Bernstein, I.S. (1966). Naturally occurring primate hybrid. *Science*, 154, 1559–60. (Chapter 3)
Bernstein, I.S. (1968). Social status of two hybrids in wild troops of *Macaca irus*. *Folia primatol.*, 8, 121–31. (Chapter 3)
Bischof, N. (1972). The biological foundations of the incest taboo. *Soc. Sci. Inf.*, 11, 7–36. (Chapter 7)
Bischof, N. (1985). *Das Rätsel Ödipus. Die biologischen Wurzeln des Urkonfliktes von Intimität und Autonomie.* Piper, München. (Chapter 7)
Bodemeyer, J. (1990). Feinanalytische Untersuchung einer Paarbildung beim Wiessbüschelaffen (*Callithrix jacchus* Erxleben, 1777) unter besonderer Berücksichtigung des Näherungsverhaltens. Unpublished diploma thesis, University of Göttingen, Göttingen. (Chapter 7)
Bodini, R. and Perez-Hernandez, R. (1987). Distribution of the species and subspecies of cebids in Venezuela. *Fieldiana, Zoology, New Series*, (39), 231–44. (Chapter 1)
Boer, L.E.M. de (1974). Cytotaxonomy of the Platyrrhini (Primates). *Genen. Phaenen.*, 17, 1–115. (Chapters 1, 2)
Boinski, S. (1987*a*). Habitat use by squirrel monkeys (*Saimiri oerstedi*) in Costa Rica. *Folia primatol.*, 49, 151–67. (Chapter 13)
Boinski, S. (1987*b*). Birth synchrony in squirrel monkeys (*Saimiri oerstedi*): a strategy to reduce neonatal predation. *Behav. Ecol. Sociobiol.*, 21, 393–400. (Chapter 8)
Boinski, S. and Fowler, N.L. (1989). Seasonal patterns in a tropical lowland forest. *Biotropica*, 21, 223–33. (Chapter 13)
Boinski, S. and Scott, P.E. (1988). Association of birds with monkeys in Costa Rica. *Biotropica*, 20, 136–43. (Chapter 15)
Box, H.O. (1977). Quantitative data on the carrying of young captive monkeys (*Callithrix jacchus*) by other members of their family groups. *Primates*, 18, 475–84. (Chapters 5, 7, 10)
Box, H.O. (1978). Social interactions in family groups of marmosets (*Callithrix jacchus*). In *The biology and conservation of the Callitrichidae* (ed. D.G. Kleiman), pp.239–49. Smithsonian Institution Press, Washington, D.C. (Chapter 4)
Box, H.O. and Morris, J.M. (1980). Behavioural observations on captive pairs of wild-caught tamarins (*Saguinus mystax*). *Primates*, 21, 53–65. (Chapter 8)
Branch, L.C. (1983). Seasonal and habitat differences in the abundance of primates in the Amazon (Tapajós) National Park, Brazil. *Primates*, 24(3), 424–31. (Chapter 1)
Brand, H.M. and Martin, R.D. (1983). The relationship between urinary excretion and mating behavior in cotton-topped tamarins, *Saguinus oedipus oedipus. Int. J. Primatol.*, 4, 275–90. (Chapters 6, 8)
Brazil, MA-IBDF and FBCN. (1979). *Plano de Manejo. Parque Nacional da Amazônia (Tapajós).* Ministério da Agricultura (MA), Instituto Brasileiro de Desenvolvimento Florestal (IBDF), Brasília, and Fundação Brasileira para a Conservação da Natureza (FBCN), Rio de Janeiro. 79pp. (Chapter 1)
Brazil, MINTER-Ibama. (1989). *Unidades de Conservação do Brasil. 1—Parques Nacionais e Reservas Biológicas.* Ministério do Interior (MINTER), Instituto Brasileiro do Meio Ambiente e dos Recursos Naturais Renováveis (Ibama), Brasília. 182 pp. (Chapter 1)
Brown, A.D. and Rumiz, D.I. (1986). Distribucion y conservacion de los Primates en Bolivia—estado actual de su conocimiento. In *A primatologia*

*no Brasil—2* (ed. M.T. de Mello), pp.335–63. Sociedade Brasileira de Primatologia, Brasília. (Chapter 1)

Brown, K. and Mack, D.S. (1978). Food sharing among captive *Leontopithecus rosalia. Folia primatol.*, 29, 268–90. (Chapters 9, 10)

Brown, R.E. and Macdonald, D.W. (ed.) (1985). *Social odours in mammals.* Oxford University Press, Oxford. (Chapter 4)

Brusek, P. (1985). Ontogenetische Untersuchungen zur Strategie des Futtererwerbs in Grossfamilien von *Callithrix jacchus* Erxleben, 1777 während der ersten fünf Lebensmonate. Unpublished diploma thesis, University of Göttingen, Göttingen. (Chapter 7)

Buchanan-Smith, H. (1984). Preliminary report on infant development of the black-tailed marmosets, *Callithrix argentata melanura*, at the Jersey Wildlife Preservation Trust. *Dodo, J. Jersey Wildl. Preserv. Trust*, 21, 57–67. (Chapters 9, 10)

Buchanan-Smith, H. (1990). Polyspecific association of two tamarin species, *Saguinus labiatus* and *Saguinus fuscicollis*, in Bolivia. *Am. J. Primatol.*, 22, 205–14. (Chapters 8, 13, 15)

Burgos, A.E.M. (1974). Contribucion al conocimiento de las especies peruanos de monos (mamíferos, primates, platirrinos). IV Congreso Nacional de Biologia, Trujillo, November 1974. Abstract. (Chapter 1)

Buskirk, R. and Buskirk, W. (1976). Changes in arthropod abundance in a highland Costa Rican forest. *Am. Midl. Nat.*, 95, 288–98. (Chapter 13)

Byrd, K.E. (1981). Sequences of dental ontogeny and callitrichid taxonomy. *Primates*, 22, 103–18. (Chapters 1, 2, 13, 15)

Cabrera, A. (1957). Catalogo de los mamíferos de America del Sur. *Rev. Mus. Argentino de Ciencias Naturales 'Bernadino Rivadavia'*, 4(1), 1–307. (Chapter 1)

Caine, N.G. (1984). Visual scanning by tamarins: a description of the behavior and test of two derived hypotheses. *Folia primatol.*, 43, 59–67. (Chapter 8)

Caine, N.G. (1986). Visual monitoring of threatening objects by captive tamarins (*Saguinus labiatus*). *Am. J. Primatol.*, 10, 1–8. (Chapter 8)

Caine, N.G. (1987). Vigilance, vocalizations, and cryptic behavior at retirement in captive groups of red-bellied tamarins (*Saguinus labiatus*). *Am. J. Primatol.*, 12, 241–50. (Chapter 8)

Caine, N.G. (1990). Unrecognized antipredator behaviour can bias observational data. *Anim. Behav.*, 39(1), 195–7. (Chapter 8)

Caine, N.G. and Stevens, C. (1990). Evidence for a 'monitoring' call in captive red-bellied tamarins. *Am. J. Primatol.*, 22, 251–62. (Chapter 8)

Caine, N.G. and Weldon, P.J. (1989). Responses by red-bellied tamarins (*Saguinus labiatus*) to fecal scents of predatory and non-predatory neotropical mammals. *Biotropica*, 21, 186–9. (Chapter 8)

Caine, N.G., Potter, M.P., and Mayer, K.E. (1992). Sleeping site selection by captive tamarins (*Saguinus labiatus*). *Ethology*, 90, 63–71. (Chapter 8)

Carpenter, C.R. (1934). A field study of the behavior and social relations of howling monkeys (*Alouatta palliata*). *Comp. Psychol. Monog.*, 10, 1–168. (Chapter 8)

Cartmill, M. (1974). Pads and claws in arboreal locomotion. In *Primate locomotion* (ed. F.A. Jenkins), pp.45–83. Academic Press, New York. (Chapter 13)

Carvalho, C.T. de (1957*a*). Alguns mamíferos do Acre Ocidental. *Boltm. Mus. Para. Emílio Goeldi, Série Zoologia*, (6), 1–22. (Chapter 1)

Carvalho, C.T. de (1957*b*). Nova subespécie de saguim da Amazônia. *Rev. Brasil. Biol.*, 17(2), 219–22. (Chapter 1)
Carvalho, C.T. de (1959*a*). Sobre a validez de *Callithrix leucippe* (Thos.) (Callithricidae. Primates). *Papeis Avulsos do Departamento de Zoologia*, Secretaria de Agricultura, São Paulo, 13(27), 317–20. (Chapter 1)
Carvalho, C.T. de (1959*b*). Lectótipos e localidades das espécies de Goeldi (Primates, Carnivora e Rodentia). *Rev. Brasil. Biol.*, 19(4), 459–61. (Chapter 1)
Carvalho, C.T. de (1965). Comentários sobre os mamíferos descritos e figurados por Alexandre Rodrigues Ferreira em 1790. *Arq. Zool.*, 12, 7–70. (Chapter 1)
Carvalho, C.T. de and Carvalho, C.F. de (1989). A organização social dos sauís-pretos, (*Leontopithecus chrysopygus* Mikan), na reserva em Teodoro Sampaio, São Paulo (Primates, Callithricidae). *Revta. bras. Zool.*, 6, 707–717. (Chapter 14)
Carvalho, C.T. de, Albernaz, A.L.K.M., and Lucca, C.A.T. de (1989). Aspectos da bionomia do mico-leão preto (*Leontopithecus chrysopygus* Mikan). (Mammalia, Callithricidae). *Rev. Inst. Flor.*, São Paulo, 1, 67–83. (Chapter 14)
Castro, R. and Soini, P. (1978). Field studies on *Saguinus mystax* and other callitrichids in Amazonian Peru. In *The biology and conservation of the Callitrichidae* (ed. D.G. Kleiman), pp.73–8. Smithsonian Institution Press, Washington, D.C. (Chapter 8)
Cebul, M.S. and Epple, G. (1984). Father—offspring relationships in laboratory families of saddle-back tamarins (*Saguinus fuscicollis*). In *Primate paternalism* (ed. D.M. Taub), pp.1–19. Van Nostrand Reinhold, New York. (Chapters 7, 9, 10)
Cebul, M.S., Alveario, M.C., and Epple, G. (1978). Odor recognition and attachment in infant marmosets. In *Biology and behaviour of marmosets* (ed. H. Rothe, H.-J. Wolters, and J.P. Hearn), pp.141–6. Eigenverlag H. Rothe, Göttingen. (Chapter 4)
Cerquera, J.R. (1985). S.O.S. for the cotton-top tamarin (*Saguinus oedipus*). *Primate Conservation*, (6), 17–19. (Chapter 1)
Charles-Dominique, P. (1977). *Ecology and behaviour of nocturnal primates.* Columbia University Press, New York. (Chapter 13)
Chase, I.D. (1980). Cooperative and non-cooperative behavior in animals. *Am. Nat.*, 115, 827–57. (Chapter 7)
Cheney, D.L. and Seyfarth, R.M. (1983). Nonrandom dispersal in free-ranging vervet monkeys: social and genetic consequences. *Am. Nat.*, 122, 392–412. (Chapter 8)
Cheney, D.L. and Wrangham, R.W. (1987). Predation. In *Primate Societies* (ed. B.B. Smuts, D.L. Cheney, R.M. Seyfarth, R.W. Wrangham, and T.T. Struhsaker), pp.227–39. Chicago University Press, Chicago. (Chapter 15)
Cleveland, J. and Snowdon, C.T. (1982). The complex vocal repertoire of the adult cotton-top tamarin (*Saguinus oedipus*). *Z. Tierpsychol.*, 58, 231–70. (Chapter 8)
Cleveland, J. and Snowdon, C.T. (1984). Social development during the first twenty weeks in the cotton-top tamarin (*Saguinus oedipus*). *Anim. Behav.*, 32, 432–44. (Chapters 5, 7, 9, 10)
Clutton-Brock, T.H. and Harvey, P.H. (1977*a*). Primate ecology and social organization. *J. Zool., Lond.*, 183, 1–39. (Chapter 15)

Clutton-Brock, T.H. and Harvey, P.H. (1977*b*). Species differences in feeding and ranging behaviour in primates. In *Primate ecology: studies in feeding and ranging behaviour in lemurs, monkeys and apes* (ed. T.H. Clutton-Brock), pp.557–84. Academic Press, London. (Chapter 13)

Coates, A. and Poole, T.B. (1983). The behavior of the callitrichid monkey, *Saguinus labiatus*, in the laboratory. *Int. J. Primatol.*, 4, 339–71. (Chapter 8)

Coimbra-Filho, A.F. (1969). Mico-leão, *Leontideus rosalia* (Linnaeus, 1766), situação atual da espécie no Brasil. *An. Acad. Brasil. Ciênc.*, 41(Supl.), 29–52. (Chapter 1)

Coimbra-Filho, A.F. (1970*a*). Considerações gerais e situação atual dos micos-leões escuros, *Leontideus chrysomelas* (Kuhl, 1820) e *Leontideus chrysopygus* (Mikan, 1823) (Callithricidae, Primates). *Rev. Brasil. Biol.*, 30, 249–68. (Chapters 1, 14, 15)

Coimbra-Filho, A.F. (1970*b*). Acerca de um caso de hibridismo entre *Callithrix jacchus* (L. 1758) × *Callithrix geoffroyi* (Humboldt, 1812). *Rev. Brasil. Biol.*, 30, 507–14. (Chapter 3)

Coimbra-Filho, A.F. (1970*c*). Acerca da redescoberta de *Leontideus chrysopygus* (Mikan, 1823) e apontamentos sobre sua ecologia (Callithricidae, Primates). *Rev. Brasil. Biol.*, 30, 609–15. (Chapter 1, 14)

Coimbra-Filho, A.F. (1971). Os saguis do gênero *Callithrix* da região oriental brasileira e um caso de duplo-hibridismo entre três de suas formas (Callithricidae, Primates). *Rev. Brasil. Biol.*, 31, 377–88. (Chapters 1, 3, 12)

Coimbra-Filho, A.F. (1972). Mamíferos ameaçados de extinção no Brasil. In: *Espécies da fauna brasileira ameaçadas de extinção*, pp.13–98. Academia Brasileira de Ciências, Rio de Janeiro. (Chapter 1)

Coimbra-Filho, A.F. (1973). Novo aspecto de duplo-hibridismo em *Callithrix* (Callitrichidae—Primates). *Rev. Brasil. Biol.*, 33, 31–8. (Chapter 3)

Coimbra-Filho, A.F. (1974). Triplo-hibridismo em *Callithrix* (Callithricidae—Primates). *An. Acad. brasil. Ciênc.*, 46, 708. Abstract. (Chapter 3).

Coimbra-Filho, A.F. (1976*a*). Os saguis do gênero *Leontopithecus* Lesson, 1840 (Callithricidae—Primates). Unpublished Master's thesis, Universidade Federal do Rio de Janeiro, Rio de Janeiro. (Chapter 14)

Coimbra-Filho, A.F. (1976*b*). *Leontopithecus rosalia chrysopygus* (Mikan, 1823), o mico-leão do Estado de São Paulo (Callitrichidae, Primates). *Silvic. S. Paulo*, 10, 1–36. (Chapters 1, 14)

Coimbra-Filho, A.F. (1978*a*). Natural shelters of *Leontopithecus rosalia* and some ecological implications (Callitrichidae: Primates). In *The biology and conservation of the Callitrichidae* (ed. D.G. Kleiman), pp.79–89. Smithsonian Institution Press, Washington, D.C. (Chapters 14, 15)

Coimbra-Filho, A.F. (1978*b*). Sobre um caso de triplo-hibridismo em *Callithrix* (Callitrichidae—Primates). *Rev. Brasil Biol.*, 38, 61–71. (Chapter 3)

Coimbra-Filho, A.F. (1981). Animais predados ou rejeitados pelo saui-piranga, *Leontopithecus r. rosalia* (L. 1766) na sua área de ocorrência primitiva (Callitrichidae, Primates). *Rev. Brasil. Biol.*, 41, 717–31. (Chapters 14, 15)

Coimbra-Filho, A.F. (1982). Distribuição geográfica, ecologia, extinção e preservação dos platirrínos. In *Genêtica comparada de primatas brasileiros* (ed. P.H. Saldanha), pp.83–103. Sociedade Brasileira de Genêtica, São Paulo. (Chapter 1)

Coimbra-Filho, A.F. (1984). Situação atual dos calitriquídeos que ocorrem no Brasil (Callitrichidae—Primates). In *A primatologia no Brasil* (ed. M.T. de Mello), pp.15–33. Sociedade Brasileira de Primatologia, Brasília. (Chapter 1)

Coimbra-Filho, A.F. (1985). Sagui-de-Wied *Callithrix kuhlii* (Wied, 1826). *FBCN/Inf.*, Rio de Janeiro, 9(4), 5. (Chapter 1)
Coimbra-Filho, A.F. (1986*a*). Sagui-da-serra *Callithrix flaviceps* (Thomas, 1903). *FBCN/Inf.*, Rio de Janeiro, 10(1), 3. (Chapter 1)
Coimbra-Filho, A.F. (1986*b*). Sagui-da-serra-escuro *Callithrix aurita* (E. Geoffroy, 1812). *FBCN/Inf.*, Rio de Janeiro, 10(2), 3. (Chapter 1)
Coimbra-Filho, A.F. (1986*c*). Sagui-de-cara-branca *Callithrix geoffroyi* (Humboldt, 1812). *FBCN/Inf.*, Rio de Janeiro, 10(3), 3. (Chapter 1)
Coimbra-Filho, A.F. (1987). Saguis-de-cara-nua *Saguinus bicolor* (Spix, 1823). *FBCN/Inf.*, Rio de Janeiro, 11(1), 3. (Chapter 1)
Coimbra-Filho, A.F. (1990). Sistemática, distribuição geográfica e situação atual dos símios brasileiros (Platyrrhini,—Primates). *Rev. Brasil. Biol.*, 50(4), 1063–79. (Chapters 1, 2, 12)
Coimbra-Filho, A.F. (1991). Apontamentos sobre *Callithrix aurita* (E. Geoffroy, 1812), um sagüi pouco conhecido. In *A primatologia no Brasil—3* (ed. A.B. Rylands and A.T. Bernardes), pp.145–58. Sociedade Brasileira de Primatalogia and Fundação Biodiversitas, Belo Horizonte (Chapter 1)
Coimbra-Filho, A.F. and Magnanini, A. (1972). On the present status of *Leontopithecus* and some data about new behavioural aspects and management of *L. rosalia rosalia.* In *Saving the lion marmoset* (ed. D.D. Bridgewater), pp.59–69. Wild Animal Propagation Trust, Wheeling. (Chapter 1)
Coimbra-Filho, A.F. and Maia, A.A. (1976). Hibridismo de *Callithrix geoffroyi* (Humboldt, 1812) × *C. jacchus* (Linnaeus, 1758) e criação artificial de filhote hibrido. *Rev. Brasil. Biol.*, 36, 665–73. (Chapter 3)
Coimbra-Filho, A.F. and Mittermeier, R.A. (1972). Taxonomy of the genus *Leontopithecus* Lesson 1840. In *Saving the lion marmoset* (ed. D.D. Bridgewater), pp.7–22. Wild Animal Propagation Trust, Wheeling. (Chapter 1)
Coimbra-Filho, A.F. and Mittermeier, R.A. (1973*a*). Distribution and ecology of the genus *Leontopithecus* Lesson, 1840 in Brazil. *Primates*, 14(1), 47–66. (Chapters 1, 12, 14)
Coimbra-Filho, A.F. and Mittermeier, R.A. (1973*b*). New data on the taxonomy of the Brazilian marmosets of the genus *Callithrix* Erxleben, 1777. *Folia primatol.*, 20, 241–64. (Chapter 3)
Coimbra-Filho, A.F. and Mittermeier, R.A. (1976). Exudate-eating and tree-gouging in marmosets. *Nature, Lond.*, 262, 630. (Chapters 12, 15)
Coimbra-Filho, A.F. and Mittermeier, R.A. (1977). Conservation of the Brazilian lion tamarins (*Leontopithecus rosalia*). In *Primate conservation* (ed. H.S.H. Prince Rainier III of Monaco and G.H. Bourne), pp.59–94. Academic Press, New York. (Chapter 1)
Coimbra-Filho, A.F. and Mittermeier, R.A. (1978). Tree-gouging, exudate-eating, and the 'short-tusked' condition in *Callithrix* and *Cebuella*. In *The biology and conservation of the Callitrichidae* (ed. D.G. Kleiman), pp.105–15. Smithsonian Institution Press, Washington, D.C. (Chapters 4, 12, 15)
Coimbra-Filho, A.F., Rocha, N. da C., and Pissinatti, A. (1976). Gestação quadrupla de tríplo-hìbridos em *Callithrix* duplo-hîbrida (Callitrichidae, Primates). *Rev. Brasil. Biol.*, 36, 675–81. (Chapter 3).
Coimbra-Filho, A.F., Rocha N. da C., and Pissinatti, A. (1980). Morfofisiologia do ceco e sua correlação com o tipo odontológico em Callitrichidae (Platyrrhini, Primates). *Rev. Brasil. Biol.*, 41, 141–7. (Chapters 12, 15)
Coimbra-Filho, A.F., Mittermeier, R.A., and Constable, I.D. (1981*a*). *Callithrix flaviceps* (Thomas, 1903) recorded from Minas Gerais, Brazil (Callitrichidae, Primates). *Rev. Brasil. Biol.*, 41(1), 141–7. (Chapters 1, 7)

Coimbra-Filho, A.F., Rocha e Silva, R. da, and Pissinatti, A. (1981*b*). Sobre a dieta de Callitrichidae em cativeiro. *Biotérios*, 1, 83–93. (Chapter 3)

Coimbra-Filho, A.F., Rocha e Silva, R. da and Pissinatti, A. (1984). Heterose em fêmea híbrida de *Callithrix* (Callitrichidae—Primates). In *A primatologia no Brasil* (ed. M.T. de Mello), pp.213–15. Sociedade Brasileira de Primatologia, Brasília. (Chapter 3)

Coimbra-Filho, A.F., Padua, C.B.V., Rocha e Silva, R. da, Pissinatti, A., and Fischer, L.R. de B. (1989). A ciência primatológica e o Centro de Primatologia do Rio de Janeiro. In *La primatologia en Latinoamerica* (ed. C.J. Saavedra, R.A. Mittermeier, and I.B. Santos), pp.259–70. World Wildlife Fund, Washington, D.C. (Chapters 1, 3)

Coimbra-Filho, A.F., Rocha e Silva, R. da, and Pissinatti, A. (1991). Acerca da distribuição geográfica original de *Cebus apella xanthosternos* Wied, 1820 (Cebidae, Primates). In *A primatologia no Brasil—3* (ed. A.B. Rylands and A.T. Bernardes), pp.215–24. Sociedade Brasileira de Primatologia and Fundação Biodiversitas, Belo Horizonte. (Chapter 1)

Colombia, MA-INDERENA. (1989). *A guide to the National Natural Parks system of Colombia.* Ministério de Agricultura (MA), Instituto Nacional de los Recursos Naturales Renovables y del Medio Ambiente—INDERENA, Bogota. (Chapter 1)

Cords, M. (1987). Forest guenons and patas monkeys: male-male competition in one-male groups. In *Primate societies* (ed. B.B. Smuts, D.L. Cheney, R.M. Seyfarth, R.W. Wrangham, and T.T. Struhsaker), pp.98–111. Chicago University Press, Chicago. (Chapter 6)

Crandlemire-Sacco, J.L. (1986). The ecology of the saddle-backed tamarin, *Saguinus fuscicollis*, of southeastern Peru. Unpublished Ph.D. thesis, University of Pittsburgh, Pittsburgh. (Chapter 13)

Crandlemire-Sacco, J.L. (1988). An ecological comparison of two sympatric primates: *Saguinus fuscicollis* and *Callicebus moloch* of Amazonian Peru. *Primates*, 29, 465–75. (Chapter 13)

Crewe, R.M., Burger, B.V., Le Roux, M., and Katsir, Z. (1979). Chemical constituents of the chest gland secretion of the thick-tailed galago (*Galago crassicaudatus*). *J. Chem. Ecol.*, 5, 861–8. (Chapter 4)

Cronin, J.E. and Sarich, V.M. (1978). Marmoset evolution: the molecular evidence. In *Primates in medicine*, vol. 10 (ed. N. Gengozian and F. Deinhardt), pp.12–19. S. Karger, Basel. (Chapters 1, 2, 15)

Crook, G.A. (1988). An incidence of breakdown of incest taboo in the common marmoset. *Australian Primatol.*, 3, 23. (Chapter 7)

Crump, D., Swigar, A.A., West, J.R., Silverstein, R.M., Muller-Schwarze, D., and Altieri, R. (1984). Urine fractions that release Flehmen in black-tailed deer (*Odocoileus hemionus columbianus*). *J. Chem. Ecol.*, 10, 203–15. (Chapter 4)

Cruz Lima, E. de (1945). *Mammals of Amazônia 1. General Introduction and Primates. Contrib. Mus. Para. Emílio Goeldi Hist. Nat. Etnogr.*, Belém. 274 pp. (Chapter 1)

Darms, K. (1987*a*). Analyse interindividueller Distanzen zwischen den Mitgliedern zweier Weissbüschelaffengruppen (*Callithrix jacchus* Erxleben, 1777). Unpublished Ph.D. thesis, University of Göttingen, Göttingen. (Chapter 7)

Darms, K. (1987*b*). Individual distances as an indicator of social relationships in groups of common marmosets. *Int. J. Primatol.*, 8, 499. Abstract. (Chapter 7)

Darms, K. (1989). Dynamics of group size in the common marmoset, *Callithrix jacchus*: social conflicts, disintegration, and expulsion of family members. Paper presented at a Meeting of the Tropical Ecology Group of the British Ecological Society, 'Behavioural Ecology of Neotropical Primates', Linnaean Society, London, 16th June 1989. (Chapter 7)

Davis, S. and Richter, C. (1980). Food transfers between family members in *Saguinus oedipus*. Unpublished abstract, Third Annual Meeting, American Society of Primatologists, Winston-Salem, North Carolina. (Chapter 9)

Dawson, G.A. (1976) . Behavioral ecology of the Panamanian tamarin, *Saguinus oedipus* (Callitrichidae, Primates). University Microfilms International, Ann Arbor. (Chapter 13)

Dawson, G.A. (1978). Composition and stability of social groups of the tamarin, *Saguinus oedipus geoffroyi*, in Panama: ecological and behavioral implications. In *The biology and conservation of the Callitrichidae* (ed. D.G. Kleiman), pp.23–37. Smithsonian Institution Press, Washington, D.C. (Chapters 7, 8, 10, 13)

Dawson, G.A. (1979). The use of time and space by the Panamanian tamarin, *Saguinus geoffroyi. Folia primatol.*, 31, 253–84. (Chapters 4, 8, 9, 13)

Dawson, G.A. and Dukelow, W.R. (1976). Reproductive characteristics of free-ranging Panamanian tamarins (*Saguinus oedipus geoffroyi*). *J. Med. Primatol.*, 5, 266–75. (Chapter 7)

Deinhardt, F., Peterson, D., Cross, G., Wolfe, L., and Holmes, A.W. (1975). Hepatitis in marmosets. *Amer. J. Med. Sci.*, 270, 73–80. (Chapter 3)

Deinhardt, F., Wolfe, L., and Ogden, J. (1976). The importance of rearing marmoset monkeys in captivity for conservation of the species and for biomedical research. In *First InterAmerican Conference on Conservation and Utilization of American Nonhuman Primates in Biomedical Research*, pp.65–71. Pan American Health Organization, Washington, D.C. (Chapter 3)

Della Serra, O. (1951). Divisão do gênero *Leontocebus* (Macacos, Platyrrhina) em dois subgêneros sob bases de caracteres dento-morfológicos. *Pap. Avulsos, São Paulo*, 10(8), 147–154. (Chapter 1)

Dietz, J. M. and Baker, A.J. (1991). O modelo limiar de poliginia e sucesso reprodutivo em *Leontopithecus rosalia* (Primates: Callitrichidae). Paper presented at the XVIII Congresso Brasileiro de Zoologia, Universidade Federal da Bahia, Salvador, 26 February to 1st March 1991. (Chapters 5, 7, 14)

Dietz, J.M. and Kleiman, D.G. (1986). Reproductive parameters in groups of free-living golden lion tamarins. *Prim. Rep.*, 14, 77. Abstract. (Chapters 7, 14)

Dietz, J.M. and Kleiman, D.G. (1987). Sexual dimorphism and alternative reproductive tactics in the golden lion tamarin. *Int. J. Primatol.*, 8, 506. Abstract. (Chapter 14)

Dietz, L.A. (1985). Captive-born lion tamarins released into the wild: a report from the field. *Primate Conservation*, (6), 21–7. (Chapter 1)

Dixson, A.F. (1987*a*). Observations on the evolution of the genitalia and copulatory behaviour in male primates. *J. Zool., Lond.*, 213, 423–43. (Chapter 6)

Dixson, A.F. (1987*b*). Effects of adrenalectomy upon proceptivity, receptivity and sexual attractiveness in ovariectimized common marmosets (*Callithrix jacchus*). *Physiol. Behav.*, 39, 495–9. (Chapter 6)

Dixson, A.F. and Lloyd, S.A.C. (1988). The hormonal and hypothalamic control of primate sexual behaviour. *Symp. zool. Soc. Lond.*, 60, 81–117. (Chapter 6)

Dixson, A.F. and Lunn, S.F. (1987). Post-partum changes in hormones and sexual behaviour in captive groups of marmosets (*Callithrix jacchus*). *Physiol. Behav.*, 41, 577–83. (Chapter 6)

Dixson, A.F., Hastie, N., Patel, I., and Jeffreys, A.J. (1988). DNA 'fingerprinting' of captive family groups of common marmosets (*Callithrix jacchus*). *Folia primatol.*, 51, 52–5. (Chapter 6)

Dollman, G. (1933). *Primates*, ser. 3. British Museum (Natural History), London. (Chapter 1)

Duvall, D., Müller-Schwarze, D., and Silverstein, R.M. (ed.) (1986). *Chemical signals in vertebrates 4: ecology, evolution and comparative biology*. Plenum Press, New York. (Chapter 4)

Easley, S.P. (1982). Ecology and behavior of *Callicebus torquatus*: Cebidae, Primates. Unpublished Ph.D. thesis, Washington University, St. Louis. (Chapter 13)

Egler, S.G. (1983). Current status of the pied bare-face tamarin in Brazilian Amazonia. *IUCN/SSC Primate Specialist Group Newsletter*, (3), 20. (Chapter 1)

Egler, S.G. (1991). Double toothed kites following tamarins. *Wilson Bull.*, 103, 510–12. (Chapter 15)

Eisenberg, J.F. (1977). The evolution of the reproductive unit in the Class Mammalia. In *Reproductive behavior and evolution* (ed. J.S. Rosenblatt and B.R. Komisaruk), pp.39–71. Plenum Press, New York. (Chapter 7)

Eisenberg, J.F. (1978). Comparative ecology and reproduction of New World monkeys. In *The biology and conservation of the Callitrichidae* (ed. D.G. Kleiman), pp.13–22. Smithsonian Institution Press, Washington, D.C. (Chapters 8, 13, 15)

Eisenberg, J.F. (1981). *The mammalian radiations. An analysis of trends in evolution, adaptation, and behavior.* University of Chicago, Chicago. (Chapter 7)

Elgar, M.A. and Catterall, C.P. (1981). Flocking and predator surveillance in house sparrows: test of an hypothesis. *Anim. Behav.*, 29, 868–72. (Chapter 8)

Emlen, S.T. (1984). Cooperative breeding in birds and mammals. In *Behavioural ecology: an evolutionary approach* (ed. J.R. Krebs and N.B. Davies), pp.305–39. Blackwell Scientific Publications, Oxford. (Chapters 5, 7)

Emlen, S.T. and Verhencamp, S.L. (1985). Cooperative breeding strategies among birds. In *Fortschritte der Zoologie*, vol. 31 (ed. B. Holldobbler and M. Lindauer), pp.359–374. G. Fischer, Stuttgart. (Chapter 7)

Encarnacion, F. (1990). Informe 01/90 Sub-proyecto: Investigacion y manejo de *Saguinus labiatus* en el sur oriente del Peru . Unpublished report, Proyecto Peruano de Primatologia 'Manuel Moro Sommo', Iquitos. 12pp. (Chapter 1)

Encarnacion, F. and Castro, R. (1978). Informe preliminar sobre censo de primatas no humanos en el sur-oriente peruano: Iberia e Inapari (Departamento de Madre de Dios) Mayo 15–Junio 14, 1978. Unpublished report, Proyecto Peruano de Primatologia, Iquitos. 16pp. (Chapter 1)

Engel, C. (1985*a*). Observations on the interaction between adult group members, group members without rearing experience and infants in the common marmoset, *Callithrix jacchus. Folia primatol.*, 45, 117–28. (Chapter 7)

Engel, C. (1985*b*). Observations on the interaction between adult infant-carrying animals and group members without rearing experience in the common marmoset, *Callithrix jacchus. Folia primatol.*, 45, 225–35. (Chapter 7)

English, W.L. (1932). Exhibition of living hybrid marmosets. *Proc. zool. Soc. Lond.*, 1932, 1079. (Chapter 3)

Epple, G. (1967). Vergleichunde Untersuchungen über Sexual- und Sozialverhalten der Krallenaffen (Hapalidae). *Folia primatol.*, 7, 37–65. (Chapters 5, 7)

Epple, G. (1970). Maintenance, breeding and development of marmoset monkeys (Callitrichidae) in captivity. *Folia primatol.*, 12, 56–76. (Chapter 7)

Epple, G. (1972). Social behavior of laboratory groups of *Saguinus fuscicollis*. In *Saving the lion marmoset* (ed. D.D. Bridgewater), pp.50–8. Wild Animal Propagation Trust, Wheeling. (Chapter 8)

Epple, G. (1974*a*). Olfactory communication in South American primates. *Ann. N.Y. Acad. Sci.*, 237, 261–78. (Chapter 6)

Epple, G. (1974*b*). Pheromones in primate reproduction and social behavior. In *Reproductive behavior* (ed. W. Montagna and W.A. Sadler), pp.131–55. Plenum Press, New York. (Chapter 4)

Epple, G. (1975*a*). The behaviour of marmoset monkeys (Callithricidae). In *Primate behaviour*, vol. 4 (ed. L.A. Rosenblum), pp.195–239. Academic Press, New York. (Chapters 4, 5, 7)

Epple, G. (1975*b*). Parental behaviour in *Saguinus fuscicollis* sspp. (Callithricidae). *Folia primatol.*, 24, 221–8. (Chapters 7, 9, 10)

Epple, G. (1978*a*). Reproductive and social behavior of marmosets with special reference to captive breeding. In *Marmosets in experimental medicine* (ed. N. Gengozian and F.W. Deinhardt), pp.50–62. S. Karger, Basel. (Chapters 7, 10)

Epple, G. (1978*b*). Notes on the establishment and maintenance of the pair-bond in *Saguinus fuscicollis*. In *The biology and conservation of the Callitrichidae* (ed. D.G. Kleiman), pp.231–237. Smithsonian Institution Press, Washington, D.C. (Chapters 7, 8)

Epple, G. (1978*c*). Studies on the nature of chemical signals in scent marks and urine of *Saguinus fuscicollis*. *J. Chem. Ecol.*, 4, 383–94. (Chapters 4, 8)

Epple, G. (1980). Relationships between aggression, scent marking and gonadal state in a primate, the tamarin *Saguinus fuscicollis*. In *Chemical signals in vertebrates 2—vertebrates and aquatic invertebrates*, (ed. D. Müller-Schwarze and R.M. Silverstein), pp.87–105. Plenum Press, New York. (Chapter 4)

Epple, G. (1981*a*). Effects of prepubertal castration on the development of the scent glands, scent marking and aggression in the saddle-back tamarin (*Saguinus fuscicollis*) (Callitrichidae, Primates). *Horm. Behav.*, 15, 54–67. (Chapter 4)

Epple, G. (1981*b*). Effects of pair-bonding with adults in the ontogenetic manifestation of aggressive behaviour in a primate *Saguinus fuscicollis*. *Behav. Ecol. Sociobiol.*, 8, 117–23. (Chapter 10)

Epple, G. (1982). Effects of prepubertal ovariectomy on the development of scent glands, scent marking and aggressive behaviors of female tamarin monkeys (*Saguinus fuscicollis*). *Horm. Behav.*, 16, 330–42. (Chapter 4)

Epple, G. (1990). Sex differences in partner preference in mated pairs of saddle-back tamarins (*Saguinus fuscicollis*). *Behav. Ecol. Sociobiol.*, 27, 455–9. (Chapter 8)

Epple, G. and Alveario, M.C. (1985). Social facilitation of agonistic responses to strangers in pairs of saddle-back tamarins (*Saguinus fuscicollis*). *Am. J. Primatol.*, 9(3), 207–18. (Chapters 8, 10)

Epple, G. and Katz, Y. (1980). Social influences on first reproductive success and related behaviors in the saddle-back tamarin (*Saguinus fuscicollis*, Callitrichidae). *Int. J. Primatol.*, 1, 171–83. (Chapter 10)
Epple, G. and Katz, Y. (1984). Social influences of estrogen excretion and ovarian cyclicity in saddle-back tamarins (*Saguinus fuscicollis*). *Am. J. Primatol.*, 6, 215–27. (Chapters 4, 5, 8)
Epple, G. and Lorenz, R. (1967). Vorkommen, Morphologie und Funktion der Sternaldruse bei den Platyrrhini. *Folia primatol.*, 7, 98–126. (Chapter 4)
Epple, G. and Smith, A.B., III (1985). The primates II: a case study of the saddle back tamarin, *Saguinus fuscicollis*. In *Social odours in mammals*, vol. 2 (ed. R.E. Brown and D.W. MacDonald), pp.770–803. Clarendon Press, Oxford. (Chapter 8)
Epple, G., Golob, N.F., and Smith, A.B., III (1979). Odor communication in the tamarin *Saguinus fuscicollis*. Behavioral and chemical studies. In *Chemical ecology: odour communication in animals* (ed. F.J. Ritter), pp.117–30. Elsevier, Amsterdam. (Chapter 4)
Epple, G., Alveario, M.C., Golob, N.F., and Smith, A.B., III (1980). Stability and attractiveness related to age of scent marks of saddle-back tamarins (*Saguinus fuscicollis*). *J. Chem. Ecol.*, 6, 735–48. (Chapter 4)
Epple, G., Belcher, A.M., and Smith, A.B., III (1986). Chemical signals in callitrichid monkeys—a comparative review. In *Chemical signals in vertebrates 4: ecology, evolution and comparative biology* (ed. D. Duvall, D. Müller-Schwarze and R.M. Silverstein), pp.653–72. Plenum Press, New York. (Chapter 4)
Epple, G., Alveario, M.C., Belcher, A.M., and Smith, A.B., III (1987). Species and subspecies specificity in urine and scent marks of saddle-back tamarins (*Saguinus fuscicollis*). *Int. J. Primatol.*, 8, 663–80. (Chapter 4)
Epple, G., Küderling, I., and Belcher, A.M. (1988). Some communicatory functions of scent marking in the cotton-top tamarin (*Saguinus oedipus oedipus*). *J. Chem. Ecol.*, 14, 503–13. (Chapter 4)
Epple, G., Belcher, A.M., Greenfield, K.L., Küderling, I., Nordstrom, K., and Smith, A.B., III. (1989). Scent mixtures used as social signals in two primate species: *Saguinus fuscicollis* and *Saguinus o.oedipus*. In *Perception of complex smells and tastes* (ed. D.G. Laing, W.S. Cain, R.L. McBridge, and B.W. Ache), pp.1–25. Academic Press, Sydney. (Chapter 4)
Epple, G., Belcher, A.M., Greenfield, K.L., Küderling, I., Smith, A.B., III (1990). Chemical signals in the social environment of a primate, *Saguinus fuscicollis*: fractionation studies of complex scent mixtures. In *ISOT X, Proc. X Int. Symp. Olf. Taste, Oslo, 1989* (ed. K.B. Doving), pp.36–44. GCS A S, Oslo. (Chapter 4)
Erb, R. (1983). Auswirkungen auf die Gruppendynamik nach Verlust des Vaters bei Weissbüscheläffchen (*Callithrix jacchus jacchus*). Unpublished Master's thesis, University of Zürich, Zürich. (Chapter 7)
Erickson, C.J. (1978). Sexual affiliation in animals. In *Biological determinants of sexual behaviour* (ed. C.J. Erickson), pp.697–725. Chichester, New York. (Chapter 7)
Evans, S. (1983). The pair-bond of the common marmoset, *Callithrix jacchus jacchus*: an experimental investigation. *Anim. Behav.*, 31, 651–8. (Chapter 7)
Evans, S. (1986). The pair-bond of the common marmoset, *Callithrix jacchus jacchus*. In *Current perspectives in primate social dynamics* (ed. D.M. Taub and F.A. King), pp.51–65. Van Nostrand Reinhold, New York. (Chapter 7)

Evans, S. and Hodges, J.K. (1984). Reproductive status of adult daughters in family groups of common marmosets (*Callithrix jacchus*). *Folia primatol.*, 42, 127–33. (Chapters 5, 7, 10)

Evans, S. and Poole, T.B. (1984). Long-term changes and maintenance of the pair bond in common marmosets (*Callithrix jacchus jacchus*). *Folia primatol.*, 42, 33–41. (Chapter 6)

Fang, T.G. (1990). La importancia de los frutos en la dieta de *Saguinus mystax* y *S. fuscicollis* (Primates, Callitrichidae), en el Río Tahuayo, Departamento de Loreto, Peru. In *La primatologia en el Peru. Investigaciones Primatologicos (1973–1985)*, pp.342–58. Proyecto Peruana de Primatologia 'Manuel Moro Sommo', Iquitos. (Chapter 15)

Faria, D.S. de (1984*a*). Aspectos gerais do comportamento de *Callithrix jacchus penicillata* em mata ciliar de cerrado. In: *A Primatologia no Brasil* (ed. M.T. de Mello), pp.55–65. Sociedade Brasileira de Primatologia, Brasília. (Chapter 12)

Faria, D.S. de (1984*b*). Uso de árvores gomíferas do cerrado por *Callithrix jacchus penicillata*. In *A Primatologia no Brasil* (ed. M.T. de Mello), pp.83–96. Sociedade Brasileira de Primatologia, Brasília. (Chapter 12)

Faria, D.S. de (1986). Tamanho, composição de um grupo social e área de vivência (home-range) do sagui *Callithrix jacchus penicillata* na mata ciliar do córrego Capetinga, Brasilia, DF. In *A Primatologia no Brasil—2* (ed. M.T. de Mello), pp.87–105. Sociedade Brasileira de Primatologia, Brasília. (Chapter 12)

Faria, D.S. de (1989). O estudo de campo com o mico-estrela no planalto central brasileiro. In *Etologia de Animais e de Homens* (ed. C. Ades), pp.109–21. EDICON/EDUSP, São Paulo. (Chapter 12)

Feistner, A.T.C. (1984). Food offering in cotton-top tamarins, *Saguinus oedipus. Int. J. Primatol.*, 5(4), 338. Abstract. (Chapter 8)

Feistner, A.T.C. (1985). Food-sharing in the cotton-top tamarin, *S. oedipus oedipus*. Unpublished Master's thesis, University of Stirling, Stirling. (Chapter 9)

Feistner, A.T.C. and Chamove, A.S. (1986). High motivation toward food increases in food-sharing in cotton-top tamarins. *Dev. Psychobiol.*, 19, 439–52. (Chapters 8, 9)

Feistner, A.T.C. and Price, E.C. (In press). Food offering in New World primates: two species added. *Folia primatol.* (Chapters 9, 10)

Fergus, C. (1991). The Florida panther verges on extinction. *Science*, 251, 1178–80. (Chapter 3)

Ferrari, S.F. (1987). Food transfer in a wild marmoset group. *Folia primatol.*, 48, 203–6. (Chapter 10)

Ferrari, S.F. (1988*a*). Social behaviour in *Callithrix flaviceps*: an ecological perspective. Paper presented at the Workshop on Monogamy in the Callitrichidae. University of Gottingen, Gottingen, 26–28th February, 1988. (Chapter 7)

Ferrari, S.F. (1988*b*). The behaviour and ecology of the buffy-headed marmoset, *Callithrix flaviceps* (O. Thomas, 1903). Unpublished Ph.D. thesis, University College, London. (Chapters 7, 12, 14)

Ferrari, S.F. (1990). A foraging association between two kite species (*Ictinea plumbea* and *Leptodon cayanensis*) and buffy-headed marmosets (*Callithrix flaviceps*) in southeastern Brazil. *Condor*, 92, 781–3. (Chapters 12, 15)

Ferrari, S.F. (In press). The care of infants in a wild marmoset, *Callithrix flaviceps*, group. *Am. J. Primatol.* (Chapters 7, 10, 12)

Ferrari, S.F. and Diego, V.H. (In press). Long-term changes in a wild marmoset group. *Folia primatol.* (Chapter 15)
Ferrari, S.F. and Lopes Ferrari, M.A. (1989). A re-evaluation of the social organisation of the Callitrichidae, with special reference to the ecological differences between genera. *Folia primatol.*, 52, 132–47. (Chapters 4, 5, 7, 8, 9, 10, 12, 14, 15)
Ferrari, S.F. and Lopes Ferrari, M.A. (1990*a*). A survey of primates in central Pará. *Bol. Mus. Para. Emílio Goeldi, ser. Zool.*, 6(2), 169–79. (Chapters 1, 15)
Ferrari, S.F. and Lopes Ferrari, M.A. (1990*b*). Predator avoidance behaviour in the buffy-headed marmoset, *Callithrix flaviceps. Primates*, 31, 323–8. (Chapters 12, 15)
Ferrari, S.F. and Lopes, M.A. (1992*a*). New data on the distribution of primates in the region of the confluence of the Jiparaná and Madeira rivers in Amazonas and Rondonia. *Goeldiana, Zoologia*, (11), 1–12. (Chapter 15)
Ferrari, S.F. and Lopes, M.A. (1992*b*). A new species of marmoset, genus *Callithrix* Erxleben 1777 (Callitrichidae, Primates) from western Brazilian Amazônia. *Goeldiana, Zoologia*, (12), 1–13. (Chapters 1, 12, 15)
Ferrari, S.F. and Martins, E.S. (In press). Gummivory and gut morphology in two sympatric callitrichids (*Callithrix emiliae* and *Saguinus fuscicollis weddelli*) from western Brazilian Amazonia. *Am. J. Phys. Anthrop.* (Chapter 15)
Ferrari, S.F. and Mendes, S.L. (1991). Buffy-headed marmosets 10 years on. *Oryx*, 25(2), 105–9. (Chapters 1, 3, 12)
Fleagle, J.G. and Mittermeier, R.A. (1980). Locomotor behavior, body size and comparative ecology of seven Surinam monkeys. *Am. J. Phys. Anthropol.*, 22, 301–14. (Chapter 13)
Fonseca, G.A.B. da (1985). The vanishing Brazilian Atlantic forest. *Biol. Conserv.*, 34, 17–34. (Chapter 1)
Fonseca, G.A.B. and Lacher, T.E., Jr. (1984). Exudate-feeding by *Callithrix jacchus penicillata* in semideciduous woodland (cerradão) in central Brazil. *Primates*, 25:441–50. (Chapter 12)
Fonseca, G.A.B. da, Lacher, T.E., Jr, Alves, C., Jr, and Magalhães-Castro, B. (1980). Some ecological aspects of free-living black tufted-ear marmosets (*Callithrix jacchus penicillata*). *Anthropol. Contemp.*, 3, 197. Abstract. (Chapters 7, 12)
Ford, S.M. (1980*a*). A systematic revision of the Platyrrhini based on features of the postcranium. Unpublished Ph.D. dissertation, University of Pittsburgh, Pittsburgh. (Chapter 2)
Ford, S.M. (1980*b*). Callitrichids as phyletic dwarfs, and the place of the Callitrichidae in Platyrrhini. *Primates*, 21, 31–43. (Chapters 12, 13, 15)
Ford, S.M. (1986*a*). Systematics of the New World monkeys. In *Comparative primate biology, Vol. 1, systematics, evolution and anatomy* (ed. D. Swindler and J. Erwin), pp.73–135. Alan R. Liss, New York. (Chapters 1, 13, 15)
Ford, S.M. (1986*b*). Comment on the evolution of claw-like nails in callitrichids (marmosets/tamarins). *Am. J. Phys. Anthrop.*, 70, 25–6. (Chapters 1, 13)
Forman, L., Kleiman, D.G., Bush, R.M., Deitz, J.M., Ballou, J.D., Phillips, L.G., Coimbra-Filho, A.F., and O'Brien, S.J. (1986). Genetic variation within and among lion tamarins. *Am. J. Phys. Anthrop.*, 71:1–11. (Chapter 1)
Frailey, C.D., Lavina, E.L., Rancy, A., and Souza-Filho, J.P. de (1988). A proposed Pleistocene/Holocene lake in the Amazon basin and its significance to Amazonian geology and biogeography. *Acta Amazonica*, 18, 119–43. (Chapter 15)

Freese, C.H., Heltne, P.G., Castro R.N., and Whitesides, G. (1982). Patterns and determinants of monkey densities in Peru and Bolivia, with notes on distributions. *Int. J. Primatol.*, 3(1), 53–90. (Chapters 1, 8)

French, J.A. (1983). Lactation and fertility: an examination of nursing and interbirth intervals in cotton-top tamarins (*Saguinus oedipus*). *Folia primatol.*, 40, 276–82. (Chapter 7)

French, J.A. (1986). Encounters with unfamiliar conspecifics in the laboratory: what do species and context differences reveal about callitrichid social structure? *Prim. Rep.*, 14, 75. Abstract. (Chapter 8)

French, J.A. (1987). Female-female reproductive competition and its manifestations in callitrichid primates. *Int. J. Primatol.*, 8, 460. Abstract. (Chapter 14).

French, J.A. and Cleveland, J. (1984). Scent-marking in the tamarin, *Saguinus oedipus*: sex differences and ontogeny. *Anim. Behav.*, 32, 615–23. (Chapters 4, 8, 10)

French, J.A. and Inglett, B. (1989). Female-female aggression and male indifference in response to unfamiliar intruders in lion tamarins. *Anim. Behav.*, 37, 487–97. (Chapters 5, 8, 10, 14)

French, J.A. and Inglett, B. (1991). Responses to novel social stimuli on callitrichid monkeys. In *Primate responses to environmental change* (ed. H.O. Box), pp.275–94. Chapman and Hall Ltd., London. (Chapters 5, 10)

French, J.A. and Snowdon, C.T. (1981). Sexual dimorphism in responses to unfamiliar intruders in the tamarin, *Saguinus oedipus. Anim. Behav.*, 29, 822–9. (Chapters 4, 5, 8)

French, J.A. and Snowdon, C.T. (1984). Reproduction and behavior in marmosets and tamarins: an introduction. *Am. J. Primatol.*, 6, 211–13. (Chapter 8)

French, J.A. and Stribley, J.A. (1985). Patterns of urinary oestrogen excretion in female golden lion tamarins (*Leontopithecus rosalia*). *J. Reprod. Fert.*, 75, 537–46. (Chapter 5)

French, J.A. and Stribley, J.A. (1987). Synchronization of ovarian cycles within and between social groups in golden lion tamarins (*Leontopithecus rosalia*). *Am. J. Primatol.*, 12, 469–78. (Chapters 5, 8, 10, 14)

French, J.A., Abbott, D.H., Scheffler, G., Robinson, J.A., and Goy, R.W. (1983). Cyclic excretions of urinary oestrogens in female tamarins (*Saguinus oedipus*). *J. Reprod. Fert.*, 67, 177–84. (Chapter 5)

French, J.A., Abbott, D.H., and Snowdon, C.T. (1984). The effect of social environment on estrogen secretion, scent marking, and socio-sexual behavior in tamarins (*Saguinus oedipus*). *Am. J. Primatol.*, 6, 155–67. (Chapters 4, 5, 8, 10)

French, J.A., Umstead, L., and Young, P. (1985). 'Selfish' food sharing in golden lion tamarins. *Am. J. Primatol.*, 8, 338. Abstract. (Chapter 9)

French, J.A., Inglett, B.J., and Dethlefs, T.M. (1989). The reproductive status of nonbreeding group members in captive golden lion tamarin social groups. *Am. J. Primatol.*, 18, 73–86. (Chapters 5, 14)

Gabow, S.A. (1975). Behavioral stabilization of a baboon hybrid zone. *Am. Nat.*, 109, 701–12. (Chapter 3)

Garber, P.A. (1980*a*). Locomotor behavior and feeding ecology of the Panamanian tamarin (*Saguinus oedipus geoffroyi*, Callitrichidae, Primates). Unpublished Ph.D. thesis, Washington University, St. Louis. (Chapters 12, 13)

Garber, P.A. (1980*b*). Locomotor behavior and feeding ecology of the Panamanian tamarin (*Saguinus oedipus geoffroyi*, Callitrichidae, Primates). *Int. J. Primatol.*, 1(2):185–201. (Chapters 1, 12, 13, 15)

Garber, P.A. (1984*a*). Use of habitat and positional behavior in a neotropical primate, *Saguinus oedipus*. In *Adaptations for foraging in nonhuman primates* (ed. P. Rodman and J. Cant), pp.112–33. Columbia University Press, New York. (Chapters 13, 14)

Garber, P.A. (1984*b*). Proposed nutritional importance of plant exudates in the diet of the Panamanian tamarin, *Saguinus oedipus geoffroyi*. *Int. J. Primatol.*, 5, 1–15. (Chapters 8, 9, 12, 13, 14, 15)

Garber, P.A. (1986). The ecology of seed dispersal in two species of callitrichid primates (*Saguinus mystax* and *Saguinus fuscicollis*). *Am. J. Primatol.*, 10, 155–77. (Chapter 13)

Garber, P.A. (1988*a*). Diet, foraging patterns, and resource defense in a mixed species troop of *Saguinus mystax* and *Saguinus fuscicollis* in Amazonian Peru. *Behaviour*, 105, 18–34. (Chapters 8, 9, 13)

Garber, P.A. (1988*b*). Foraging decisions during nectar feeding in tamarin monkeys (*Saguinus mystax* and *Saguinus fuscicollis*, Callitrichidae, Primates) in Amazonian Peru. *Biotropica*, 20, 100–6. (Chapters 13, 15)

Garber, P.A. (1989). Role of spatial memory in primate foraging patterns: *Saguinus mystax* and *Saguinus fuscicollis*. *Am. J. Primatol.*, 19, 203–16. (Chapter 13)

Garber, P.A. (1991*a*). Seasonal variation in diet and ranging patterns in 2 species of tamarin monkeys. *Am. J. Phys. Anthropol.*, Suppl. 12, 75. Abstract. (Chapter 13)

Garber, P.A. (1991*b*). A comparative study of positional behavior in three species of tamarin monkeys. *Primates*, 32. In press. (Chapter 13)

Garber, P.A. (In press). Vertical clinging, small body size and the evolution of feeding adaptations in the Callitrichinae. *Am. J. Phys. Anthropol.* (Chapter 13)

Garber, P.A. and Teaford, M.F. (1986). Body weights in mixed species troops of *Saguinus mystax mystax* and *Saguinus fuscicollis nigrifrons* in Amazonian Peru. *Am. J. Phys. Anthrop.*, 71, 331–6. (Chapters 6, 13, 15)

Garber, P.A., Moya, L., and Malaga, C. (1984). A preliminary field study of the moustached tamarin monkey (*Saguinus mystax*) in northeastern Peru: questions concerned with the evolution of a communal breeding system. *Folia primatol.*, 42, 17–32. (Chapters 5, 7, 8, 9)

Garber, P.A., Moya, L., Encarnacion, F., and Pruetz, J. (1991). Demography, patterns of dispersal, and polyandrous matings in an island population of moustached tamarin monkeys. *Am. J. Primatol.*, 24, 102. Abstract. (Chapter 8)

Gautier, J.-P. and Gautier, A. (1977). Communication in Old World monkeys. In *How animals communicate* (ed. T. Sebeok), pp.890–964. Indiana University Press, Bloomington. (Chapter 2)

Goldizen, A.W. (1986). Dynamics and proximate causes of facultative polyandry in wild saddle-backed tamarins (*Saguinus fuscicollis*). *Primate Report*, 14, 77. Abstract. (Chapter 14)

Goldizen, A.W. (1987*a*). Facultative polyandry and the role of infant-carrying in wild saddle-back tamarins (*Saguinus fuscicollis*). *Behav. Ecol. Sociobiol.*, 20, 99–109. (Chapters 6, 7, 8, 9, 10, 13, 14)

Goldizen, A.W. (1987*b*). Tamarins and marmosets: communal care of offspring. In *Primate societies* (ed. B.B. Smuts, D.L. Cheney, R.M. Seyfarth, R.W. Wrangham, and T.T. Struhsaker), pp.34–43. University of Chicago Press, Chicago. (Chapters 7, 8, 13, 15)

Goldizen, A.W. (1988). Tamarin and marmoset mating systems: unusual flexibility. *Trends Ecol. Evol.*, 3(2), 36–40. (Chapters 8, 10)

Goldizen, A.W. (1989). Social relationships in a cooperatively polyandrous group of tamarins (*Saguinus fuscicollis*). *Behav. Ecol. Sociobiol.*, 24, 79–89. (Chapter 8)

Goldizen, A.W. (1990). A comparative perspective on the evolution of tamarin and marmoset social systems. *Int. J. Primatol.*, 11, 63–83. (Chapter 8, 13)

Goldizen, A.W. and Terborgh, J. (1986). Cooperative polyandry and helping behavior in saddle-backed tamarins (*Saguinus fuscicollis*). In *Primate ecology and conservation* (ed. J.G. Else and P.C. Lee), pp.191–8. Cambridge University Press, New York. (Chapters 8, 10)

Goldizen, A.W. and Terborgh, J. (1989). Demography and dispersal patterns of a tamarin population: possible causes of delayed breeding. *Am. Nat.*, 134, 208–224. (Chapters 7, 8)

Goldizen, A.W., Terborgh, J., Cornejo, F., Porras, D.T., and Evans, R. (1988). Seasonal food shortage, weight loss, and the timing of births in saddle-back tamarins (*Saguinus fuscicollis*). *J. Anim. Ecol.*, 57, 893–901. (Chapters 14, 15)

Gottschling, B. (1984). Interaktionsstrukturen im Zusammenhang mit der Jungenaufzucht in einer Gruppe Weissgesichtsseidenaffen, *Callithrix jacchus geoffroyi*. Unpublished diploma thesis, University of Bielefeld, Bielefeld. (Chapter 7)

Gray, A.P. (1954). *Mammalian hybrids. A check-list with bibliography*. Technical Communication No. 10. C.A.B., Edinburgh. (Chapter 3)

Green, K. (1980). An assessment of the Poço das Antas Reserve, Brazil, and prospects for survival of the golden lion tamarin, *Leontopithecus rosalia rosalia*. Unpublished report to World Wildlife Fund—US, Washington, D.C. (Chapter 14)

Groeger, D. (1988). Untersuchungen über die Position des ältesten Sohnes in einer vaterlosen Familiengruppe von Weissbüschelaffen (*Callithrix jacchus*). Unpublished diploma thesis, University of Göttingen, Göttingen. (Chapter 7)

Hamilton, W.D. (1964). The genetical evolution of social behaviour. *J. theor. Biol.*, 7, 1–52. (Chapter 7)

Hampton, J.K., Jr, Hampton, S.H., and Landwehr, B.T. (1966). Observations on a succesful breeding colony of the marmoset *Oedipomidas oedipus. Folia primatol.*, 4, 265–87. (Chapters 7, 8, 10, 13)

Hampton, J.K., Jr, Hampton, S.H., and Levy, B.M. (1971). Reproductive physiology and pregnancy in marmosets. In *Medical primatology* (ed. E.I. Goldsmith and J. Moor-Jankowski), pp.527–35. S. Karger, Basel. (Chapter 3)

Handley, C.O., Jr. (1976). Mammals of the Smithsonian Venezuelan Project. *Brigham Young Univ. Sci. Bull. Biological Series*, 20(5), 1–99. (Chapter 1)

Hanihara, T. and Natori, M. (1987). Preliminary analysis of numerical taxonomy of the genus *Saguinus* based on dental measurements. *Primates*, 28(4), 517–23. (Chapter 1)

Hanihara, T. and Natori, M. (1989). Evolutionary trends of the hairy-face *Saguinus* in terms of dental and cranial morphology. *Primates*, 30(4), 531–41. (Chapter 1)

Harcourt, A.H. (1987). Dominance and fertility among female primates. *J. Zool., Lond.*, 213, 471–87. (Chapter 5)

Harcourt, A.H., Harvey, P.H., Larson, S.G., and Short, R.V. (1981). Testis weight, body weight and breeding system in primates. *Nature, Lond.*, 293, 55–7. (Chapter 6)

Harlow, C.R., Hillier, S.G., and Hodges, J.K. (1986). Androgen modulation of follicle stimulating hormone-induced granulosa cell steroidogenesis in the primate ovary. *Endrocinology*, 119, 1403–5. (Chapter 5)
Harrison, M.L. and Tardif, S.D. (1988). Kin preference in marmosets and tamarins: *Saguinus oedipus* and *Callithrix jacchus* (Callitrichidae, Primates). *Am. J. Phys. Anthrop.*, 77, 377–84. (Chapters 7, 8)
Harrison, M.L. and Tardif, S.D. (1989). Species differences in response to intruders in *Callithrix jacchus* and *Saguinus oedipus*. *Int. J. Primatol.*, 10, 343–62. (Chapter 8)
Hearn, J.P. (1983). The common marmoset (*Callithrix jacchus*). In *Reproduction in New World primates: new models in medical science* (ed. J.P. Hearn), pp.181–215. MTP Press Ltd., Lancaster. (Chapters 5, 9, 10)
Heistermann, M., Kleis, E., Proeve, E., and Wolters, H.-J. (1989). Fertility status, dominance, and scent marking behavior of family-housed female cotton-top tamarins (*Saguinus oedipus*) in absence of their mothers. *Am. J. Primatol.*, 18, 177–89. (Chapter 7)
Heltne, P.G. (1978). Demography and wildlife management of tamarins and marmosets. In *Marmosets in experimental medicine* (ed. N. Gengozian and F.W. Deinhardt), pp.30–6. S. Karger, Basel. (Chapter 7)
Heltne, P.G., Freese, C.H., and Whitesides, G. (1976). Field survey of nonhuman primates in Bolivia. Unpublished report, Panamerican Health Organization (PAHO), Washington, D.C. (Chapter 1)
Hernandez-Camacho, J. and Cooper, R.W. (1976). The nonhuman primates of Colombia. In *Neotropical primates, field studies and conservation* (ed. R.W. Thorington Jr and P.G. Heltne), pp.35–69. National Academy of Sciences, Washington, D.C. (Chapter 1)
Hernandez-Camacho, J. and Defler, T.R. (1985). Some aspects of the conservation of non-human primates in Colombia. *Primate Conservation*, (6), 42–50. (Chapter 1)
Hernandez-Camacho, J. and Defler, T.R. (1989). Algunos aspectos de la conservacion de primates no humanos en Colombia. In *La primatologia en Latinoamerica* (ed. C.J. Saavedra, R.A. Mittermeier, and I.B. Santos), pp.68–100. World Wildlife Fund, Washington, D.C. (Chapter 1)
Hershkovitz, P. (1949). Mammals of northern Colombia. Preliminary report No. 4: monkeys (Primates), with taxonomic revisions of some forms. *Proc. U.S. Nat. Mus.*, 98, 323–427. (Chapter 1)
Hershkovitz, P. (1966*a*). On the identification of some marmosets, family Callithricidae (Primates). *Mammalia*, 30(2), 327–32. (Chapter 1)
Hershkovitz, P. (1966*b*). Taxonomic notes on tamarins, genus *Saguinus* (Callithricidae, Primates) with descriptions of four new forms. *Folia primatol.*, 4(5), 381–95. (Chapter 1)
Hershkovitz, P. (1968). Metachromism or the principle of evolutionary change in mammalian tegumentary colors. *Evolution*, 22, 556–75. (Chapters 1, 12)
Hershkovitz, P. (1975). Comments on the taxonomy of Brazilian marmosets (*Callithrix*, Callitrichidae). *Folia primatol.*, 24, 137–72. (Chapter 1)
Hershkovitz, P. (1977). *Living New World Monkeys, Part 1. (Platyrrhini), with an Introduction to Primates*. Chicago University Press, Chicago. (Chapters 1, 2, 3, 4, 6, 8, 9, 12, 13, 14, 15)
Hershkovitz, P. (1979). Races of the emperor tamarin, *Saguinus imperator* Goeldi (Callitrichidae, Primates). *Primates*, 20(2), 277–87. (Chapter 1)
Hershkovitz, P. (1982). Subspecies and geographic distribution of black-mantle

tamarins *Saguinus nigricollis* Spix (Primates: Callitrichidae). *Proc. Biol. Soc. Wash.*, 95(4), 647–56. (Chapter 1)

Heymann, E.W. (1987). A field observation of predation on a moustached tamarin (*Saguinus mystax*) by an anaconda. *Int. J. Primatol.*, 8, 193–5. (Chapter 8)

Heymann, E.W. (1990*a*). Interspecific relations in a mixed-species troop of moustached tamarins, *Saguinus mystax*, and saddle-back tamarins, *Saguinus fuscicollis* (Platyrrhini: Callitrichidae), at the Río Blanco, Peruvian Amazon. *Am. J. Primatol.*, 21, 115–27. (Chapters 8, 15)

Heymann, E.W. (1990*b*). Reactions of wild tamarins, *Saguinus mystax* and *Saguinus fuscicollis*, to avian predators. *Int. J. Primatol.*, 11, 327–37. (Chapter 8)

Heymann, E.W. (1991). Field observations on the scent marking behaviour of moustached tamarins, *Saguinus mystax* (Primates: Platyrrhini), in north-eastern Peru. *Abstracts—Symposium on Chemical Signals in Vertebrates VI*, Philadelphia, p.67. (Chapter 4)

Heymann, E.W., Zeller, U., and Schwibbe, M.H. (1989). Muzzle-rubbing in the moustached tamarin, *Saguinus mystax* (Primates, Callitrichidae)—behavioural and histological aspects. *Z. Säugetierk.*, 54, 265–75. (Chapter 4)

Hill, W.C.O. (1957). *Primates. Comparative anatomy and taxonomy, III. Pithecoidea, Platyrrhini (Families Hapalidae and Callimiconidae).* Edinburgh University Press, Edinburgh. (Chapter 1)

Hill, W.C.O. (1959). The anatomy of *Callimico goeldii* (Thomas), a primitive American primate. *Trans. Am. Phil. Soc.*, 49, 1–116. (Chapter 1)

Hill, W.C.O. (1960). *Primates. Comparative anatomy and taxonomy, IV. Cebidae, Part A*. Edinburgh University Press, Edinburgh. (Chapter 1)

Hill, W.C.O. (1961). Hybridization in marmosets. *Proc. zool. Soc. London*, 137(2), 321–2. (Chapter 3)

Hill, W.C.O. (1962). *Primates. Comparative anatomy and taxonomy, V. Cebidae, Part B*. Edinburgh University Press, Edinburgh. (Chapter 1)

Hinde, R.A. and Atkinson, S. (1970). Assessing the roles of social partners in maintaining mutual proximity, as exemplified by mother-infant interactions in rhesus monkeys. *Anim. Behav.*, 18, 169–79. (Chapter 10)

Hladik, A. and Hladik, C.M. (1969). Rapports trophique entre vegetation et primates dans la forêt de Barro Colorado (Panama). *Terre et Vie*, 23, 25–117. (Chapter 13)

Hoage, R.J. (1978). Parental care in *Leontopithecus rosalia*: sex and age differences in carrying behavior and the role of prior experience. In *The biology and conservation of the Callitrichidae* (ed. D.G. Kleiman), pp.293–305. Smithsonian Institution Press, Washington, D.C. (Chapters 5, 7, 9, 10)

Hoage, R.J. (1982). Social and physical maturation in captive lion tamarins, *Leontopithecus rosalia rosalia* (Primates: Callitrichidae). *Smithson. Contrib. Zool.*, 364, 1–56. (Chapters 4, 9, 10)

Hodges, J.K., Gulick, B.A., Czekala, N.M., and Lasley, B.L. (1981). Comparison of urinary oestrogen excretion in South American primates. *J. Reprod. Fert.*, 61, 83–90. (Chapter 5)

Hodun, A., Snowdon, C.T., and Soini, P. (1981). Subspecific variation in the long calls of the tamarin, *Saguinus fuscicollis*. *Z. Tierpsychol.*, 57, 97–110. (Chapters 1, 2)

Hubrecht, R.C. (1984). Field observations on group size and composition of the common marmoset (*Callithrix jacchus jacchus*), at Tapacurá, Brazil. *Primates*, 25, 13–21. (Chapters 7, 10, 12)

Hubrecht, R.C. (1985). Home-range size and use and territorial behavior in the common marmoset, *Callithrix jacchus jacchus*, at the Tapacurá Field Station, Recife, Brazil. *Int. J. Primatol.*, 6, 533–50. (Chapters 7, 9, 12)

Hubrecht, R.C. (1989). The fertility of daughters in common marmoset (*Callithrix jacchus jacchus*) family groups. *Primates*, 30, 423–32. (Chapter 7)

Inglett, B.J., French, J.A., Simmons, L.G., and Vires, K. (1989). Dynamics of intrafamily aggression and social reintegration in lion tamarins. *Zoo Biol.*, 8, 67–78. (Chapter 14).

Ingram, J.C. (1977). Interactions between parents and infants, and the development of independence in the common marmoset (*Callithrix jacchus*). *Anim. Behav.*, 25, 811–27. (Chapters 5, 7, 9, 10)

Ingram, J.C. (1978*a*). Parent-infant interactions in the common marmoset (*Callithrix jacchus*). In *The biology and conservation of the Callitrichidae* (ed. D.G. Kleiman), pp.281–91. Smithsonian Institution Press, Washington, D.C. (Chapter 7)

Ingram, J.C. (1978*b*). Preliminary comparisons of parental care of wild caught and captive born common marmosets. In *Biology and behaviour of marmosets* (ed. H. Rothe, H.-J. Wolters, and J.P. Hearn), pp.219–22. Eigenverlag H. Rothe, Göttingen. (Chapter 10)

Ingram, J.C. (1978*c*). Infant socialization within common marmoset family groups. In *Biology and behaviour of marmosets* (ed. H. Rothe, H.-J. Wolters and J.P. Hearn), pp.223–5. Eigenverlag H. Rothe, Göttingen. (Chapter 10)

IUCN. (1982*a*). *The IUCN mammal red data book, part 1: threatened mammalian taxa of the Americas and Australasian zoogeographic regions (excluding Cetacea)* (compilers J. Thornback and M. Jenkins). International Union for Conservation of Nature and Natural Resources (IUCN), Gland. (Chapter 1)

IUCN (1982*b*). *IUCN directory of Neotropical protected areas*. Commission on National Parks and Protected Areas (CNPPA), International Union for Conservation of Nature and Natural Resources (IUCN), Gland. Tycooly International Publishing Ltd., Dublin. (Chapter 1)

IUCN (1988). *1988 IUCN red list of threatened animals*. International Union for Conservation of Nature and Natural Resources (IUCN) Conservation Monitoring Centre, Cambridge. (Chapter 1)

Izawa, K. (1975). Foods and feeding behaviour of monkeys in the upper Amazon basin. *Primates*, 16(3), 295–316. (Chapters 1, 13)

Izawa, K. (1976). Group sizes and composition of monkeys in the upper Amazon basin. *Primates*, 17(3), 367–99. (Chapters 1, 8)

Izawa, K. (1978). A field study of the ecology and behaviour of the black-mantle tamarin (*Saguinus nigricollis*). *Primates*, 19(2), 241–74. (Chapters 1, 7, 8, 10, 13, 15)

Izawa, K. (1979). Studies on peculiar distribution pattern of *Callimico*. *Kyoto University Overseas Research Reports of New World Monkeys*, (1979), 1–19. (Chapter 1)

Izawa, K. and Bejarano, G. (1981). Distribution ranges and patterns of nonhuman primates in western Pando, Bolivia. *Kyoto University Overseas Research Reports of New World Monkeys*, (1981), 1–12. (Chapters 1, 3, 13)

Izawa, K. and Yoneda, M. (1981). Habitat utilization of nonhuman primates in a forest of western Pando, Bolivia. *Kyoto University Overseas Research Reports of New World Monkeys*, (1981), 13–22. (Chapter 13)

Jaemmrich, S. (1985). Zum Fortpflanzungsverhalten des Weissbüscheläffchens, *Callithrix jacchus* (L., 1758). *Zool. Garten N.F.*, 55, 177. (Chapter 7)

Janson, C.H. (1985). Aggressive competition and individual food consumption in wild brown capuchin monkeys (*Cebus apella*). *Behav. Ecol. Sociobiol.*, 18, 125–38. (Chapter 13)

Janson, C.H., Terborgh, J., and Emmons, L.H. (1981). Non-flying mammals as pollinating agents in the Amazonian forest. *Biotropica (suppl.)*, 14, 1–6. (Chapters 12, 13)

Janzen, D.H. (1973). Sweep samples of tropical foliage insects: description of study sites, with data on species abundances and size distributions. *Ecology*, 54, 659–86. (Chapter 13)

Janzen, D.H. and Schoener, T.W. (1968). Differences in insect abundances and diversity between wetter and drier sites during a tropical dry season. *Ecology*, 49, 96–110. (Chapter 13)

Jeffreys, A.J., Wilson, V., and Thein, S.L. (1985*a*). Hypervariable 'minisatellite' regions in human D.N.A. *Nature, Lond.*, 314, 67–73. (Chapter 6)

Jeffreys, A.J., Wilson, V., and Thein, S.L. (1985*b*). Individual specific 'fingerprints' of human D.N.A. *Nature, Lond.*, 316, 76–9. (Chapter 6)

Johns, A.D. (1985). Primates and forest exploitation at Tefé, Brazilian Amazonia. *Primate Conservation*, (6), 27–9. (Chapter 1)

Johns, A.D. (1986). Effects of habitat disturbance on rainforest wildlife in Brazilian Amazonia. Final Report. Unpublished report, World Wildlife Fund, Washington, D.C. 111pp. (Chapter 1)

Johnson, I.T. and Gee, J.M. (1986). Gastrointestinal adaptation in response to soluble non-available polysaccharides in the rat. *Brit. J. Nutrit.*, 55, 497–505. (Chapter 15)

Juennemann, B. (1990). Analyse der räumlichen Struktur einer Grossfamilie des Weissbüschellaffen (*Callithrix jacchus* Erxleben, 1777) unter besonderer Berücksichtigung möglicher Regulationsmechanismem. Unpublished diploma thesis, University of Göttingen, Göttingen. (Chapter 7)

Kaspereit, B. (1977). Untersuchungen zur Entwicklung von Interaktionsmustern zwischen aufzuchterfahrenen *Callithrix jacchus* Erxleben, 1777 und infantilen Gruppenmitgliedern in den ersten Lebenswochen der Jungtiere. Unpublished Master's thesis, University of Göttingen, Göttingen. (Chapter 7)

Kay, R.F. (1990). The phyletic relationships of extant and fossil Pitheciinae (Playtrrhini, Anthropoidea). *J. Hum. Evol.*, 19, 175–208. (Chapter 15)

Kenargy, G.J. and Trombulak, G.C. (1986). Size and function of mammalian testes in relation to body size. *J. Mammal.*, 67, 1–22. (Chapter 6)

Kendrick, K.M. and Dixson, A.F. (1983). The effect of the ovarian cycle on the sexual behaviour of the common marmoset (*Callithrix jacchus*). *Physiol. Behav.*, 30, 735–42. (Chapter 6)

Keuroghlian, A. (1990). Observations on the behavioral ecology of the black lion tamarin (*Leontopithecus chrysopygus*) at Caetetus Reserve, São Paulo, Brazil. Unpublished Master's degree in Wildlife Management, West Virginia University, Morgantown. (Chapter 14).

Keverne, E.B. (1978). Olfactory cues in mammalian behavior. In *Biological determinants of sexual behavior* (ed. J.B. Hutchison), pp.727–63. John Wiley and Sons, New York. (Chapter 4)

Kierulff, M.C.M. and Stallings, J.R. (1991). Levantamento das populações de mico-leão-dourado (*Leontopithecus rosalia*) no estado do Rio de Janeiro. In *Resumos. XVIII Congresso Brasileiro de Zoologia*, p.393. Universidade Federal da Bahia, Salvador, 24th February to 1st March, 1991. Abstract. (Chapter 1)

Kinzey, W.G. (1977). Diet and feeding behaviour of *Callicebus torquatus*. In *Primate ecology: studies of feeding and ranging behaviour in lemurs, monkeys and apes* (ed. T.H. Clutton-Brock), pp.127–51. Academic Press, London. (Chapter 13)

Kinzey, W.G. (1981). The titi monkeys, genus *Callicebus*. In *Ecology and behavior of Neotropical primates* (ed. A.F. Coimbra-Filho and R.A. Mittermeier), pp.241–76. Academia Brasileira de Ciências, Rio de Janeiro. (Chapter 13)

Kinzey, W.G. (1982). Distribution of primates and forest refuges. In *Biological diversification in the tropics* (ed. G.T. Prance), pp.455–82. Columbia University Press, New York. (Chapters 3, 13, 15)

Kinzey, W.G. (1986). New World primate field studies: what's in it for anthropology? *An. Rev. Anthropol.*, 15, 121–48. (Chapter 8)

Kinzey, W.G. and Becker, M. (1983). Activity pattern of the masked titi monkey, *Callicebus personatus. Primates*, 24, 337–43. (Chapter 13)

Kirkwood, J.K. (1985). Patterns of growth in primates. *J. Zool., Lond.*, 205, 123–36. (Chapter 9)

Kirkwood, J.K. and Underwood, S.J. (1984). Energy requirements of captive cotton-top tamarins *(Saguinus oedipus oedipus*). *Folia primatol.*, 42, 180–7. (Chapter 8)

Kleiber, M. (1961). *The fire of life*. John Wiley and Sons, New York. (Chapter 15)

Kleiman, D.G. (1977). Monogamy in mammals. *Q. Rev. Biol.*, 52, 36–69. (Chapters 4, 6, 7, 8, 9)

Kleiman, D.G. (1978*a*). Characteristics of reproduction and socio-sexual interactions in pairs of lion tamarins (*Leontopithecus rosalia*) during the reproductive cycle. In *The biology and conservation of the Callitrichidae* (ed. D.G. Kleiman), pp.181–90. Smithsonian Institution Press, Washington, D.C. (Chapter 4)

Kleiman, D.G. (1978*b*). The development of pair preferences in the lion tamarin (*Leontopithecus rosalia*): male competition or female choice? In *Biology and behaviour of marmosets* (ed. H. Rothe, H.-J. Wolters and J.P. Hearn), pp.203–7. Eigenverlag H. Rothe, Göttingen. (Chapter 4)

Kleiman, D.G. (1979). Parent-offspring conflict and sibling competition in a monogamous primate. *Am. Nat.*, 114, 753–60. (Chapters 7, 10, 14)

Kleiman, D.G. (1984). The behaviour and conservation of the golden lion tamarin, *Leontopithecus r. rosalia*. In *A primatologia no Brasil* (ed. M.T. de Mello), pp.35–53. Sociedade Brasileira de Primatologia, Brasília. (Chapter 1)

Kleiman, D.G. (1985). Paternal care in New World primates. *Am. Zool.*, 25, 857–9. (Chapters 7, 8)

Kleiman, D.G. (ed.) (In press). *A case study in conservation biology: the golden lion tamarin* (ed. D.G. Kleiman). Smithsonian Institution Press, Washington, D.C. (Chapter 1)

Kleiman, D.G. and Mack, D.S. (1977). A peak in sexual activity during midpregnancy in the golden lion tamarin, *Leontopithecus rosalia. J. Mammal.*, 58, 657–60. (Chapter 6)

Kleiman, D.G. and Mack, D.S. (1980). Effects of age, sex and reproductive status on scent marking frequencies in the golden lion tamarin, *Leontopithecus rosalia. Folia primatol.*, 33, 1–14. (Chapters 4, 10)

Kleiman, D.G., Beck, B.B., Deitz, J. M., Deitz, L.A., Ballou, J.D., and Coimbra-Filho, A.F. (1986). Conservation program for the golden lion tamarin: captive research and management, ecological studies, educational strategies and reintroduction. (1986). In *Primates. The road to self-sustaining populations* (ed. K. Benirschke), pp.959–79. Springer-Verlag, New York. (Chapters 1, 14)

Kleiman, D.G., Hoage, R.T., and Green, K.M. (1988). The lion tamarins, genus *Leontopithecus*. In *Ecology and behavior of Neotropical primates*, vol. 1 (ed. R.A. Mittermeier, A.B. Rylands, A.F. Coimbra-Filho, and G.A.B. da Fonseca), pp.299–347. World Wildlife Fund, Washington, D.C. (Chapters 5, 9, 10, 14)

Knox, K.L. (1990). Observations on dominance relations among *Saguinus imperator*, the emperor tamarin (Family: Callitrichidae). *Dissertation Abstracts International*, B51(1), 470. (Chapter 8)

Koenig, A. (1987). Zur Stellung des alpha-Männchens in Kleingruppen bei *Callithrix jacchus* Erxleben, 1777, in Gefangenschaft unter besonderer Berücksichtigung der Aufzucht eigener ind genetisch nicht verwandter Jungtiere. Unpublished diploma thesis, University of Göttingen, Göttingen. (Chapter 7)

Koenig, A. and Rothe, H. (1990). Gruppenstruktur von *Callithrix jacchus* während und nach der Immigration genetisch fremder Artgenossen. Unpublished report to DFG, Az: Ro 356/9–1. (Chapter 7)

Koenig, A. and Rothe, H. (1991*a*). Social relationships and individual contribution to cooperative behaviour in captive common marmosets (*Callithrix jacchus*). *Primates*, 32, 183–95. (Chapters 7, 14)

Koenig, A. and Rothe, H. (1991*b*). Infant carrying in a polygynous group of the common marmoset (*Callithrix jacchus*). *Am. J. Primatol.* In press. (Chapter 7)

Koenig, A. and Siess, M. (1986). Zwei Fälle von vaterloser Zwillingsaufzucht bei *Callithrix penicillata* × *jacchus* und *Callithrix geoffroyi*. *Zool. Garten*, N.F., 56, 438–40. (Chapter 7)

Koenig, A., Rothe, H., Darms, K., Siess, M., Radespiel, U., and Rock, J. (1987). Bildung vin Mehr-Männchen-Gruppen bei Krallenaffen. *Zool. Garten, N.F.*, 57, 200–1. (Chapter 7)

Koenig, A., Rothe, H., Siess, M., Darms, K., Groeger, D., Radespiel, U., and Rock, J. (1988). Reproductive reorganization in incomplete groups of the common marmoset (*Callithrix jacchus*) under laboratory conditions. *Z. Säugetierk.*, 53, 1–6. (Chapter 7)

Koenig, A., Radespiel, U., Siess, M., Rothe, H., and Darms, K. (1990). Analysis of gestation length and interbirth-intervals in a colony of common marmosets (*Callithrix jacchus*). *Z. Säugetierk.*, 55, 308–14. (Chapter 7)

Kuester, J. (1978). Preliminary investigations on the integration/reintroduction of hand-reared common marmosets (*Callithrix jacchus*) into family groups. In *Biology and behavior of marmosets* (ed. H. Rothe, H.-J. Wolters, and J.P. Hearn), pp.149–52. Eigenverlag H. Rothe, Göttingen. (Chapter 7)

Kuester, J. (1982). Analyse der sozialen Beziehungen in Gruppen gleichaltriger, künstlich aufgezogener Weissbüscheläffchen (*Callithrix jacchus* Erxleben 1977). Unpublished Ph.D. thesis, University of Göttingen, Göttingen. (Chapter 7)

Lacher, T.E., Jr, Fonseca, G.A.B. da, Alves, C., Jr, and Magalhães-Castro, B. (1981). Exudate-eating, scent marking and territoriality in a wild population of marmosets. *Anim. Behav.*, 29, 306–7. (Chapter 4)

Lacher, T.E., Jr, Fonseca, G.A.B. da, Alves, C., Jr, and Magalhães-Castro, B.

(1984). Parasitism of trees by marmosets in a central Brazilian gallery forest. *Biotropica*, 16, 202–9. (Chapters 12, 15)
Laemmert, H.W. Jr, Ferreira, L. de C., and Taylor, R.M. (1946). An epidemiological study of jungle yellow fever in an endemic area in Brazil. Part II—Investigations of vertebrate hosts and arthropod vectors. *Am. J. Trop. Med.*, 126, 23–60. (Chapter 1)
Lange, K. (1977). Untersuchungen zur Entwicklung von Interaktionsmustern zwischen aufzuchtnaiven *Callithrix jacchus* Erxleben 1977 und infantilen Gruppenmitgliedern in den ersten Lebenwochen der Jungtiere. Unpublished Master's thesis, University of Göttingen, Göttingen. (Chapter 7)
Larsson, H., Hagelin, M., and Hjern, M. (1982). Observations on a breeding group of pigmy marmosets, *Cebuella pygmaea*, at Skansen Aquarium. *Int. Zoo Yrbk.*, 22, 88–93. (Chapter 10)
Leighton, D.R. 1987. Gibbons: territoriality and monogamy. In *Primate societies* (ed. B.B. Smuts, D.L. Cheney, R.M. Seyfarth, R.W. Wrangham, and T.T. Struhsaker), pp.135–45. University of Chicago Press, Chicago. (Chapter 8)
Leutenegger, W. (1973). Maternal-fetal weight relationships in primates. *Folia primatol.*, 20, 280–94. (Chapters 10, 13, 15)
Leutenegger, W. (1979). Evolution of litter-size in primates. *Am. Nat.*, 114, 525–31. (Chapter 15)
Leutenegger, W. (1980). Monogamy in callitrichids: a consequence of phyletic dwarfism. *Int. J. Primatol.*, 1, 95–8. (Chapters 7, 8, 9, 13)
Leutenegger, W. and Larson, S. (1985). Sexual dimorphism in the postcranial skeleton of New World primates. *Folia primatol.*, 44, 82–95. (Chapter 13)
Lindsay, N.B.D. (1980). A report on a field study of Geoffroy's tamarin, *Saguinus geoffroyi. Dodo, J. Jersey Wildl. Preserv. Trust*, 17, 27–51. (Chapters 4, 8)
Lisboa, P.L.B., Maciel, U.N., and Prance, G.T. (1987). Perdendo Rondônia. *Ciência Hoje*, 6(36), 48–56. (Chapter 1)
Locke-Haydon, J. and Chalmers, N.R. (1983). The development of infant-caregiver relationships in captive common marmosets (*Callithrix jacchus*). *Int. J. Primatol.*, 4, 63–81. (Chapters 9, 10)
Lorini, M.L. and Persson, V.G. (1990). Nova espécie de *Leontopithecus* Lesson, 1840, do sul do Brasil (Primates, Callitrichidae). *Bol. Mus. Nac., Nova Série, Zoologia*, Rio de Janeiro, (338), 1–14. (Chapters 1, 15)
Lucas, N.S., Hume, E.M., and Smith, H.H. (1927). On the breeding of the common marmoset (*Hapale jacchus* Linn.) in captivity when irradiated with ultra-violet rays. *Proc. zool. Soc. Lond.*, 30, 447–50. (Chapter 7)
Macdonald, D.W. and Carr, G.M. (1989). Food security and the rewards of tolerance. In *Comparative socioecology: the behavioural ecology of humans and other mammals* (ed. V. Standon and R.A. Foley), pp.75–99. Blackwell Scientific Publications, Oxford. (Chapter 5)
Mace, G. (compiler). (1990). *1989 international studbook golden-headed lion tamarin, number three.* Jersey Wildlife Preservation Trust, on behalf of the International Recovery and Management Committee for the Golden-Headed Lion Tamarin, Jersey. (Chapter 1)
Mack, D.S. and Kleiman, D.G. (1978). Distribution of scent marks in different contexts in captive lion tamarins, *Leontopithecus rosalia* (Primates). In *Biology and behavior of marmosets* (ed. H. Rothe, H.-J. Wolters, and J.P. Hearn), pp.181–8. Eigenverlag H. Rothe, Göttingen. (Chapter 4)
Mack, D.S. and Mittermeier, R.A. (1984). *The international primate trade, 1,*

*legislation, trade and captive breeding*. Traffic (U.S.A.), World Wildlife Fund-U.S. Primate Program and IUCN/SSC Primate Specialist Group, Washington, D.C. 185pp. (Chapter 1)

Magnanini, A. (1978). Progress in the development of Poço das Antas Biological Reserve for *Leontopithecus rosalia rosalia* in Brazil. In *The biology and conservation of the Callitrichidae* (ed. D.G. Kleiman), pp.131–6. Smithsonian Institution Press, Washington, D.C. (Chapter 1)

Mahar, D.J. (1989). *Government policies and deforestation in Brazil's Amazon region*. The World Bank, in cooperation with World Wildlife Fund (WWF) and Conservation International (CI), Washington, D.C. 56pp. (Chapter 1)

Maier, W. (1978). Die bilophodonten Molaren in der Indriidae (Primates) ein evolutionmorphologischer Modellfall. *Z. Morphol. Anthropol.*, 68, 307–44. (Chapter 15)

Maier, W. (1982). Nasal structures in Old and New World primates. In *Evolutionary biology of New World monkeys and continental drift* (ed. A.L. Ciochon and A.B. Chiarelli), pp.219–41. Plenum Press, New York. (Chapter 4)

Maier, W., Alonso, C., and Langguth, A. (1982). Field observations on *Callithrix jacchus jacchus* L. *Z. Säugetierk.*, 47, 334–46. (Chapters 7, 9, 12)

Malaga, C. (1985). Nonstandard mating systems for *Saguinus mystax*. *Am. J.Phys. Anthrop.*, 66, 201. Abstract. (Chapter 8)

Mallinson, J.J.C. (1971). The breeding and maintenance of marmosets at Jersey Zoo. *Jersey Wildl. Preserv. Trust, 6th Ann. Rep.*, (1969), 5–10. (Chapter 3)

Mallinson, J. J. C. (1986). The Wildlife Preservation Trusts' (J.W.P.T./W.P.T.I.) support for the conservation of the genus *Leontopithecus*. *Dodo, J. Jersey Wildl. Preserv. Trust*, 23, 6–18. (Chapter 1)

Mallinson, J.J.C. (1987). International efforts to secure a viable population of the golden-headed lion tamarin. *Primate Conservation*, (8), 124–5. (Chapter 1)

Mallinson, J.J.C. (1989). A summary of the work of the International Recovery and Management Committee for Golden-headed Lion Tamarin *Leontopithecus chrysomelas* 1985–1990. *Dodo, J. Jersey Wildl. Preserv. Trust*, 26, 77–86. (Chapter 1)

Marchlewska-Koj, A. (1977). Pregnancy block elicited by urinary proteins of male mice. *Biol. Reprod.*, 17, 729–32. (Chapter 4)

Marler, P. (1972). Vocalizations of East African monkeys. II. Black and white colobus. *Behaviour*, 42, 175–97. (Chapter 2)

Marsh, C.W. and Mittermeier, R.A. (ed.) (1987). *Primate conservation in the tropical rain forest*. Alan R. Liss, New York. (Chapter 1)

Marshall, J.T., Jr and Marshall, E.R. (1976). Gibbons and their territorial songs. *Science*, 193, 235–7. (Chapter 2)

Martin, R.D. (1972). A preliminary field study of the lesser mouse lemur (*Microcebus murinus* J.F. Miller 1777). *Advances in Ethology*, 9, 43–89. (Chapter 15)

Martin, R.D. (1990). Goeldi and the Dwarfs: the evolutionary biology of the small New World monkeys. Osman Hill Memorial Lecture presented at the 'Primates in Evolution: John Napier Memorial Symposium', Primate Society of Great Britain, Geological Museum, London, 17th December 1990. Abstract. *Primate Eye*, (42), 8–9. (Chapter 15)

Martins, E.S., Schneider, H., and Leão, V.F. (1987). Syntopy and troops association between *Callithrix* and *Saguinus* from Rondonia, Brazil. *Int. J.Primatol.*, 8, 527. Abstract. (Chapter 15)

Martins, E.S., Ayres, J.M., and Valle, M.B.R. do (1988). On the status of *Ateles belzebuth marginatus* with notes on other primates of the Irirí river basin. *Primate Conservation*, (9), 87–91. (Chapter 1)

Masataka, N. (1987). The perception of sex-specificity in long calls of the tamarin (*Saguinus labiatus*). *Ethology*, 76, 56–64. (Chapter 8)

Mayer, K.E., Blum, C.A., and Caine, N.G. (1992). Comparative foraging styles of tamarins and squirrel monkeys. *Am. J. Primatol.* 27, 46. (Chapter 8)

Mayr, E. (1970). *Populations, species and evolution*. Harvard University Press, Cambridge. (Chapter 3)

McGrew, W.C. (1986). Kinship terms and callitrichid mating patterns: a discussion note. *Primate Eye*, 30, 25–6. (Chapter 7)

McGrew, W.C. (1988). Parental division of infant caretaking varies with family composition in cotton-top tamarins. *Anim. Behav.*, 36, 285–6. (Chapters 9, 10)

McGrew, W.C. and McLuckie, E.C. (1986). Philopatry and dispersion in the cotton top tamarin, *Saguinus oedipus oedipus*: an attempted laboratory simulation. *Int. J. Primatol.*, 7, 401–22. (Chapters 7, 8)

Melrose, D.R., Reed, H.C.B., and Patterson, R.L.S. (1971). Androgen steroids associated with boar odour as an aid to the detection of oestrus in pig artificial insemination. *Br. vet. J.*, 127, 497–502. (Chapter 4)

Mendes, S.L. (1989). Sintopia e hibridização entre dois taxons de *Callithrix* (Primates, Callitrichidae) do grupo *jacchus. In Resumos. XVI Congresso Brasileiro de Zoologia*, João Pessoa, pp.110–11. 22nd–27th January 1989. Abstract. (Chapter 1)

Mendes, S.L. (1991*a*). Situação atual dos primatas em reservas florestais do estado do Espírito Santo. In: *A primatologia no Brasil—3* (ed. A.B. Rylands and A.T. Bernardes), pp.347–56. Sociedade Brasileira de Primatologia, Fundação Biodiversitas, Belo Horizonte. (Chapter 1)

Mendes, S.L. (1991*b*). Diferenças vocais entre *Callithrix geoffroyi e Callithrix flaviceps* (Primates: Callitrichidae). In *Resumos. XVIII Congresso Brasileiro de Zoologia*, p.391. Universidade Federal da Bahia, Salvador, 24th February to 1st March, 1991. Abstract. (Chapters 1, 2, 3)

Menzel, E.W. and Menzel, C.R. (1979). Cognitive, developmental, and social aspects of responsiveness to novel objects in a family group of marmosets. *Behaviour*, 70, 251–79. (Chapter 8)

Milton, K. and May, M.L. (1976). Body weight, diet and home range area in primates. *Nature, Lond.*, 259, 459–62. (Chapter 15)

Milton, K. and Lucca, C. de (1984). Population estimate for *Brachyteles* at Fazenda Barreiro Rico, São Paulo State, Brazil. *IUCN/SSC Primate Specialist Group Newsletter*, (4), 27–8. (Chapter 1)

Miranda Ribeiro, A. (1912). Dois novos simios da nossa fauna. *Brasil. Rundschau*, 2(1), 21–3. (Reprinted in 1955, *Arq. Mus. Nacional*, Rio de Janeiro, 42, 414). (Chapter 1)

Mittermeier, R.A. (1973). Recommendations for the creation of National Parks and Biological Reserves in the Amazonian region of Brazil, based on a four-month primate survey in the upper Amazon, Rio Negro and Rio Tapajós. Unpublished report, Instituto Brasileiro de Desenvolvimento Florestal (IBDF), Brasília. 4pp. (Chapter 1)

Mittermeier, R.A. (1977). The distribution, synecology and conservation of Surinam monkeys. Unpublished Ph.D. dissertation, University of Harvard, Cambridge. (Chapter 15)
Mittermeier, R.A. and Cheney, D.L. (1987). Conservation of primates and their habitats. In *Primate societies* (ed. D.L. Cheney, R.M. Seyfarth, B. Smuts, R.W. Wrangham, and T.T. Struhsaker), pp.477–90. Chicago University Press, Chicago. (Chapter 1)
Mittermeier, R.A. and Coimbra-Filho, A.F. (1977). Primate conservation in Brazilian Amazonia. In *Primate conservation* (ed. H.S.H. Prince Rainier III of Monaco and G.H. Bourne), pp.117–66. Academic Press, New York. (Chapter 1)
Mittermeier, R.A. and Coimbra-Filho, A.F. (1981). Systematics: species and subspecies. In *Ecology and behaviour of Neotropical primates*, vol. 1 (ed. A.F. Coimbra-Filho and R.A. Mittermeier), pp.29–109. Academia Brasileira de Ciências, Rio de Janeiro. (Chapters 1, 5, 13)
Mittermeier, R.A. and Coimbra-Filho, A.F. (1983). Distribution and conservation of New World primate species used in biomedical research. In *Reproduction in New World primates: new models in medical science* (ed. J.P. Hearn), pp.1–37. MTP Press, Lancaster. (Chapter 1)
Mittermeier, R.A. and Roosmalen, M.G.M. van (1981). Preliminary observations on habitat utilization and diet in eight Surinam monkeys, *Folia primatol.*, 44, 82–95. (Chapter 13)
Mittermeier, R.A. and Roosmalen, M.G.M. van (1982). Conservation of primates in Surinam. *Int. Zoo Yrbk.*, 22, 59–68. (Chapter 1)
Mittermeier, R.A., Bailey, R.C., and Coimbra-Filho, A.F. (1978). Conservation status of the Callitrichidae in Brazilian Amazonia, Surinam and French Guiana. In *The biology and conservation of the Callitrichidae* (ed. D.G. Kleiman), pp.137–46. Smithsonian Institution, Washington, D.C. (Chapter 1)
Mittermeier, R.A., Coimbra-Filho, A.F., and Constable, I. (1981). Conservation of eastern Brazilian primates. Unpublished report for the period 1979/1980, Project No. 1614, World Wildlife Fund, Washington, D.C. 39 pp. (Chapter 1)
Mittermeier, R.A., Coimbra-Filho, A.F., Constable, I.D., Rylands, A.B., and Valle, C. (1982). Conservation of primates in the Atlantic forest region of eastern Brazil. *Int. Zoo Yrbk.*, 22, 2–17. (Chapter 1)
Mittermeier, R.A., Padua, C.V., Valle, C.M.C., and Coimbra-Filho, A.F. (1985). Major program underway to save the black lion tamarin in São Paulo, Brazil. *Primate Conservation*, (6), 19–21. (Chapter 1)
Mittermeier, R.A., Oates, J.F., Eudey, A.E., and Thornback, J. (1986). Primate conservation. In: *Comparative primate biology, volume 2A: behavior, conservation and ecology*, pp.3–72. Alan R. Liss Inc., New York. (Chapter 1)
Mittermeier, R.A., Rylands, A.B., Coimbra-Filho, A.F., and Fonseca, G.A.B. da (ed.) (1988*a*). *Ecology and behavior of Neotropical primates*, vol. 2. World Wildlife Fund, Washington, D.C. (Chapter 5)
Mittermeier, R.A., Rylands, A.B., and Coimbra-Filho, A.F. (1988*b*). Systematics: Species and subspecies—un update. In *Ecology and behavior of Neotropical primates*, vol. 2, (ed. R.A. Mittermeier, A.B. Rylands, A.F. Coimbra-Filho, and G.A.B. da Fonseca), p.13–75. World Wildlife Fund, Washington, D.C. (Chapters 1, 2, 7, 12, 13, 15)
Mittermeier, R.A., Kinzey, W.G., and Mast, R.B. (1989). Neotropical primate conservation. *J. Hum. Evol.*, 18, 597–610. (Chapter 1)

Monteiro da Cruz, M.A.O. and Silva, G.S. (1989). Mudanças na composição de um grupo social de *Callithrix jacchus* na mata de Dois Irmãos, PE. In *Resumos. XVI Congresso Brasileiro de Zoologia*, p.128. Federal University of Paraíba, João Pessoa. Abstract. (Chapter 7)

Moody, M.I. and Menzel, E.W. (1976). Vocalizations and their behavioral contexts in the tamarin, *Saguinus fuscicollis. Folia primatol.*, 25, 73–94. (Chapter 8)

Moore, W.S. (1977). An evaluation of narrow hybrid zones in vertebrates. *Q. Rev. Biol.*, 52, 263–77. (Chapter 3)

Morosetti, G. (1986). Beobachtungen zur Entwicklung der Eltern-Jungtier-Beziehung in den ersten funf Lebensmonaten bei Grossfamilien des Weissbüscheläffchens *Callithrix jacchus* Erxleben, 1777. Unpublished diploma thesis, University of Göttingen, Göttingen. (Chapter 7)

Moya, L., Trogoso, P., and Heltne, P.G. (1980). Manejo de fauna silvestre em semicautiverio en la isla de Iquitos y Padre Isla—Ano 1980. Unpublished technical report, Pan American Health Organisation, Washington, D.C. (Chapter 7)

Moynihan, M. (1970). Some behavior patterns of platyrrhine monkeys II. *Saguinus geoffroyi* and some other tamarins. *Smithson. contr. Zool.*, (28), 1–77. (Chapters 2, 4, 8, 13)

Moynihan, M. (1976*a*). Notes on the ecology and behaviour of the pigmy marmoset (*Cebuella pygmaea*) in Amazonian Colombia. In *Neotropical primates, field studies and conservation* (ed. R.W. Thorington Jr and P.G. Heltne), pp.79–84. National Academy of Sciences, Washington, D.C. (Chapter 1)

Moynihan, M. (1976*b*). *The New World primates.* Princeton University Press, Princeton, N.J. (Chapter 1)

Muckenhirn, N.A. (1967). The behavior and vocal repertoire of *Saguinus oedipus* (Hershkovitz, 1966) (Callitrichidae, Primates). Unpublished Master's thesis, University of Maryland. (Chapter 4)

Müller-Schwarze, D., Morehouse, L., Corradi, R., Cheng-hua, Z., and Silverstein, R.M. (1986). Odor images: responses of beaver to castoreum fractions. In *Chemical signals in vertebrates 4: ecology, evolution and comparative biology* (ed. D. Duvall, D. Müller-Schwarze, and R.M. Silverstein), pp.561–70. Plenum Press, New York. (Chapter 4)

Muskin, A. (1984*a*). Preliminary field observations of *Callithrix aurita* (Callitrichinae, Cebidae). In *A primatologia no Brasil* (ed. M.T. de Mello), pp.79–82. Sociedade Brasileira de Primatologia, Brasília. (Chapters 7, 12)

Muskin, A. (1984*b*). Field notes and geographic distribution of *Callithrix aurita* in eastern Brazil. *Am. J. Primatol.*, 7, 377–80. (Chapters 7, 12)

Nagel, U. (1971). Social organization in a baboon hybrid zone. *Proc. 3rd int. Congr. Primat., Zurich, 1970*, 3, 48–57. S. Karger, Basel. (Chapter 3)

Napier, J.S. and Napier, P.H. (1967). *A handbook of living primates.* Academic Press, London. (Chapter 1)

Napier, P.H. (1976). *Catalogue of primates in the British Museum (Natural History). Part 1: Families Callitrichidae and Cebidae.* British Museum (Natural History), London. (Chapter 1)

Nash, L.T. (1986). Dietary, behavioral, and morphological aspects of gummivory in primates. *Yrbk. Phys. Anthropol.*, 29, 113–37. (Chapters 12, 13, 15)

Natori, M. (1986). Interspecific relationships of *Callithrix* based on the dental characters. *Primates*, 27(3), 321–6. (Chapters 1, 12)

Natori, M. (1988). A cladisitic analysis of interspecific relationships of *Saguinus*. *Primates*, 29, 263–76. (Chapter 1)

Natori, M. (1989). An analysis of cladistic relationships of *Leontopithecus* based on dental and cranial characters. *J. Anthrop. Soc. Nippon*, 97, 157–167. (Chapter 1)

Natori, M. (1990). A numerical analysis of the taxonomical status of *Callithrix kuhli* based on measurements of the postcanine dentition. *Primates*, 31(4), 555–62. (Chapters 1, 12)

Natori, M. and Hanihara, T. (1992). Variations in dental measurements between *Saguinus* species and their systematic relationships. *Folia primatol.*, **58**, 84–92. (Chapter 1)

Neville, M.K., Castro, N., Marmol, A., and Revilla, J. (1976). Censusing primate populations in the reserved area of Pacaya and Samiria rivers, Department of Loreto, Peru. *Primates*, 17(2), 151–81. (Chapter 1)

Neyman, P.F. (1978). Aspects of the ecology and social organization of free-ranging cotton-top tamarins (*Saguinus oedipus*) and the conservation status of the species. In *The biology and conservation of the Callitrichidae* (ed. D.G. Kleiman), pp.39–71. Smithsonian Institution Press, Washington, D.C. (Chapters 1, 7, 8, 9, 10, 13, 14)

Nicolson, N.A. (1987). Infants, mothers and other females. In *Primate societies* (ed. B.B. Smuts, D.L. Cheney, R.M. Seyfarth, R.W. Wrangham, and T.T. Struhsaker), pp.330–42. University of Chicago Press, Chicago. (Chapter 10)

Nogami, Y. and Natori, M. (1986). Fine structure of the dental enamel in the Family Callitrichidae (Ceboidea, Primates). *Primates*, 27, 245–58. (Chapters 1, 12)

Noll, S., Zeller, U., Epple, G. and Küderling, I. (1989). Zur Ontogenese des circumgenitalen Hautdrusenorgans vin *Saguinus fuscicollis* (Callitrichidae). *Abstracts. Deutsche Ges. Saugetierk, 63rd Meeting*. Paul Parey, Berlin. (Chapter 4)

Norconk, M.A. (1986). Interactions between primate species in a neotropical forest: mixed species troops of *Saguinus mystax* and *S. fuscicollis* (Callitrichidae). Unpublished Ph.D. thesis, University of California, Los Angeles. (Chapter 13)

Norconk, M.A. (1990). Mechanisms promoting stability in mixed *Saguinus mystax* and *Saguinus fuscicollis* troops. *Am. J. Primatol.*, 21, 159–70. (Chapter 8, 15)

Nordstrom, K.M., Belcher, A.M., Epple, G., Greenfield, K.L., and Smith, A.B., III (1989). Microbial flora of the skin of the saddle-back tamarin monkey, *Saguinus fuscicollis*. *J. Chem. Ecol.*, 15, 629–39. (Chapter 4)

Novotny, M., Jorgensen, J.W., Carmack, M., Wilson, S.R., Boyse, E.A., Yamazaki, K., Wilson, M., Beamer, W., and Whitten, W.K. (1980). Chemical studies of the primer mouse pheromones. In *Chemical signals in vertebrates 2: vertebrates and aquatic invertebrates* (ed. D. Müller-Schwarze and R.M. Silverstein), pp.377–90. Plenum Press, New York. (Chapter 4)

Novotny, M., Harvey, S., Jemiolo, B., and Alberts, A. (1985). Synthetic pheromones that promote inter-male aggression in mice. *Proc. Natn. Acad. Sci. U.S.A.*, 82, 2059–61. (Chapter 4)

Oates, J.F. (1987). Food distribution and foraging behavior. In *Primate societies* (ed. B.B. Smuts, D.L. Cheney, R.M. Seyfarth, R.W. Wrangham, and T.T. Struhsaker), pp.197–209. University of Chicago Press, Chicago. (Chapter 13)

Oates, J.F. and Trocco, T.F. (1983). Taxonomy and phylogeny of black and white colobus monkeys: inferences from an analysis of loud call variations. *Folia primatol.*, 40, 83–113. (Chapter 5)

Oliver, W.L.R. and Santos, I.B. (1991). Threatened endemic mammals of the Atlantic forest region of south-east Brazil. *Wildl. Preserv. Trust, Special Scientific Report*, (4), 1–126. (Chapters 1, 3)

Omedes A. and Carroll, J.B. (1980). A comparative study of pair behaviour of four callitrichid species and the Goeldi's monkey at Jersey Wildlife Preservation Trust. *Dodo, J. Jersey Wildl. Preserv. Trust*, 17, 51–62. (Chapter 8)

Opler, P.A., Baker, H.G., and Frankie, G.W. (1980). Plant reproductive characteristics during secondary succession in neotropical lowland forest ecosystems. *Biotropica*, 12 (suppl.), 40–6. (Chapter 13)

Oppenheimer, J.R. (1982). *Cebus capucinus*: home range, population dynamics, and interspecific relationships. In *The ecology of a tropical rainforest: seasonal rhythms and long-term changes* (ed. E.G. Leigh Jr, A.S. Rand, and D. M. Windsor), pp.253–73. Smithsonian Institution Press, Washington, D.C. (Chapter 13)

Paccagnella, S.G. (1986). Relatório sobre o censo da população de monos-carvoeiros do Parque Estadual de 'Carlos Botelho'. Unpublished report, Instituto de Florestas (IF), São Paulo. 39pp. (Chapter 1)

Padua, C.V., Simon, F., Padua, S., Keuroghlian, A., Faria, H., Max, J., and Sério, F. Black lion tamarin working group report. (1990). In *Leontopithecus*: population viability analysis workshop report (ed. U.S. Seal, J.D. Ballou and C.V. Padua), pp.43–7. Captive Breeding Specialist Group (IUCN/SSC/CBSG), Apple Valley, Minnesota. (Chapter 14)

Parker, G.A. (1970). Sperm competition and its evolutionary consequences in insects. *Biol. Rev.*, 45, 525–67. (Chapter 6)

Passos, F.C. and Carvalho, C.T. de (1991). Importância de exsudatos na alimentação do mico-leão preto, *Leontopithecus chrysopygus* (Callithricidae, Primates). In *Resumos. XVIII Congresso Brasileiro de Zoologia*, p.392. Universidade Federal da Bahia, Salvador, 24th February to 1st March 1999. Abstract. (Chapters 12, 14)

Peres, C.A. (1986*a*). Costs and benefits of territorial defense in the golden lion tamarin, *Leontopithecus rosalia.* Unpublished Master's thesis, University of Florida, Gainesville. (Chapters 9, 12, 14)

Peres, C.A. (1986*b*). Golden lion tamarin project. II. Ranging patterns and habitat selection in golden lion tamarins *Leontopithecus rosalia* (Linnaeus, 1766) (Callitrichidae, Primates. In *A primatologia no Brasil—2* (ed. M.T. de Mello), pp.223–41. Sociedade Brasileira de Primatologia, Brasília. (Chapters 12, 14)

Peres, C.A. (1987). Conservation of primates in western Brazilian Amazonia. Unpublished report to World Wildlife Fund-US, Washington, D.C. (Chapter 1)

Peres, C.A. (1989). Exudate-eating by wild golden lion tamarins, *Leontopithecus rosalia. Biotropica*, 21, 287–8. (Chapters 12, 14, 15)

Perkins, E.M. (1966). The skin of the black-collared tamarin (*Tamarinus nigricollis*). *Am. J. Phys. Anthrop.*, 25, 41–69. (Chapter 4)

Perkins, E.M. (1968). The skin of the pygmy marmoset, *Callithrix* (= *Cebuella*) *pygmaea. Am. J. Phys. Anthrop.*, 29, 349–64. (Chapter 4)

Perkins, E.M. (1969*a*). The skin of the cotton-top pinché, *Saguinus* (= *Oedipomidas*) *oedipus. Am. J. Phys. Anthrop.*, 30, 13–27. (Chapter 4)

Perkins, E.M. (1969*b*). The skin of the silver marmoset (*Callithrix* (= *Mico*) *argentata. Am. J. Phys. Anthrop.*, 30, 361–87. (Chapter 4)

Persson, V.G. and Lorini, M.L. (1991). Notas sobre o mico-leão-de-cara-preta, *Leontopithecus caissara* Lorini e Persson, 1990, no sul do Brasil (Primates, Callitrichidae). In *Resumos. XVIII Congresso Brasileiro de Zoologia*, p.385. Universidade Federal da Bahia, Salvador, 24th February to 1st March, 1999. Abstract. (Chapter 1)

Persson, V.G. and Lorini, M.L (In press). Notas sobre o mico-leão-de-cara-preta, *Leontopithecus caissara* Lorini ê Persson, 1990, no sul do Brasil (Primates, Callitrichidae). In *A primatologia no Brasil—4*, M.E. Yamamoto and M.B. Cordeiro de Sousa (ed.). Sociedade Brasileira de Primatologia and Fundação Biodiversitas, Belo Horizonte. (Chapters 1, 14)

Pocock, R.I. (1925). Additional notes on the external characters of some platyrrhine monkeys. *Proc. zool. Soc. Lond.*, 1925, 91–113. (Chapter 1)

Pook, A.G. (1978*a*). A comparison between the reproduction and parental behaviour of the Goeldi's monkey (*Callimico goeldii*) and of the true marmosets (Callitrichidae). In *Biology and behaviour of marmosets* (ed. H. Rothe, H.-J. Wolters, and J.P. Hearn), pp.1–14. Eigenverlag H. Rothe, Göttingen. (Chapters 8, 10, 15)

Pook, A.G. (1978*b*). Some notes on the re-introduction into groups of six hand-reared marmosets of different species. In *Biology and behaviour of marmosets* (ed. H. Rothe, H.-J. Wolters, and J.P. Hearn), pp.155–159. Eigenverlag H. Rothe, Göttingen. (Chapters 7, 8)

Pook, A.G. (1984). The evolutionary role of soci-ecological factors in the development of paternal care in the New World Family Callitrichidae. In *Primate paternalism* (ed. D.M. Taub), pp.336–45. Van Nostrand Reinhold, New York. (Chapters 7, 8)

Pook, A.G. and Pook, G. (1979). A field study of the status and socioecology of Goeldi's monkey (*Callimico goeldii*) and other primates in northern Bolivia. Unpublished report, New York Zoological Society, New York. (Chapter 1)

Pook, A.G. and Pook, G. (1981). A field study of the socioecology of the Goeldi's monkey (*Callimico goeldii*) in northern Bolivia. *Folia primatol.*, 35, 288–312. (Chapter 13)

Pook, A.G. and Pook, G. (1982). Polyspecific association between *Saguinus fuscicollis*, *S. labiatus* and *Callimico goeldi*, and other primates in northwestern Bolivia. *Folia primatol.*, 38, 196–216. (Chapters 8, 9, 13, 14, 15)

Poole, T.B. and Evans, R.G. (1982). Reproduction, infant survival and productivity of a colony of common marmosets (*Callithrix jacchus jacchus*). *Lab. Anim.*, 16, 88–97. (Chapter 7)

Powell, G.V.N. (1974). Experimental analysis of the social value of flocking in starlings (*Sturnus vulgaris*) in relation to predation and foraging. *Anim. Behav.*, 22, 501–5. (Chapter 8)

Power, M.L., Milton, K., and Oftedal, O.T. (1990). An examination of the relative abilities of marmosets and tamarins to digest a gum. *Am. J. Primatol.*, 20, 222. Abstract. (Chapter 15)

Prance, G.T. (ed.) (1983). *Biological diversification in the tropics*. Columbia University Press, New York. (Chapter 1)

Price, E.C. (1990*a*). Parturition and perinatal behaviour in captive cotton-top tamarins (*Saguinus oedipus*). *Primates*, 31, 523–35. (Chapter 10)

Price, E.C. (1990*b*). Infant-carrying as a courtship strategy of breeding male cotton-top tamarins. *Anim. Behav.*, 40, 784–6. (Chapters 8, 10)

Price, E.C. (In press *a*). The costs of infant carrying in captive cotton-top tamarins. *Am. J. Primatol.* (Chapters 7, 8, 9, 10)

Price, E.C. (In press *b*). The benefits of helpers: effects of group and litter size on infant care in tamarins (*Saguinus oedipus*). *Am. J. Primatol.* (Chapter 7)
Price, E.C. (In press *c*). Stability of wild callitrichid groups. *Folia primatol.* (Chapters 7, 14)
Price, E.C. and Evans, S. (In press). Terminology in the study of callitrichid reproductive strategies. *Anim. Behav.* (Chapter 7)
Price, E.C. and Hannah, A.C. (1983). A preliminary comparison of group structure in the golden lion tamarin, *Leontopithecus r. rosalia* and the cotton-topped tamarin, *Saguinus oedipus. Dodo, J. Jersey Wildl. Preserv. Trust*, 20, 36–48. (Chapters 8, 10)
Price, E.C. and McGrew, W.C. (1990). Cotton-top tamarins (*Saguinus* (*o.*) *oedipus*) in a semi-naturalistic captive colony. *Am. J. Primatol.*, 20, 1–12. (Chapter 7)
Price, E.C. and McGrew, W.C. (1991). Departures from monogamy in colonies of captive cotton-top tamarins. *Folia primatol.* 57, 16–27. (Chapters 7, 14)
Pruetz, J.D. and Garber, P.A. (1991). Patterns of resource utilization, home range overlap, and intergroup encounters in moustached tamarin monkeys. *Am. J. Phys. Anthrop.*, suppl. 12, 146. Abstract. (Chapter 13)
Pryce, C.R. (1988). Individual and group effects on early caregiver-infant relationships in red-bellied tamarin monkeys. *Anim. Behav.*, 36, 1455–64. (Chapters 7, 8)
Pulliam, H.R. (1973). On the advantages of flocking. *J. theor. Biol.*, 38, 419–22. (Chapter 8)
Radespiel, U. (1990). Die räumlichen und sozialen Strukturen innerhalb einer Familie von Weissbüschelaffen (*Callithrix jacchus* Erxleben, 1777) unter dem experimentallen Einfluss der zeitweiligen Abwesenheit der Eltern. Unpublished diploma thesis, University of Göttingen, Göttingen. (Chapter 7)
Ramirez, M. (1984). Population recovery in the moustached tamarin (*Saguinus mystax*): management strategies and mechanisms of recovery. *Am. J. Primatol.*, 7, 245–59. (Chapters 7, 8)
Ramirez, M. (1986). Feeding ecology of the moustached tamarin, *Saguinus mystax*. In *A primatologia no Brasil—2* (ed. M.T. de Mello), pp.211–12. Sociedade Brasileira de Primatologia, Brasília, D.F. Abstract. (Chapter 11)
Ramirez, M. (1989). Feeding ecology and demography of the moustached tamarin, *Saguinus mystax*, in northeastern Peru. Unpublished Ph.D thesis, City University of New York, New York. (Chapter 13)
Ramirez, M., Freese, C., and Revilla, J. (1978). Feeding ecology of the pygmy marmoset, *Cebuella pygmaea*, in north-eastern Peru. In *The biology and conservation of the Callitrichidae* (ed. D.G. Kleiman), pp.91–104. Smithsonian Institution Press, Washington, D.C. (Chapters 12, 13)
Rasa, A.E. (1989). Helping in dwarf mongoose societies: an alternative reproductive strategy. In *The sociobiology of sexual and reproductive strategies* (ed. A.E. Rasa, C. Vogel, and E. Voland), pp.61–73. Chapman and Hall Ltd., London. (Chapter 5)
Rathbun, C.D. (1979). Description and analysis of the arch display in the golden lion tamarin, *Leontopithecus rosalia rosalia. Folia primatol.*, 32, 124–44. (Chapter 10)
Raymer, J., Wiesler, D., Novotny, M., Asa, C., Seal, U.S., and Mack, L.D. (1986). Chemical scent constituents in the urine of wolf (*Canis lupus*) and their dependence on reproductive hormones. *J. Chem. Ecol.*, 12, 291–314. (Chapter 4)

Rhine, R. and Westlund, B.J. (1981). Adult male positioning in baboon progressions: order and chaos revisited. *Folia primatol.*, 35, 77–116. (Chapter 8)

Riedman, M.L. (1982). The evolution of alloparental care and adoption in mammals and birds. *Q. Rev. Biol.*, 57, 405–35. (Chapter 7)

Rizzini, C.T. (1963). Nota previa sobre a divisão fitogeográfica do Brasil. *Rev. Brasil. Geogr.*, Rio de Janeiro, 25, 1–64. (Chapter 14)

Rizzini, C.T. and Coimbra-Filho, A.F. (1981). Lesões produzidas pelo sagui, *Callithrix p. penicillata* (E. Geoffroy, 1812), em árvores do cerrado (Callitrichidae, Primates). *Rev. Brasil. Biol.*, 41, 579–83. (Chapter 12)

Robinson, J.G. (1986). Seasonal variation in use of time and space by the wedge-capped capuchin monkey, *Cebus olivaceus*: implications for a foraging theory. *Smithson. contrib. Zool.*, 431, 1–60. (Chapter 13)

Rosenberger, A.L. (1977). *Xenothrix* and ceboid phylogeny. *J. Hum. Evol.*, 6, 461–81. (Chapters 13, 15)

Rosenberger, A.L. (1978). Loss of incisor enamel in marmosets. *J. Mammal.*, 59, 207–8. (Chapter 12)

Rosenberger, A.L. (1980). Gradistic views and adaptive radiation of platyrrhine primates. *Z. Morph. Anthrop.*, 71(2), 157–63. (Chapter 1)

Rosenberger, A.L. (1981). Systematics: the higher taxa. In: *Ecology and behavior of Neotropical primates*, vol. 1 (ed. A.F. Coimbra-Filho and R.A. Mittermeier), pp.9–27. Academia Brasileira de Ciências, Rio de Janeiro. (Chapters 1, 13, 15)

Rosenberger, A.L. (1984). Aspects of the systematics and evolution of the marmosets. In *A primatologia no Brasil* (ed. M.T. de Mello), pp.159–80. Sociedade Brasileira de Primatologia, Brasília. (Chapters 1, 2, 15)

Rosenberger, A.L. and Coimbra-Filho, A.F. (1984). Morphology, taxonomic status and affinities of the lion tamarins, *Leontopithecus* (Callitrichinae, Cebidae). *Folia primatol.*, 42, 149–79. (Chapters 1, 2, 14, 15)

Rosenberger, A.L. and Strier, K.B. (1989). Adaptive radiation of the ateline primates. *J. Hum. Evol.*, 18, 717–50. (Chapter 1)

Rosenberger, A.L., Setoguchi, T., and Shigehara, N. (1990). The fossil record of callitrichine primates. *J. Hum. Evol.*, 19, 209–36. (Chapters 1, 15)

Rothe, H. (1973). Beobachtungen zur Geburt beim Weisbüscheläffchen (*Callithrix jacchus* Erxleben, 1777). *Folia primatol.*, 19, 257–85. (Chapter 7)

Rothe, H. (1974). Further observations on the delivery behaviour of the common marmoset (*Callithrix jacchus*). *Z. Saugetierk.*, 39, 135–42. (Chapters 7, 10)

Rothe, H. (1975*a*). Influence of newborn marmosets' (*Callithrix jacchus*) behaviour on expression and efficiency of maternal and paternal care. In *Contemporary primatology, Proceedings of the 5th International Congress of Primatology* (ed. S. Kondo, M. Kawai and A. Ehara), pp.315–20. S. Karger, Basel. (Chapters 7, 10)

Rothe, H. (1975*b*). Some aspects of sexuality and reproduction in groups of captive marmosets (*Callithrix jacchus*). *Z. Tierpsychol.*, 37, 255–73. (Chapters 5, 7)

Rothe, H. (1978*a*). Parturition and related behaviour in *Callithrix jacchus* (Ceboidea, Callitrichidae). In *The biology and conservation of the Callitrichidae* (ed. D.G. Kleiman), pp.193–206. Smithsonian Institution Press, Washington, D.C. (Chapters 7, 10)

Rothe, H. (1978*b*). Sub-grouping behaviour in captive *Callithrix jacchus* families:

a preliminary investigation. In *Biology and behaviour of marmosets* (ed. H. Rothe, H.-J. Wolters and J.P. Hearn), pp.233–57. Eigenverlag H. Rothe, Göttingen. (Chapter 7)

Rothe, H. (1979). Das Ethogramm con *Callithrix jacchus* Erxleben, 1777 (Primates, Ceboidea, Callitrichidae). Eine morphaktische Analyse des Verhaltens mit besonderer Berücksichtigung des sozialen Umfeldes. Unpublished Habilitation thesis, University of Göttingen, Göttingen. (Chapter 7)

Rothe, H. and Koenig, A. (1987). Polygynes Verhalten in einer Weissbüschelaffen-Gruppe (*Callithrix jacchus*). *Zool. Garten, N.F.*, 57, 368–9. (Chapter 7)

Rothe, H. and Koenig, A. (1988). Polygynous mating in the common marmoset. Paper presented at the International Workshop on Monogamy in the Callitrichidae, 26th–28th February, 1988. University of Göttingen, Göttingen. (Chapter 7)

Rothe, H. and Koenig, A. (1989). Variability of social organisation in the common marmoset (*Callithrix jacchus*). Paper presented at a Workshop on Callitrichid Behaviour, Scottish Primate Research Group, University of Stirling, Stirling. (Chapter 7)

Rothe, H. and Koenig, A. (1991). Variability of social organization in captive common marmosets (*Callithrix jacchus*). *Folia primatol.*, 57, 28–33. (Chapters 7, 14)

Rothe, H. and Radespiel, U. (1988). Relationships of expelled common marmosets to their native families and unrelated conspecifics. *Int. J. Primatol.*, 8, 499. Abstract. (Chapter 7)

Rothe, H., Darms, K., Koenig, A. Siess, M., and Brusek, P. (1986). Dynamics of group size in the common marmoset (*Callithrix jacchus*) under laboratory conditions. Paper presented at the XIth Congress of the International Primatological Society, 20th to 25th July 1986. University of Göttingen, Göttingen. (Chapter 7)

Rothe, H., Koenig, A., Siess, M., Radespiel, U., and Darms, K. (1987). Integration adulter *Callithrix jacchus*—Männchen in Rumpfgruppen. *Zool. Garten, N.F.*, 57, 202–3. (Chapter 7)

Rothe, H., Koenig, A., Radespiel, U., Darms, K., and Siess, M. (1988). Occurrence and frequency of twin-fights in the common marmoset (*Callithrix jacchus*). *Z. Säugetierk.*, 53, 325–32. (Chapter 8)

Rothe, H., Darms, K., and Koenig, A. (In press). Sex ratio and mortality in a laboratory colony of the common marmoset (*Callithrix jacchus*). *Lab. Anim.* (Chapter 7)

Rothe, H., Darms, K., Koenig, A., Radespiel, U., and Juenemann, B. (In review a). Long term study of infant-carrying behaviour in captive common marmosets, *Callithrix jacchus*: effects of non-reproductive helpers on the parents' carrying performance. *Int. J. Primatol.* (Chapter 7)

Rothe, H., Koenig, A., and Darms, K. (In review b). Infant survival and number of helpers in captive groups of common marmosets. *Am. J. Primatol.* (Chapter 7)

Rothe, H., Koenig, A., and Achilles, L. (Submitted a). Effects of age and sex of non-reproductive helpers on infant-carrying behaviour in groups of captive common marmosets (*Callithrix jacchus*). *Primates.* (Chapter 7)

Rothe, H., Koenig, A., and Achilles, L. (Submitted b). Infant-carrying in captive groups of black tufted-ear marmosets (*Callithrix penicillata*): the parents' benefit from non-reproductive helpers. *Primates.* (Chapter 7)

Roussilhon, C. (1988). The general status of monkeys in French Guiana. *Primate Conservation*, (9), 70–4. (Chapter 1)

Rowell, T.E. (1988). Beyond the one-male group. *Behaviour*, 104, 189–201. (Chapter 3)

Rutberg, A.T. (1983). The evolution of monogamy in primates. *J. theor. Biol.*, 104, 93–112. (Chapter 7)

Ruth, B. (1987). Social behavior of free-living tamarin monkeys (*Saguinus mystax*) on Padre Isla, Peru. *Int. J. Primatol.*, 8, 475. Abstract. (Chapter 8)

Rylands, A.B. (1981). Preliminary field observations on the marmoset, *Callithrix humeralifer intermedius* (Hershkovitz, 1977), at Dardanelos, Rio Aripuanã, Mato Grosso. *Primates*, 22(1), 46–59. (Chapters 1, 4, 10, 12)

Rylands, A.B. (1982). The behaviour and ecology of three species of marmosets and tamarins (Callitrichidae, Primates) in Brazil. Unpublished Ph.D thesis, University of Cambridge, Cambridge. (Chapters 6, 7, 12, 14, 15)

Rylands, A.B. (1984). Exudate-eating and tree-gouging by marmosets (Callitrichidae, Primates). In *Tropical rain forest: The Leeds Symposium* (ed. A.C. Chadwick and S.L. Sutton), pp.155–68. Leeds Philosophical and Literary Society, Leeds. (Chapters 12, 14, 15)

Rylands, A.B. (1985*a*). Conservation areas protecting primates in Brazilian Amazonia. *Primate Conservation*, (5), 24–7. (Chapter 1)

Rylands, A.B. (1985*b*). Tree-gouging and scent-marking by marmosets. *Anim. Behav.*, 33, 1365–7. (Chapter 4)

Rylands, A.B. (1986*a*). Infant-carrying in a wild marmoset group, *Callithrix humeralifer*: evidence for a polyandrous mating system. In *A primatologia no Brasil—2* (ed. M.T. de Mello), pp.131–44. Sociedade Brasileira de Primatologia, Brasília. (Chapters 6, 7, 10, 14, 15)

Rylands, A.B. (1986*b*). Ranging behaviour and habitat preference of a wild marmoset group, *Callithrix humeralifer* (Callitrichidae, Primates). *J. Zool, Lond.*, (A), 210, 489–514. (Chapters 8, 12, 13, 14, 15)

Rylands, A.B. (1987). Primate communities in Amazonian forests: their habitats and food resources. *Experientia*, 43(3), 265–79. (Chapters 1, 8, 12, 14)

Rylands, A.B. (1989*a*). Evolução do sistema de acasalamento em Callitrichidae. In *Etologia de animais e de homens* (ed. C. Ades), pp.87–108. Edicon, University of São Paulo Press, São Paulo. (Chapters 7, 8, 10, 12, 14)

Rylands, A.B. (1989*b*). Sympatric Brazilian callitrichids: the black tufted-ear marmoset, *Callithrix kuhli*, and the golden-headed lion tamarin, *Leontopithecus chrysomelas*. *J. Hum. Evol.*, 18, 679–95. (Chapters 9, 12, 14, 15)

Rylands, A.B. (1990). Scent marking behaviour of wild marmosets, *Callithrix humeralifer* (Callitrichidae, Primates). In *Chemical signals in vertebrates 5* (ed. D.W. Macdonald, D. Müller-Schwarze, and S.E. Natynczuk), pp.415–29. Oxford University Press, Oxford. (Chapter 4)

Rylands, A.B. (1991). The status of conservation areas in the Brazilian Amazon. World Wildlife Publications, Washington, D.C. 146pp. (Chapter 1)

Rylands, A.B. and Bernardes, A.T. (1989). Two priority regions for conservation in Brazilian Amazonia. *Primate Conservation*, (10), 56–62. (Chapter 1)

Rylands, A.B. and Costa, C.M.R. (1988). Population Density and habitats of marmosets. Some observations on a hybrid population of *Callithrix geoffroyi* × *Callithrix penicillata* at the Research and Environmental Development Station of Peti, Minas Gerais, Brazil. Report to the Companhia Energética de Minas Gerais (CEMIG) and the Program for Studies in Tropical Conservation (PSTC), University of Florida, Gainesville. 15pp. (Chapter 3)

Rylands, A.B. and Mittermeier, R.A. (1982). Conservation of primates in Brazilian Amazonia. *Int. Zoo Yrbk.*, 22, 17–37. (Chapter 1)

Rylands, A.B., Spironelo, W.R., Tornisielo, V.L., Sa, R.L. de, Kierulff, M.C.M., and Santos, I.B. (1988). Primates of the Rio Jequitinhonha valley, Minas Gerais, Brazil. *Primate Conservation*, (9), 100–9. (Chapters 1, 3)

Rylands, A.B., Monteiro da Cruz, M.A.O., and Ferrari, S.F. (1989). An association between marmosets and army ants in Brazil. *J. Trop. Ecol.*, 5, 113–16. (Chapters 12, 15)

Rylands, A.B., Santos, I.B., and Mittermeier, R.A. (In press). Distribution and status of *Leontopithecus chrysomelas* in the wild. In *A case study in conservation biology: the golden lion tamarin* (ed. D.G. Kleiman). Smithsonian Institution Press, Washington, D.C. (Chapter 1)

Santos, I.B., Mittermeier, R.A., Rylands, A.B., and Valle, C.M.C. (1987). The distribution and conservation status of primates in southern Bahia, Brazil. *Primate Conservation*, (8), 126–42. (Chapter 1)

Savage, A., Ziegler, T.E., and Snowdon, C.T. (1988). Socio-sexual development, pair bond formation, and mechanisms of fertility suppression in female cotton-top tamarins (*Saguinus oedipus oedipus*). *Am. J. Primatol.*, 14, 345–9. (Chapters 4, 5, 8, 10)

Savage, A., Snowdon, C.T., and Giraldo, H. (1990). The ecology of the cotton-top tamarin in Colombia. *Am. J. Primatol.*, 20, 230. Abstract. (Chapter 8)

Scanlon, C.E., Chalmers, N.R., and Monteiro da Cruz, M.A.O. (1988). Changes in the size, composition and reproductive condition of wild marmoset groups (*Callithrix jacchus jacchus*) in north east Brazil. *Primates*, 29, 295–305. (Chapters 5, 6, 7, 12)

Scanlon, C.E, Chalmers, N.R., and Monteiro da Cruz, M.A.O. (1989). Home range use and the exploitation of gum in the marmoset *Callithrix jacchus jacchus*. *Int. J. Primatol.*, 19, 123–36. (Chapter 12)

Schaffer, J. (1940). *Die Hautdrusenorgane der Säugetiere.* Urban and Schwarzenberg, Berlin. (Chapter 4)

Schaffner, C. (1991). Aggression and post-conflict behavior in red-bellied tamarins. Unpublished Master's thesis, Bucknell University, Lewisburg. (Chapter 8)

Schilling, A. and Perret, M. (1987). Chemical signals and reproductive capacity in male prosimian primates (*Microcebus murinus*). *Chem. Senses*, 12, 143–58. (Chapter 4)

Schoener, T.W. (1971). Theory of feeding strategies. *Ann. Rev. Ecol. Syst.*, 2, 369–404. (Chapter 15)

Seal, U.S., Ballou J.D., and Padua, C.V. (ed.) (1990). *Leontopithecus*: population viability analysis workshop report. Captive Breeding Specialist Group (IUCN/SSC/CBSG), Apple Valley, Minnesota. (Chapter 1)

Sick, H. and Teixeira, D.M. (1979). Notas sobre aves brasileiras raras ou ameaçadas de extinção. *Publ. Avulsas do Museu Nacional, Rio de Janeiro,* (62), 1–39. (Chapter 1)

Siebels, F. (1991). Experimentelle Untersuchungen zur möglichen wechselseitigen Beeinflussung bei der Wahl des Aufenthaltsortes beim Weissbüschelaffen (*Callithrix jacchus* Erxleben, 1777). Unpublished diploma thesis, University of Göttingen, Göttingen. (Chapter 7)

Siegel, C.E., Hamilton, J.M., and Castro, N.R. (1989). Observations of the red-billed ground-cuckoo (*Neomorphus pucheranii*) in association with tamarins (*Saguinus*) in northeastern Peru. *Condor*, 91, 720–2. (Chapter 15)

Siess, M. (1988). Experimente zur kooperativen Polyandrie beim Weissbüscheläffen (*Callithrix jacchus* Erxleben, 1777). Unpublished diploma thesis, University of Göttingen, Göttingen. (Chapter 7)

Silva, E.M.B., Yamamoto, M.E., and Arruda, M.F. (1991). Independência e socialização do *Callithrix jacchus* na ausência do gêmeo: um estudo piloto. In *A primatologia no Brasil—3* (ed. A.B. Rylands and A.T. Bernardes), pp.47–55. Sociedade Brasileira de Primatologia e Fundação Biodiversitas, Belo Horizonte. (Chapter 10)

Silva, N.G, Yamamoto, M.E., and Arruda, M.F. (1987). O cuidado com a prole no sagui comum (*Callithrix jacchus*): uma comparação entre animais capturados e nascidos em cativeiro. *Resumos. 39ª. Reunião Anual da SBPC*, p.926. Sociedade Brasileira para o Progressa da Ciência (SBPC), Rio de Janeiro. (Chapter 10)

Simek, M.A. (1988). Food provisioning of infants in captive social groups of common marmosets (*Callithrix jacchus*) and cotton-top tamarins (*Saguinus oedipus*). Unpublished Master's thesis, University of Tennessee, Knoxville. (Chapter 9)

Simon, F. (1989). Livro de linhagens (oficial) studbook 1989 *Leontopithecus chrysopygus*. International Committee for the Preservation and Management of the Black lion tamarin (*Leontopithecus chrysopygus*), São Paulo. (Chapter 1)

Simons, E.L. (1972). *Primate evolution: an introduction to man's place in nature.* Macmillan, New York. 322pp. (Chapter 1)

Simpson, G.G. (1945). The principles of classification and classification of mammals. *Bull. Am. Mus. nat. Hist.*, 85, 1–350. (Chapter 1)

Singer, A.G., Macrides, F., Clancy, A.N., and Agosta, W.C. (1986). Purification and analysis of proteinaeous aphrodisiac pheromone from hamster vaginal discharge. *J. Biol. Chem.*, 261, 13323–6. (Chapter 4)

Singer, A.G., Clancy, A.N., Macrides, F., Agosta, W.C., and Bronson, F.H. (1988). Chemical properties of a female mouse pheromone that stimulates gonadotropin secretion in males. *Biol. Reprod.*, 38, 193–9. (Chapter 4)

Skinner, C. (1985). Report on a field study of Geoffroy's tamarin in Panama. *Primate Conservation*, (5), 22–4. (Chapter 1)

Skinner, C. (1986). A life history study of the Geoffroy's tamarin, *Saguinus geoffroyi*, with emphasis on male-female relationships in captive animals. Dissertation Abstracts International. B47(5), 1844. (Chapter 8)

Smith, A,B., III, Belcher, A.M., Epple, G., Jurs, P.C., and Levine, B. (1985). Computerized pattern recognition: a new technique for the analysis of chemical communication. *Science*, 228, 175–7. (Chapter 4)

Smith, W.J. (1966). Communication and relationships in the genus *Tyrannis. Monog. Nuttall Ornith. Club.*, 6, 1–250. (Chapter 2)

Snowdon, C.T. (1986). Language parallels in the vocal communication of callitrichids. In *A primatologia no Brasil* (ed. M.T. de Mello), pp.221–32. Sociedade Brasileira de Primatologia, Brasília. (Chapter 8)

Snowdon, C.T. (1990). Mechanisms maintaining monogamy in monkeys. In *Contemporary issues in comparative psychology* (ed. D.A. Dewsbury), pp.225–51. Sinauer Associates, Sunderland. (Chapter 10)

Snowdon, C.T. and Hodun, A. (1981). Acoustic adaptations in pygmy marmoset contact calls: locational cues vary with distance between conspecifics. *Behav. Ecol. Sociobiol.*, 9, 295–300. (Chapter 2)

Snowdon, C.T. and Hodun, A. (1985). Troop specific responses to long calls

of isolated tamarins (*Saguinus mystax*). *Am. J. Primatol.*, 8, 205–13. (Chapter 2)
Snowdon, C.T. and Soini, P. (1988). The tamarins, genus *Saguinus*. In *Ecology and behavior of Neotropical primates*, vol. 2 (ed. R.A. Mittermeier, A.B. Rylands, A.F. Coimbra-Filho, G.A.B. da Fonseca), pp.223–98. World Wildlife Fund, Washington, D.C. (Chapters 2, 5, 8, 10, 11, 12, 13, 14, 15)
Snowdon, C.T., Hodun, A., Rosenberger, A.L., and Coimbra-Filho, A.F. (1986). Long call structure and its relation to taxonomy in lion tamarins. *Am. J. Primatol.*, 11, 253–61. (Chapter 2)
Soini, P. (1982*a*). Primate conservation in Peruvian Amazonia. *Int. Zoo Yrbk.*, (22), 37–47. (Chapter 1)
Soini, P. (1982*b*). Ecology and population dynamics of the pygmy marmoset, *Cebuella pygmaea. Folia primatol.*, 39, 1–21. (Chapters 7, 11)
Soini, P. (1982*c*). Informe de Pacaya No. 6: distribucion geografica y ecologia poblacional de *S. mystax* (Primates, Callitrichidae). Unpublished report, Direccion Regional de Agricultura y Alimentacion, Iquitos, Peru. (Chapter 13)
Soini, P. (1987*a*). Sociosexual behavior of a free-ranging *Cebuella pygmaea* (Callitrichidae, Platyrrhini) troop during postpartum estrus of its reproductive female. *Am. J. Primatol.*, 13, 223–30. (Chapters 6, 11)
Soini, P. (1987*b*). Ecology of the saddleback tamarin *Saguinus fuscicollis illigeri* on the Río Pacaya, northeastern Peru. *Folia primatol.*, 49, 11–32. (Chapters 7, 8, 11, 12, 13, 14, 15)
Soini, P. (1988). The pygmy marmoset, genus *Cebuella*. In *Ecology and behavior of Neotropical primates*, vol. 2 (ed. R.A. Mittermeier, A.B. Rylands, A.F. Coimbra-Filho, and G.A.B. da Fonseca), pp.79–129. World Wildlife Fund, Washington, D.C. (Chapters 1, 5, 7, 10, 11, 15)
Soini, P. and Cóppula, M. (1981). Ecología y dinamica poblacional del pichico, *Saguinus fuscicollis* (Primates: Callitrichidae). Informe de Pacaya, No. 4. Ordeloreto, Dirección Regional de Agricultura y Alimentación, Iquitos. (Chapters 11, 13)
Soini, P. and Soini, M. (1982). Distribución geográfica y ecología poblacional de *Saguinus mystax* (Primates: Callitrichidae). Informe de Pacaya, No.6. Ordeloreto, Dirección Regional de Agricultura y Alimentación, Iquitos. (Chapters 11, 13)
Soini, P., Aquino, R., Encarnacion, F., Moya, L., and Tapia, J. (1989). Situacion de los primates en la Amazonia Peruana. In *La primatologia en Latinoamerica* (ed. C.J. Saavedra, R.A. Mittermeier, and I.B. Santos), pp.13–19. World Wildlife Fund, Washington, D.C. (Chapter 1)
Sokal, R.R. and Rohlf, F.J. (1981). *Biometry*. W.H. Freeman, New York. (Chapter 9)
Sommer, V. (1980). Zur Gruppenstuktur und deren Dynamik in einer Familie von *Callithrix jacchus* (Ceboidea, Callitrichidae). Unpublished diploma thesis, University of Göttingen, Göttingen. (Chapter 7)
Sommer, V. (1984). Dynamics of group structure in a family of the common marmoset *Callithrix jacchus* (Callitrichidae). In *Current primate researches* (ed. M.L. Roonwal, S.M. Mohnot, and N.S. Rathore), pp.315–42. Jodhpur University, Jodhpur. (Chapter 10)
Sousa e Silva, J. de, Jr (1988). A range extension for *Saguinus labiatus thomasi. Primate Conservation*, (9), 23–4. (Chapter 1)
Spichiger-Carlsson, P. (1982). Beziehungen unter Familienmitgliedern bei Weissbuschelaffen. Unpublished Master's thesis, University of Zurich, Zurich. (Chapter 7)

Stallings, J.R. (1985). Distribution and status of primates in Paraguay. *Primate Conservation*, (6), 51–8. (Chapters 1, 12)

Stallings, J.R. (1988). Small mammal communities in an eastern Brazilian park. Unpublished Master's thesis, University of Florida, Gainesville. (Chapter 12)

Stallings, J.R. and Mittermeier, R.A. (1983). The black-tailed marmoset (*Callithrix argentata melanura*) recorded from Paraguay. *Am. J. Primatol.*, 4, 159–63. (Chapters 1, 12, 15)

Stallings, J.R. and Robinson, J.G. (1991). Disturbance, forest heterogeneity and primate communities in a Brazilian Atlantic forest park. In: *A Primatologia no Brasil—3* (ed. A.B. Rylands and A.T. Bernardes), pp.357–68. Sociedade Brasileira de Primatologia and Fundação Biodiversitas, Belo Horizonte. (Chapter 12)

Stallings, J.R., West, L., Hahn, W., and Gamarra, I. (1989). Primates and their relation to habitat in the Paraguayan chaco. In: *Advances in Neotropical mammalogy* (ed. K.H. Redford and J.F. Eisenberg), pp.425–42. Sandhill Crane Press, Inc., Gainesville. (Chapter 12)

Starck, D. (1969). Die circumgenitalen Drüsenorgane von *Callithrix* (*Cebuella*) *pygmaea* (Spix, 1923). *Zool. Garten, Lpz.*, 36, 312–26. (Chapter 4)

Stephens, D.W. and Krebs, J.R. *Foraging theory*. Princeton University, Princeton, N.J. (Chapter 8)

Stevenson, M.F. (1976*a*). Behavioural observations on groups of Callitrichidae with an emphasis on playful behaviour. *Jersey Wildl. Preserv. Trust, Ann. Rep.*, 13, 47–52. (Chapter 7)

Stevenson, M.F. (1976*b*). Birth and perinatal behaviour in family groups of the common marmoset (*Callithrix jacchus jacchus*), compared to other primates. *J. Hum. Evol.*, 5, 365–81. (Chapters 5, 7)

Stevenson, M.F. (1978). Ontogeny of playful behaviour in family groups of the common marmoset. In *Recent advances in primatology*, vol. 1, behaviour (ed. D.J. Chivers and J. Herbert), pp.139–43. Academic Press, London. (Chapter 10)

Stevenson, M.F. and Poole, T.B. (1976). An ethogram of the common marmoset, *Callithrix jacchus jacchus*: general behavioural repertoire. *Anim. Behav.*, 24, 428–51. (Chapter 10)

Stevenson, M.F. and Rylands, A.B. (1988). The marmosets, genus *Callithrix*. In *Ecology and behavior of Neotropical primates*, vol. 2 (ed. R.A. Mittermeier, A.B. Rylands, A.F. Coimbra-Filho, and G.A.B. da Fonseca), pp.131–222. World Wildlife Fund, Washington, D.C. (Chapters 1, 4, 5, 7, 10, 12, 13, 14, 15)

Stevenson, M.F. and Sutcliffe, A.G. (1978). Breeding a second generation of common marmosets (*Callithrix jacchus jacchus*). *Int. Zoo. Yrbk.*, 18, 109–14. (Chapter 5)

Stribley, J.A., French, J.A., and Inglett, B.J. (1987). Mating patterns in the golden lion tamarin (*Leontopithecus rosalia*): continuous receptivity and concealed estrus. *Folia primatol.*, 49, 137–50. (Chapter 6)

Strier, K.B. (1990). New World primates, new frontiers: insights from the woolly spider monkey, or muriqui (*Brachyteles arachnoides*). *Int. J. Primatol.*, 11, 7–19. (Chapter 8)

Struhsaker, T.T. (1970). Phylogenetic implications of some vocalizations of *Cercopithecus* monkeys. In *Old World monkeys: evolution, systematics and behaviour* (ed. J.R. Napier and P.H. Napier), pp.365–444. Academic Press, London. (Chapter 2)

Struhsaker, T.T. (1984). Hybrid monkeys of the Kibale forest. Successful mating between redtail and blue monkeys. In *The Encyclopaedia of Mammals*, vol. 1 (ed. D.W. Macdonald), pp.396–7. George Allen and Unwin, London. (Chapter 3)

Sussman, R.W. and Garber, P.A. (1987). A new interpretation of the social organization and mating system of the Callitrichidae. *Int. J. Primatol.*, 8, 73–92. (Chapters 4, 5, 6, 7, 8, 9, 12, 13, 15)

Sussman, R.W. and Kinzey, W.G. (1984). The ecological role of the Callitrichidae. *Am. J. Phys. Anthropol.*, 64, 419–49. (Chapters 4, 5, 8, 12, 13, 14, 15)

Sutcliffe, A.G. and Poole, T.B. (1978). Scent marking and associated behaviour in captive common marmosets (*Callithrix jacchus jacchus*) with a description of the histology of scent glands. *J. Zool., Lond.*, 185, 41–56. (Chapter 4)

Sutcliffe, A.G. and Poole, T.B. (1984). An experimental analysis of social interaction in the common marmoset (*Callithrix jacchus jacchus*). *Int. J. Primatol.*, 5, 591–607. (Chapters 7, 10)

Tanaka, L. and Tanaka, S. (1982). Rainfall and seasonal changes in arthropod abundance on a tropical ocean island. *Biotropica*, 14, 114–23. (Chapter 13)

Tardif, S.D. (1983). Relationship between social interactions and sexual maturation in female *Saguinus oedipus. Folia primatol.*, 40, 268–75. (Chapters 8, 10)

Tardif, S.D. (1984). Social influences on sexual maturation of female *Saguinus oedipus. Am. J. Primatol.*, 6, 199–209. (Chapters 5, 8, 14)

Tardif, S.D. and Harrison, M.L. (1986). Energetic demands of infant care in the callitrichids: species comparison. *Prim. Rep.*, 14, 78. Abstract. (Chapter 7)

Tardif, S.D., Richter, C.B., and Carson, R.L. (1984). Effects of sibling rearing experience on future: reproductive success in two species of Callitrichidae. *Am. J. Primatol.*, 6, 377–80. (Chapters 5, 7, 10)

Tardif, S.D., Carson, R.L., and Gangaware, B.L. (1986). Comparison of infant care in family groups of the common marmoset (*Callithrix jacchus*) and the cotton-top tamarin (*Saguinus oedipus*). *Am. J. Primatol.*, 11, 103–10. (Chapters 7, 8, 9, 10)

Tardif, S.D., Carson, R.L., and Gangaware, B.L. (1990). Infant-care behavior of mothers and fathers in a communal-care primate, the cotton-top tamarin, *Saguinus oedipus. Am. J. Primatol.*, 22, 73–85. (Chapters 7, 9)

Tardif, S.D., Carson, R.L., and Gangaware, B.L. (In review). Examination of factors controlling infant care in a communal care primate, the cotton-top tamarin (*Saguinus oedipus*). Submitted to *Ethology*. (Chapter 9)

Terborgh, J. (1983). *Five New World primates: a study in comparative ecology*. Princeton University Press, Princeton, N.J. (Chapters 1, 4, 8, 12, 13, 14, 15)

Terborgh, J. (1985). The ecology of Amazonian primates. In *Amazonia* (ed. G.T. Prance and T.E. Lovejoy), pp.284–304. Pergamon Press, New York. (Chapters 12, 13)

Terborgh, J. (1986). The social systems of New World primates: an adaptionist view. In *Primate ecology and conservation* (ed. J.G. Else and P.C. Lee), pp.199–211. Cambridge University Press, Cambridge. (Chapter 13)

Terborgh, J. and Goldizen, A.W. (1985). On the mating system of the cooperatively breeding saddle-backed tamarin (*Saguinus fuscicollis*). *Behav. Ecol. Sociobiol.*, 16, 293–9. (Chapters 4, 5, 7, 8, 9, 14, 15)

Terborgh, J. and Janson, C.H. (1986). The socioecology of primate groups. *Ann. Rev. Ecol. Syst.*, 75, 111–36. (Chapter 8)

Terborgh, J. and Stern, M. (1987). The surreptitious life of the saddle-backed tamarin. *Am. Sci.*, 75, 260–69. (Chapters 8, 14, 15)

Terborgh. J., Robinson, S.K., Parker, T.A., III., Munn, C.A. and Pierpont, N. (1990). Structure and organization of an Amazonian bird community. *Ecol. Monog.*, 60, 213–38. (Chapter 15)

Thomas, O. (1920). On mammals from the lower Rio Amazonas in the Goeldi Museum, Pará. *Ann. Mag. nat. Hist.*, 9(6): 266–83. (Chapter 1)

Thorington, R.W., Jr (1968). Observations of the tamarin, *Saguinus midas. Folia primatol.*, 9, 85–98. (Chapters 8, 13)

Thorington, R.W., Jr (1976). The systematics of New World monkeys. In *First Inter-American Conference on Conservation and Utilization of American Nonhuman Primates in Biomedical Research, Pan American Health Organization Sci. Publ. 317*, pp.8–18. Pan American Health Organization (PAHO), World Health Organization (WHO), Washington, D.C. (Chapter 1)

Thorington, R.W., Jr. (1988). Taxonomic status of *Saguinus tripartitus* (Milne-Edwards, 1878). *Am. J. Primatol.*, 15, 367–71. (Chapter 1)

Torres de Assumpção, C. (1983a). Conservation of primates in Brazil: Atlantic forest primates. In *Proceedings of the Symposium on the Conservation of Primates and Their Habitats, Vol. 1, Primate Conservation in the Wild* (ed. D. Harper), pp.34–48. University of Leicester, Leicester. (Chapter 1)

Torres de Assumpção, C. (1983b). An ecological study of primates in southeastern Brazil, with a reappraisal of *Cebus apella* races. Unpublished Ph.D. thesis, University of Edinburgh, Edinburgh. (Chapters 1, 12)

Trivers, R.L. (1971). The evolution of recripocal altruism. *Q. Rev. Biol.*, 46, 35–57. (Chapter 7)

Vandenbergh, J.G., Whitsett, J.M., and Lombardi, J.R. (1975). Partial isolation of a pheromone accelerating puberty in female mice. *J. Reprod. Fert.*, 43, 515–23. (Chapter 4)

Vandenbergh, J.G., Finlayson, S.J., Dobrogosz, W.J., Dills, S.S., and Kost, T.A. (1976) Chromatographic separation of puberty accelerating pheromone from male mouse urine. *Biol. Reprod.*, 15, 260–5. (Chapter 4)

Vieira, C.O.da C. (1944). Os símios do estado de São Paulo. *Pap. Avulsos, Dept. Zool. Sec. Agric., São Paulo*, 4, 1–31. (Chapter 1)

Vivo, M. de (1979). Primatas do Parque Nacional do Tapajós. Unpublished report, Instituto Brasileiro de Desenvolvimento Florestal (IBDF), Brasília. 22pp. (Chapter 1)

Vivo, M. de (1985). On some monkeys from Rondônia, Brasil (Primates: Callitrichidae, Cebidae). *Papeis Avulsos Zool.*, São Paulo, 36(11), 103–10. (Chapter 1)

Vivo, M. de (1988). Sistemática de *Callithrix* Erxleben, 1777 (Callitrichidae, Primates). Unpublished doctoral thesis, Instituto de Biociências, Universidade de São Paulo, São Paulo. (Chapters 1, 15)

Vivo, M. de (1991). *Taxonomia de* Callithrix *Erxleben, 1777 (Callitrichidae, Primates)*. Fundação Biodiversitas, Belo Horizonte. (Chapter 1)

Vogt, J.L., Carlson, H., and Menzel, E.W., Jr (1978). Social behavior of a marmoset (*Saguinus fuscicollis*) group. I. Parental care and infant development. *Primates*, 19, 715–26. (Chapters 7, 9)

Voland, E. (1977). Social play behaviour of the common marmoset (*Callithrix jacchus* Erxl., 1777) in captivity. *Primates*, 18, 883–902. (Chapter 10)

Waal, F.B.M. de (1986). Conflict resolution in monkeys and apes. In *Primates: the road to self-sustaining populations* (ed. K. Benirschke), pp.341–50. Springer-Verlag, New York. (Chapter 8)

Walek, M.L. (1978). Vocalizations of the black and white colobus monkey (*Colobus polykomos* Zimmermann, 1780). *Am. J. Phys. Anthropol.*, 49, 227–40. (Chapter 2)

Wamboldt, M.Z., Gelhard, R.E., and Insel, T.R. (1988). Gender differences in caring for infant *Cebuella pygmaea*: the role of infant age and relatedness. *Develop. Psychobiol.*, 21, 187–202. (Chapter 9)

Waser, P.M. (1988). Resources, philopatry, and social interaction among mammals. In *The ecology of social behavior* (ed. C.N. Slobodchikoff), pp.109–30. Academic Press, New York. (Chapter 7)

Waser, P.M. and Waser, M.S. (1977). Experimental studies of primate vocalizations: specializations for long distance propagation. *Z. Tierpsychol.*, 43, 239–63. (Chapter 2)

Wasser, S.K. and Barash, D.P. (1983). Reproductive suppression among female mammals: implications for biomedicine and sexual selection theory. *Q. Rev. Biol.*, 58, 513–38. (Chapter 5)

Webley, G.E., Abbott, D.H., George, L.M., Hearn, J.P., and Mell, H. (1989). The circadian pattern of plasma melatonin concentrations in the marmoset monkey (*Callithrix jacchus*). *Am. J. Primatol.*, 17, 73–9. (Chapter 5)

West Eberhard, M.J. (1975). The evolution of social behavior by kin selection. *Q. Rev. Biol.*, 50, 1–33. (Chapter 7)

Wetton, J.H., Carter, R.E., Parker, D.T., and Walters, D. (1987). Demographic study of a wild house sparrow population by D.N.A. fingerprinting. *Nature, Lond.*, 327, 147–9. (Chapter 6)

Wheeler, J.W., Blum, M.S., and Clark, A. (1977). β-(p-hydroxyphenyl) ethanol in the chest gland secretion of a galago (*Galago crassicaudatus*). *Experientia*, 33, 988–9. (Chapter 4)

Whitmore, T.C. and Prance, G.T. (ed.) (1987). *Biogeography and Quaternary history in tropical America*. Clarendon Press, Oxford. (Chapter 1)

Whitten, P.L. (1987). Infants and adult males. In *Primate societies* (ed. B.B. Smuts, D.L. Cheney, R.M. Seyfarth, R.W. Wrangham, and T.T. Struhsaker), pp.343–57. Chicago University Press, Chicago. (Chapters 7, 10)

Wickler, W. and Seibt, U. (1983). Monogamy: an ambiguous concept. In *Mate choice* (ed. P. Bateson), pp.33–50. Cambridge University Press, Cambridge. (Chapter 7)

Wilson, W.L. and Wilson, C.C. (1975). Species-specific vocalizations and the determination of phylogenetic affinities of the *Presbytis aygula-melalophus* group in Sumatra. In *Contemporary primatology* (ed. S. Kondo, M. Kawai, and A. Ehara), pp.459–63. S. Karger, Basel. (Chapter 2)

Wislocki, G.B. (1930). A study of the scent glands in the marmosets, especially *Oedipomidas geoffroyi*. *J. Mammal.*, 11, 475–83. (Chapter 4)

Wislocki, G.B. (1939). Observations on twinning in marmosets. *Am. J. Anat.*, 664, 445–83. (Chapter 6)

Wolfe, L.G., Deinhardt, F., Ogden, J.D., Adams, M.R., and Fischer, L.E. (1975). Reproduction of wild-caught and laboratory-born marmoset species used in biomedical research (*Saguinus* sp., *Callithrix jacchus*). *Lab. Anim. Care.*, 25, 802–13. (Chapter 3)

Wojcik, J. and Heltne, P.G. (1978). Tail marking in *Callimico goeldi*. In *Recent advances in primatology. Behaviour*, vol. 1 (ed. D.J. Chivers and J. Herbert), pp.507–9. Academic Press, London. (Chapter 4)

Wolters, H.-J. (1978). Some aspects of role-taking behaviour in captive family groups of the cotton-top tamarin *Saguinus oedipus oedipus*. In *Biology and*

*behaviour of marmosets* (ed. H. Rothe, H.-J. Wolters, and J.P. Hearn), pp.259–78. Eigenverlag H. Rothe, Göttingen. (Chapters 4, 9, 10)
Wrangham, R.W. (1980). An ecological model of female primate groups. *Behaviour*, 75, 262–300. (Chapter 8)
Wright, P.C. (1984). Biparental care in *Aotus trivirgatus* and *Callicebus moloch*. In *Female primates: studies by women primatologists* (ed. M. F. Small), pp.59–75. Alan R. Liss, New York. (Chapter 13)
Wright, P.C. (1986). Ecological correlates of monogamy in *Aotus* and *Callicebus*. In *Primate ecology and conservation* (ed. J.G. Else and P.C. Lee), pp.159–67. Cambridge University Press, Cambridge. (Chapter 13)
Ximenes, M.F.F.M. and Sousa, M.B.C. (In press). Composição do grupo familiar e características da amamentação (contato com o mamilo) no *Callithrix jacchus*. In *A primatologia no Brasil—4* (ed. M.E. Yamamoto and M.B.C. de Sousa). Sociedade Brasileira de Primatologia e Fundação Biodiversitas, Belo Horizonte. (Chapter 10)
Yamamoto, M.E. (1990). Ontogenese das relações sociais e dinâmica do cuidado com a prole no sagui comum (*Callithrix jacchus*). Unpublished doctoral dissertation, Escola Paulista de Medicina, São Paulo. (Chapter 10)
Yamamoto, M.E., Arruda, M.F. de, and Bueno, O.F.A. (1987). Compensation in abnormal conditions of infant care in the common marmoset (*Callithrix jacchus*). *Int. J. Comp. Psychol.*, 1, 97–106. (Chapters 7, 10)
Yarger, T.G., Smith, A.B., III, Preti, G., and Epple, G. (1977). The major volatile constituents of the scent mark of a South American primate, *Saguinus fuscicollis*, Callithricidae. *J. Chem. Ecol.*, 3, 45–56. (Chapter 4)
Yoneda, M. (1981). Ecological studies of *Saguinus fuscicollis* and *S. labiatus* with reference to habitat segregation and height preference. *Kyoto University Overseas Research Reports*, 2, 43–50. (Chapters 8, 13, 14, 15)
Yoneda, M. (1984*a*). Comparative studies on vertical separation, foraging behavior and travelling mode of saddle-backed tamarins (*Saguinus fuscicollis*) and red-chested moustached tamarins (*Saguinus labiatus*) in northern Bolivia. *Primates*, 25, 414–42. (Chapters 13, 14, 15)
Yoneda, M. (1984*b*). Ecological study of the saddle-backed tamarin (*Saguinus fuscicollis*) in northern Bolivia. *Primates*, 25, 1–12. (Chapters 4, 8, 13)
Zeller, U., Epple, G., Küderling, I., and Kühn, H.-J. (1988). The anatomy of the circumgenital scent gland of *Saguinus fuscicollis* (Callitrichidae, Primates). *J. Zool., Lond.*, 214, 141–56. (Chapter 4)
Zeller, U., Richter, J., Epple, G., and Kühn, H.-J. (1989). Die circumgenitaldruse von Krallenaffen der Gattung *Saguinus*. *Verh. anat. Ges., Jena*, 82, 171–4. (Chapter 4)
Ziegler, T.E., Bridson, W.E., Snowdon, C.T., and Eman, S. (1987*a*). Urinary gonadotropin and estrogen excretion during the postpartum estrus, conception and pregnancy in the cotton-top tamarin (*Saguinus oedipus oedipus*). *Am. J. Primatol.*, 12, 127–40. (Chapter 6)
Ziegler, T.E., Savage, A., Scheffler, G., and Snowdon, C.T. (1987*b*). The endocrinology of puberty and reproductive functioning in female cotton-top tamarins (*Saguinus oedipus*) under varying social conditions. *Biol. Reprod.*, 37, 618–27. (Chapters 5, 10)
Ziegler, T.E., Scholl, S.A., Scheffler, G., Haggerty, M.A., and Lasley, B.L. (1989). Excretion of oestrone, oestradiol and progesterone in the urine and faeces of the female cotton-top tamarin (*Saguinus oedipus oedipus*). *Am. J. Primatol.*, 17, 185–95. (Chapter 5)

Ziegler, T.E., Widowski, T.M., Larson, M.L., and Snowdon, C.T. (1990). Nursing does affect the duration of the post-partum to ovulation interval in cotton-top tamarins (*Saguinus oedipus*). *J. Reprod. Fert.*, 90, 563–571. (Chapter 9)

Zimmerman, E. (1990). Differentiation of vocalizations in bushbabies (Galaginae, Prosimiae, Primates) and the significance for assessing phylogenetic relationships. *Zeitschrift fuer Zoologische Systematiks und Evolutionsforschung*, 28, 217–39. (Chapter 2)

Zullo, J. and Caine, N.G. (1988). The use of sentinels in captive groups of red-bellied tamarins. *Am. J. Primatol.*, 14, 455. Abstract. (Chapter 8)

# Author index

# Subject index

# Index by species